KU-400-526

Theory of Stellar Pulsation

PRINCETON SERIES IN ASTROPHYSICS

Edited by Jeremiah P. Ostriker

Theory of Rotating Stars, *by J. L. Tassoul*
Theory of Stellar Pulsation, *by J. P. Cox*

John P. Cox

THEORY OF STELLAR PULSATION

PRINCETON UNIVERSITY PRESS
Princeton, New Jersey

Published by Princeton University Press, Princeton, New Jersey
In the United Kingdom: Princeton University Press, Guildford, Surrey

Library of Congress Cataloging in Publication Data will be found on the last printed page of this book

This book has been composed in Times Roman

Clothbound editions of Princeton University Press books are printed on acid-free paper, and binding materials are chosen for strength and durability

Printed in the United States of America by Princeton University Press, Princeton, New Jersey

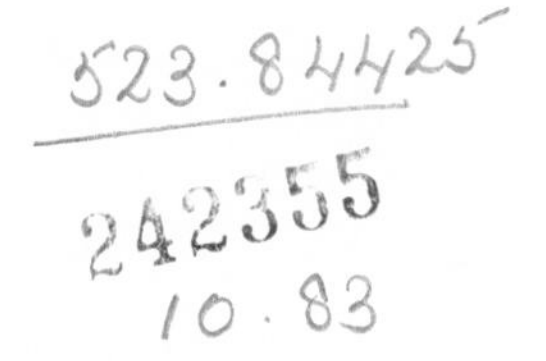

To Richard N. Thomas

who provided me with the opportunity
to develop extensive lecture notes,
on which this book is partly based.

Contents

Preface

It is now about thirty years since a new text or monograph dealing mainly with the theory of stellar pulsation has been published in the English language. The last such book, to the best of my knowledge, was S. Rosseland's classic, *The Pulsation Theory of Variable Stars* (Oxford University Press, 1949), which dealt almost exclusively with the theory of purely radial oscillations.* Even the monumental and remarkably comprehensive encyclopedia article by P. Ledoux and Th. Walraven *(Handb. d. Phys.,* Vol. **51,** 1958) is by now more than twenty years out of date. This article has been the standard reference in the field for many years and will no doubt continue in this role; indeed, the work is referred to in almost every section of this book. However, some of the most important astrophysical problems having to do with variable stars have been solved since the Ledoux-Walraven article was published. Also, much of the basic theory of nonradial oscillations was not developed until the mid and late 1960's, and new developments are still occasionally coming in. Some of these recent developments have been described in various review articles (see the references given in Chapter 1). These articles, however, along with most of the background material needed for a detailed understanding of the theory of stellar pulsation, are scattered throughout the physical and astrophysical literature. In this book I have collected much of this material into one place, and attempted to fill the need in the astrophysical literature for an up-to-date, reasonably comprehensive, and sufficiently detailed treatment of these matters.

The theory of both radial and nonradial oscillations is discussed in this book. However, the recent (mid 1960's and later) extensions of the theory into general relativity are considered for the most part outside the scope of the present work. Thus, except as mentioned otherwise, the treatment throughout most of this book is based on nonrelativistic, Newtonian physics.

The book is divided into three main parts. Part I, consisting of five chapters, is devoted to fundamentals. It contains a brief summary of the main observations (Chap. 3); a brief summary of the basic equations of hydrodynamics and heat flow, couched in forms suitable for later astrophysical applications (Chap. 4); and a fairly thorough discussion of the linear theory (Chap. 5). I have taken considerable pains to elucidate the

*Recently, a monograph dealing with nonradial stellar oscillations has been published by Unno, Osaki, Ando, and Shibahashi (1979).

differences between Eulerian and Lagrangian variations, and to write the linearized equations for the more general case in which there is a velocity field present. Much of this material is based on a paper published in 1967. I also refer to an unpublished proof worked out as recently as 1974.

Part II, consisting of eight chapters, is concerned with purely radial oscillations. The theory of linear, adiabatic, radial oscillations is presented in considerable detail (Chapter 8), as is the theory and calculations of nonadiabatic and nonlinear radial oscillations (Chapters 9 through 12). Some simple models of stellar pulsation (essentially radial) are described in Chapter 13. Some of the important recent developments in our understanding of variable stars are also summarized.

Part III, made up of the final six chapters, is concerned primarily with the theory of nonradial stellar oscillations. Most of the conventional notation and terminology associated with nonradial oscillations is contained in Chapter 17. The bulk of this chapter was written in 1976–1977; therefore, the level of sophistication contained in, for example, the papers of Christensen-Dalsgaard (1979), Shibahashi (1979), and Wolff (1979), and in the monograph by Unno, Osaki, Shibahashi, and Ando (1979), is not reflected in this chapter. This fact, though regrettable, is also inevitable in a rapidly developing field, especially when delays in publication are taken into account. In this Part I have included not only the topics that might logically be considered a part of this subject, but also some of the newer developments referred to above. Chapter 19, in particular, is devoted to "miscellaneous" topics. These are primarily characterized by a relaxation of one or more of the assumptions usually adopted—assumptions which are held through most of the rest of the book.

The book is aimed at about the level of the first-year graduate student. A knowledge of calculus, differential equations, vector analysis, and matrix algebra is assumed. Because of space limitations, I have not been able to include detailed proofs and derivations in most cases. Therefore, considerable demands may be made on the reader.

Although the book was not written as a text (for example, no problems are included), it may nevertheless be useful in that capacity. There is probably more than enough material to comprise a one- or two-semester graduate course in variable stars. The book will probably be of greatest use to students and research workers in this and related fields.

Theory of Stellar Pulsation has grown, for the most part, out of lecture notes developed for a graduate course in variable stars that the author has given at the University of Colorado several times over the past fifteen years. Some of the contents of several of the author's review papers on variable stars have also been incorporated (of course, some of the contents of some of these review papers had their first origin in these same lecture notes!).

I take great pleasure in acknowledging the many people who have contributed, directly or indirectly, to this project. I would particularly like to express my gratitude to the University of Colorado for granting me a sabbatical leave during the Spring semester of 1975, when the actual transformation from lecture notes into a book was begun and partially completed. I am also grateful to Professor V. H. Regener for making available the facilities of the Department of Physics and Astronomy of the University of New Mexico during part of this period, and to Professor D. S. King of that Department for much moral support and for many useful discussions.

Particular thanks go to Dr. Morris Aizenman for suggestions regarding publication and for many discussions, and to Professor S. Chandrasekhar for reading and making useful comments on some of the chapters, as well as for much general assistance and advice. I am grateful to Professor Carl Hansen, who read and commented on many chapters, and offered useful general suggestions; to Professor N. Baker, who also read and commented on several chapters; and to Professor M. Smith, who made helpful comments and suggestions, mostly about Chapter 17. The efforts and interest of Dr. C. G. Davis, who kindly brought to the writer's attention a number of references and conference proceedings which might otherwise have been missed, are also much appreciated. On the other hand, I offer sincere apologies in advance to those persons whose work has been insufficiently cited or referenced.

Useful discussions with Professor A. Weigert, Dr. M. Aizenman, and Professor P. Smeyers have resulted in a clarification of many concepts for the writer. I am especially grateful to Professor Smeyers for some expert help on some difficult points. I also appreciate the very proficient assistance of D. Schwank, R. Gross, and B. Carroll in attending to the seemingly infinite amount of work required in assembling a book of this kind. My gratitude also to Allen Wynne and Myrle Crouch for their kind assistance with some of the bibliographical material. Professor R. McCray was also most helpful, particularly with respect to some of the references on X-ray bursters.

I wish to thank my wife, Dr. J. B. Blizard-Cox, for much patience, encouragement, and moral support over the nearly four years and many absences required to write this book.

Discussions with numerous persons, in addition to those mentioned above, have influenced, either directly or indirectly, the final form of the book. Among these persons are T. Adams, G. L. Berggren, W. Brittin, T. Brown, R. Buchler, J. Castor, J. Christensen-Daalsgard, R. Christy, L. Cloutman, B. Cogan, A. Cox, E. F. Cox, W. R. Davey, D. Davison, R. Deupree, B. L. Dickerson, W. Dziembowski, D. Eilers, J. Faller, D. Fischel, W. Fitch, P. Flower, R. H. Garstang, M. Goossens, D. Gough, H.

Hill, S. Hill, N. Hoffman, D. Hummer, C. Keller, G. E. Langer, J. Latour, P. Ledoux, J. Lesh, D. Lind, J. M. Malville, J. McGraw, P. Melvin, B. Mihalis, D. Mihalis, G. Nelson, Y. Osaki, A. Phelps, R. Ross, H. Saio, E. Schmidt, M. Schwarzschild, R. Scuflaire, N. Simon, A. Skumanich, W. Spangenburg, W. Sparks, S. Starrfield, R. Stellingwerf, C. Sterken, R. Stobie, P. Stry, J. Toomre, H. Van Horn, G. Wallerstein, B. Warner, J. C. Wheeler, C. Whitney, D. Winget, C. Wolff, C. Zafiratos, and K. Ziebarth.

I am also especially grateful to the many students, too numerous to mention by name, who have suffered through my lectures, and whose questions have resulted in a sharpening of my understanding of certain points. This enhanced degree of understanding has presumably resulted in a clarification of a number of discussions in the book.

Sincere thanks are also extended to the numerous persons who have sent me preprints of their work over the past few years.

For their skillful preparation of the typescript and assistance with numerous other editorial matters, I am greatly indebted to Lorraine Volsky, Leslie Haas, and Gwendy Romey.

For their kind permission to reproduce material extensively, I am grateful to The Institute of Physics, The Astronomical Society of Japan, Astronomy and Astrophysics, The Astrophysical Journal, and Annual Reviews, Inc. I would also like to thank the Princeton University Press and its editorial staff for much patience and forbearance. Particular thanks are extended to Mr. Edward Tenner.

Much of the support for the time required to write this book has been provided by National Science Foundation Grants MPS72-05309, AST72-05039 A04, AST77-23183, AST76-01586, and AST78-42115, all through the University of Colorado. I am grateful also to Dr. Charles F. Keller, who made the facilities of the Los Alamos Scientific Laboratory available to the writer (support provided by the Energy Research and Development Administration), and I appreciate the interest of and many discussions with Dr. Arthur N. Cox, of that Laboratory, over a number of years.

August 15, 1979

I

PRELIMINARIES

1

Introduction

Pulsating stars are stars in which large-scale dynamical motions, usually including the entire star, and usually more or less rhythmic, are present. The simplest kind of such motion is a purely radial pulsation, in which the star maintains a spherical shape at all times, but changes its volume, as if it were breathing.

The study of pulsating stars constitutes a relatively small, but highly important, area of modern stellar astrophysics. The idea that certain types of variable stars owe their variability to periodic or cyclic expansions and contractions dates from the work of Shapley (1914), and was given a firm mathematical foundation by Eddington (1918a,b). Since then the "pulsation hypothesis" has gained wide acceptance. The study of pulsating stars, both theoretical and observational, has proved a powerful tool in the study of stellar structure and in other aspects of modern astrophysics. (Summaries of some of the early history of the pulsation theory have been given by Rosseland 1949, Chap. 1; Eddington 1926, Chap. 8; and Ledoux and Walraven 1958.) One of the more spectacular and far-reaching fruits of the observational study of one of the best-known types of pulsating stars, the *classical Cepheids,* is the famous *period-luminosity* relation (see §3.1). This relation provides the astronomer with one of the most basic "yardsticks" for the measurement of truly great astronomical distances, of the order of the mean separation between galaxies, and has played a crucial role in the establishment of the basic distance scale of the universe. In addition, the attempts to understand the cause and nature of stellar pulsations have served as a challenge to the theorist, and have provided some fascinating and, in some ways, unique applications of physical theory. Further discussion of the importance and significance of the study of pulsating stars will be found in §19.7.

Pulsating stars comprise only a subset of the wider class of *intrinsic variable stars.* These are stars whose variability arises from causes entirely *within* themselves, and not from geometric effects such as eclipses in binary stars; or to some external agency such as interaction with the interstellar medium or with circumstellar matter. The whole class of intrinsic variables includes many different kinds of objects, some of which, such as the *quasi-stellar objects,* are probably not stars in the usual sense of the term. (For recent reviews of these objects, see, for example, Burbidge and Burbidge 1967; Perry, Burbidge, and Burbidge 1978; Schmidt 1969; an updated, semi-popular account is given in H. Smith

1978). The intrinsic variables are usually divided into two broad groups, the *pulsating variables* and the *eruptive variables*. This monograph is concerned primarily with the former group: a brief survey of the types of stars included therein will be presented in Chapter 3. Among the eruptive variables are the spectacular novae and supernovae, which will not be discussed in detail in this book. Recent reviews of certain types of eruptive variables may be found, however, in Shklovsky (1968) and in Oke and Searle (1974) (supernovae); Payne-Gaposchkin (1957) (novae); Kukarkin and Parenago (1963), Payne-Gaposchkin (1954), Ledoux and Walraven (1958) (the whole class of intrinsic variables, including recurrent novae and nova-like stars); Mumford (1967) (dwarf novae); Robinson (1976), Warner (1976a) (cataclysmic variables in general); and Herbig (1962) (T Tauri stars). See also many of the papers in Kippenhahn, Rahe, and Strohmeier (1977). Unless we explicitly state otherwise, we shall always in this book mean "intrinsic variable star" when we use the term "variable star."

The most general definition of a variable star is that it is a star whose physical properties change with time. However, a more restricted definition is implied in normal usage: by variable stars is usually meant stars whose properties change appreciably at a rate fairly easily detectable by astronomers—during, say, a few seconds or fractions of seconds to a few years or decades.

The most obvious and most easily detectable distinguishing feature of a variable star is its apparent brightness: most such stars, in fact, are detected by their light variations. Other observable properties, such as spectral type or color, and radial velocity, usually also vary during the light variations. In the case of *pulsars* (for recent reviews, see, for example, Hewish 1970; Ruderman 1972, 1975; Ginzburg and Zheleznyakov 1975; F. G. Smith 1977; Taylor and Manchester 1977), it is the variable radio radiation, on time scales of a few seconds to a few hundredths of a second, by which these objects are generally detected. However, light variations, synchronized with the radio variations, have been detected in the Crab pulsar, NP-0532 (Cocke, Disney, and Taylor 1969). The light variations are synchronized with the X-ray pulses in some X-ray "pulsars" (e.g., Hiltner and Mook 1970; Lamb and Sorvari 1972; Davidson, Henry, Middleditch, and Smith 1972; Forman, Jones, and Liller 1972). On the other hand, in some cases, such as in spectrum or magnetic variables or in the "line profile variable B stars" (the "53 Persei stars") (M. Smith 1977; Smith and McCall 1978; M. Smith 1978, 1979a,b), the brightness may be almost constant in time, and some other property, such as spectral details or magnetic field strength, may betray the variability of the star (for example, Deutsch 1958; Sargent 1964; Ledoux and Renson 1966). For example, the eleven-year solar cycle makes the sun, strictly speaking, a variable star (not to mention the small-scale oscillations recently reported by Hill and collaborators [Hill, Stebbins, and Brown 1975; Brown, Stebbins, and Hill 1976; Hill 1978; and numerous papers in Hill and Dziembowski 1979]; see also the numerous references in Gough 1977c and, in relation to the whole question of the possible variability of the sun, White 1977 and Eddy 1978). Also, the "X-ray bursters" exhibit variations on scales of minutes (see, e.g., Gursky 1977; Lewin 1977; Lewin *et al.* 1977; Lewis and van Paradijs 1979), while "γ-ray bursts" are characterized by time scales for variability of ~0.1–100s (e.g., Strong, Klebesadel, and Evans 1975; Klebesadel and Strong 1976; Cline and Desai 1976; Fishman, Watts, and Derrickson 1978). (For a recent review of X-ray sources in general, see Ostriker 1977.)

The variations associated with the pulsating variables may be periodic or cyclic, semi-regular, or irregular. The corresponding time scales range all the way from a few tens of seconds to a few years. It is, of course, possible that time scales lying outside this range exist; but then the problem of detection might become somewhat difficult.

The discovery of periodic or cyclic variables came relatively late in the whole history of astronomy. Apparently, the first authenticated discovery of such a variable star was that of *o* Ceti (Mira), a Long Period Variable (see Chap. 3), by Fabricius in 1596 (Ledoux and Walraven 1958). A few supernovae, such as the Crab supernova of 1054, Tycho's supernova of 1572, and Kepler's supernova of 1604, had been recorded, but these belong to the class of eruptive variables. Before the end of the eighteenth century, only sixteen variable stars had been discovered, two of which were later found to be eclipsing binaries and five of which were novae (Campbell and Jacchia 1941). Two of these were classical Cepheids: δ Cephei, the prototype of this kind of star (see Chap. 3), discovered by John Goodricke in 1784; and η Aquilae, discovered by Edward Pigott also in 1784 (Campbell and Jacchia 1941). The total number of intrinsic variable stars now known in the Galaxy is some 25,000, of which over 20,000 are listed in the catalog of Kukarkin *et al.* (1969). Over ninety per cent of these are pulsating variables. The total number of such variables in the entire Galaxy is estimated to be $\sim 2 \times 10^6$ (Kukarkin and Parenago 1963). However, since the total number of stars in the whole Galaxy is some 10^{11}–10^{12}, it follows that only about one star in 10^5–10^6 is a pulsating star. Stellar pulsation is therefore quite rare, on the whole, among stars. Nevertheless, it is highly important in astrophysics, as will be seen in later portions of this book. (The recent discovery of the variable white dwarfs, or "ZZ Ceti stars" [McGraw 1977; Robinson and McGraw 1976a,b; Robinson, Nather, and McGraw 1976; Nather 1978], may cause the above numbers to be revised somewhat.)

In Chapter 2 we introduce, primarily for orientation, some important time scales for stars. In Chapter 3 we present a brief survey of empirical information on pulsating variables. Since the observational literature on variable stars is quite extensive, and since a number of good reviews of this subject exist (to be referenced there), we shall only mention the particularly important points.

In Chapter 4 we shall summarize some basic theoretical information which will frequently be referred to in later parts of the book. We shall here and throughout, except when explicitly stated otherwise, employ nonrelativistic mechanics and Newtonian gravitation theory. The neglect of special relativity in the consideration of pulsating stars is well justified in most cases because the relevant velocities are generally small compared to the speed of light. The neglect of general relativistic gravitation theory is also well justified for most pulsating stars because the gravitational fields are usually very weak; equivalently, the mean radii of most kinds of pulsating stars are much larger than their Schwarzschild radii $R_s = 2GM/c^2$, where G is the gravitation constant, M is the mass of the star, and c is the velocity of light. Examples of stellar objects in which these approximations are not justified are dense white dwarfs (see, e.g., Misner, Thorne, and Wheeler 1973); neutron stars; "supermassive stars," if they exist (see, e.g., Wagoner 1969); and collapsed stars, or "black holes" (see, e.g., Ruffini and Wheeler 1971; Penrose 1972; Zeldovitch and Novikov 1971, Chap. 11; Thorne 1967a; Eardley and Press 1975). These developments in the general relativistic theory of pulsating stars will not be considered in detail in this book (however, see §19.5 and Cox 1974a).

The linear theory of stellar oscillations is discussed in Chapter 5. This theory has played a vital role in the development of our present understanding of pulsating stars. Until recent years this theory formed the basis of nearly all theoretical discussions of pulsating stars, even though it was well known that pulsations of actual stars are generally of a large enough amplitude that nonlinear effects are certainly important. The linear theory is nevertheless extremely useful, in part because its relative mathematical simplicity facilitates understanding in physical terms of some of the complicated phenomena involved. This theory is also useful if we believe that at least some types of actual stellar oscillations arose because the star was at one time unstable against infinitely small oscillations. The fact that most of the recognized types of pulsating stars occupy more-or-less well defined regions on the Hertzsprung-Russell (H-R) diagram (see Fig. 3.1 below) suggests a relation between linear instability, which depends (presumably) on the "static" characteristics of a star, and actual stellar oscillations.

Part II will be devoted exclusively to purely radial motion, which will receive a fair amount of emphasis in this book. There are two main reasons for this relatively heavy emphasis. First, this is the simplest kind of motion for spherical stars. It is therefore relatively tractable mathematically, and many of its aspects can be understood physically. Second, most actual pulsating stars appear, fortunately, to be undergoing predominantly just this simple kind of motion.

Part III will be devoted primarily to the theory of nonradial stellar oscillations.

In Chapter 19, certain complicating factors, such as rotation, viscosity, magnetic fields, thermal imbalance, and general relativity in stellar pulsations (both radial and nonradial) will either be dealt with briefly, or at least mentioned, with appropriate references to the literature. A few other miscellaneous topics, such as secular stability of stars, will be referred to, and some comments will be made about the significance of stellar oscillation theory to other areas of astrophysics.

Other recent reviews of pulsating stars and pulsation theory have been provided by Payne-Gaposchkin (1951, 1954); Ledoux and Walraven (1958); Ledoux and Whitney (1961); Ledoux 1963, 1965, 1974, 1978); Zhevakin (1963); Christy (1966a, 1967, 1968, 1969a,b, 1970); J. P. Cox (1967, 1974a, 1975, 1976a, 1979); A. N. Cox and J. P. Cox (1967); King and Cox (1968); J. P. Cox and Giuli (1968; Chap. 27); Iben (1971a); Hoffmeister (1971); Percy (1975); Glasby (1975); and Kukarkin (1976). Earlier reviews are those of Eddington (1926, Chap. 8); and Rosseland (1949). Other useful recent collections of papers on pulsating stars are the proceedings of the Third I.A.U. Colloquium on Variable Stars (Bamberg, Germany, 1965), *The Position of Variable Stars on the* H-R *Diagram;* the proceedings of the Fifth I.A.U. Colloquium on Variable Stars (Bamberg, Germany, 1971), *New Directions and New Frontiers in Variable Star Research;* Detre (1968); Philip (1972); Strohmeier and Knigge (1972); Demarque (1973); Ledoux, Noels, and Rodgers (1974); Fischel and Sparks (1975); Fitch (1976a); A. N. Cox and Deupree (1976); Kippenhahn, Rahe, and Strohmeier (1977); Fischel, Lesh, and Sparks (1978); and Hill and Dziembowski (1979). A monograph on nonradial stellar oscillations has recently been published by Unno, Osaki, Ando, and Shibahashi (1979).

Some works dealing with wave phenomena in general have been found very helpful to the author, and may also be helpful to the reader. Among these are Morse (1936), Greenspan (1968), Tolstoy (1973), Lighthill (1978), and Main (1978).

2

Some Important Time Scales

In this chapter we shall consider, for orientation, some important stellar time scales and their rough orders of magnitude. Time scales lying in the general range of a few seconds to a few years will be of particular interest in connection with pulsating stars.

2.1. THE PULSATION PERIOD

The first and most relevant of these time scales for pulsating stars is the pulsation period Π of the fundamental mode of purely radial oscillations. While accurate methods of calculating Π will be considered in later chapters, it is instructive to consider first some simple, approximate methods of estimating its value that give nearly the same results as do the more elaborate methods. To order of magnitude, this value of Π also applies to the lower pressure and gravity modes of nonradial oscillations of somewhat realistic stellar models (see Chapter 17).

Perhaps the most general of these simple methods is that described by Cox (1967). This method uses the fact that stellar pulsations (at least those of low modes) can be regarded, approximately, as a kind of "long-wave" acoustics (wavelength of the "sound wave" of the order of or larger than the dimensions of the system), as is shown by Ledoux and Walraven (1968, Sect. 60; see also §8.9 of this book). The pulsation period Π then ought to be of the order of the time required for a sound wave to propagate through the mean or equilibrium diameter of the star. A general expression for the Laplacian (adiabatic) sound speed (see, e.g., §5.5), averaged in some suitable manner over the entire star in its equilibrium state, can be obtained from the virial theorem (see, e.g., Cox and Giuli 1968, Chap. 17). This expression is essentially independent of the material properties of the star.

A crude but roughly equivalent method of obtaining an expression for the mean sound speed is the following. The equation of hydrostatic equilibrium is used (see, e.g., Chapter 4), and all quantities therein are regarded as average, or representative, values throughout the star. Substituting these values into the expression (eq. [5.38]) for the sound speed then yields the desired result.

This procedure shows that, very nearly, $\Pi(\bar{\rho})^{1/2}$ = *constant*. This is the famous period-mean density relation, which seems to be satisfied by most types of pulsating stars (see, e.g., Chapter 3). According to this relation, a large, tenuous star will have a longer period than will a small, compact star.

The details of the above considerations will reveal that the *constant* in the above expression contains the factor $\Gamma_1^{-1/2}$, where Γ_1 is one of the adiabatic exponents (see Chapter 4), here assumed constant. A more careful derivation of the above expression shows that this factor should actually be replaced by the factor $(3\Gamma_1 - 4)^{-1/2}$, which arises from the spherical symmetry and the variations of gravity which are not fully taken into account in the above considerations. This latter factor may cause Π to be considerably larger than the value given in the above expression if Γ_1 is close to 4/3; this will be the case in relativistic white dwarfs or neutron stars, or in very massive stars where radiation pressure is more important than gas pressure. For $\Gamma_1 = 4/3$, $\Pi = \infty$; and the star is dynamically unstable if $\Gamma_1 < 4/3$.

Another simple, approximate expression for the pulsation period of a star, which, moreover, yields the correct factor $(3\Gamma_1 - 4)^{-1/2}$, is the following. Suppose that the entire mass M of the star is concentrated in a point at the center, and that the stellar surface, lying at a mean distance R from the center, is represented by a thin, spherical shell of this radius, having a mass m small compared with M, and offering no resistance, other than inertia, to changes in its radius (that is, the shell is completely compressible, inviscid, and has zero surface tension). The entire volume within the shell is filled with a uniform, massless gas whose only function is to supply pressure to support the shell against gravity, and the shell is surrounded by vacuum (pressure $P = 0$). If r is the instantaneous radius of the membrane, its equation of motion is

$$m\ddot{r} = 4\pi r^2 P - \frac{GMm}{r^2}, \tag{2.1}$$

where a dot denotes the time derivative, P denotes the (spatially constant) gas pressure inside the membrane, and G is the constant of gravitation. We now assume small, adiabatic oscillations about the hydrostatic equilibrium state ($\ddot{r} = 0$); that is, $\delta P/P = \Gamma_1\, \delta\rho/\rho$, where δP, for example, denotes the departure of the pressure from its equilibrium value. Linearizing eq. (2.1) (further details regarding linearization may be found in Chap. 5) and assuming a time dependence of the form $e^{i\sigma t}$, it is a simple matter to show that the angular pulsation frequency σ is given by the relation

$$\sigma^2 = (3\Gamma_1 - 4) \cdot \frac{GM}{R^3} = (3\Gamma_1 - 4) \cdot \tfrac{4}{3}\pi G\bar{\rho}, \tag{2.2}$$

which defines the mean density $\bar{\rho}$. We then obtain the following expression for the pulsation period $\Pi = 2\pi/\sigma$:

$$\Pi = 2\pi/[(3\Gamma_1 - 4) \cdot \tfrac{4}{3}\pi G\bar{\rho}]^{1/2} . \tag{2.3}$$

Interestingly enough, this is precisely the expression for the fundamental pulsation period of purely radial pulsations of the homogeneous (constant-density) model of given Γ_1 and $\bar{\rho}$.

By carefully following through the derivation of eq. (2.2), it is easy to discover the origin of the "magic" critical number 4/3 (see above remarks concerning dynamical instability). Write the number as (2 + 2)/3. One of the 2's comes from the inverse square character of Newtonian gravitation (which is the only kind we consider in this book unless we specifically state otherwise); the other 2 comes from the fact that the total pressure force on a sphere of radius r varies as r^2. The 3 comes from the three-dimensionality of physical space: the volume of a sphere of radius r varies as r^3.

It is customary to write the period-mean density relation in the form

$$\Pi(\bar{\rho}/\bar{\rho}_\odot)^{1/2} = Q, \tag{2.4}$$

where $Q \sim (G\bar{\rho}_\odot)^{-1/2}$ ($\bar{\rho}_\odot = 1.41$ gm cm^{-3} = mean density of the sun) is the "pulsation constant." It is not actually a constant, as its value depends, generally only weakly, on Γ_1 and on the structure of the star. Accurate calculations show that, for the fundamental radial mode and for $\Gamma_1 = 5/3$,

$$0\overset{d}{.}03 \lesssim Q \lesssim 0\overset{d}{.}12, \tag{2.5}$$

while a representative value is $Q \approx 0\overset{d}{.}04$. Fitting formulae, giving Q as a function of stellar parameters (mostly mass and equilibrium radius), have been provided by Cox, King, and Stellingwerf (1972) and Faulkner (1977b). Since Q is the period that the sun would have if it were pulsating, we see that its period would be of the order of an hour. Observations (uncertain as they are!) of many variable stars yield values in the general range

$$0\overset{d}{.}02 \lesssim Q \lesssim 0\overset{d}{.}11, \tag{2.6}$$

in reasonable agreement with theory.

The pulsation periods to be expected of known kinds of stars can be estimated on the basis of the period-mean density relation (2.4). Considering stars of mean densities lying between those of moderately dense white dwarfs, $\bar{\rho} \sim 10^6$ gm cm^{-3}, and those of tenuous red supergiants, $\bar{\rho} \sim 10^{-9}$ gm cm^{-3}, we obtain periods lying in the approximate range

$$3 \text{ seconds} \lesssim \Pi \lesssim 1000 \text{ days}, \tag{2.7}$$

which nicely spans the range of periods observed for most types of periodic or cyclic intrinsic variables (see Chapter 3). This rough agreement provides good general support for the pulsation theory of variable stars. There are stronger and more specific arguments in favor of this theory that are, however, outside the scope of this book (see, e.g., Eddington 1926, Chap. 8). We may note that, had neutron stars (which are probably represented by the pulsars), with mean densities $\bar{\rho} \sim 10^{15}$ gm cm^{-3}, been included in the above selection of stars, the lower limit of the above period range would have been a few milliseconds. While there is as yet no direct evidence that neutron stars are pulsating, some of the finer details, with time scales of a few milliseconds, observed in the pulsed radio radiation from pulsars (e.g., Taylor and Huguenin 1971), could well be a result of pulsations.

2.2 THE "FREE-FALL" TIME

The "free-fall," or "dynamical," time scale, t_{ff}, is the characteristic time associated with dynamical collapse, or with the orbital motion of a satellite circling the parent body very close to its surface; t_{ff} is also the characteristic time for a significant departure from hydrostatic equilibrium to alter the state of a star appreciably.

A simple estimate of the order of magnitude of t_{ff} can be obtained by calculating the time required for a unit mass to fall freely through a distance of the order of R (stellar radius) under the influence of a (constant) gravitational acceleration equal to the surface gravity GM/R^2 of a star of mass M. This procedure yields

$$t_{ff} \sim (G\bar{\rho})^{-1/2} \tag{2.8}$$

(other approximate methods of obtaining eq. [2.8] are presented in, e.g., Cox and Giuli 1968, Chap. 1). Equation (2.8) shows that, aside from numerical factors generally of order unity, t_{ff} is of the order of the pulsation period Π. This well-known result is a consequence of the fact that the characteristic velocities associated with low-order, largely radial pulsations (the sound speed) and with dynamical processes (e.g., free fall or orbital speeds) or low-order, nonradial gravity oscillations are all determined, via the virial theorem, essentially by the gravitational energy of the star.

2.3 THE KELVIN TIME

The "Kelvin time," t_K, is essentially the "relaxation time" for departures of a star from *thermal* equilibrium, that is from balance between energy generated by thermonuclear reactions in the stellar interior and energy lost by radiation, both photonic and neutrinic, through the stellar surface. The order of magnitude of t_K can be estimated as follows. Let $E_{\rm th}$ be the total internal (thermal) energy of a star and L the *luminosity* (net rate of loss of energy through the surface) of the star. Then we have, to order of magnitude,

$$t_K \sim E_{\rm th}/L. \tag{2.9}$$

However, $E_{\rm th}$ can be related to Ω, the gravitational energy of the star, by the virial theorem. This theorem can be written in the following general form, for a self-gravitating system in hydrostatic equilibrium that possesses no mass motions (for example, turbulence, rotation, pulsation) and no magnetic fields, and for which the pressure vanishes on the surface:

$$3 \int_V P dV = -\Omega, \tag{2.10}$$

where P is the total pressure and the integration is extended over the entire volume V of the star, and

$$\Omega \equiv - \int_M \frac{Gmdm}{r} \equiv -q \frac{GM^2}{R} \tag{2.11}$$

is the gravitational potential energy of a spherical star. Here q is a dimensionless constant whose value depends on the mass concentration of the star but is of order unity for chemically homogeneous stars, and the integration is extended over the entire stellar mass M. If we assume that the pressure is supplied by a simple, perfect, nonrelativistic gas, then we have $E_{\rm th} = (3/2) \int_V PdV$, which yields the simple form $E_{\rm th} = -(1/2)\,\Omega$ of the virial theorem. Using this last result in eqs. (2.9) and (2.11) for Ω, and taking $q \approx 3/2$, we obtain

$$t_k \sim 3/4 \frac{GM^2}{LR} \sim 2 \times 10^7 \frac{M^2}{LR} \text{ years}, \tag{2.12}$$

where L, M, and R are in solar units. The Kelvin time t_K is also the time that would be required for a star to contract from infinite dispersion to its present radius if L were to remain constant during the entire contraction.

The Kelvin time is normally not of immediate concern as far as the *periods* of pulsating stars are concerned. However, as we shall see (Chapter 9), it is relevant in connection with growth rates, or e-folding times, for the growth or decay of pulsations.

A useful dimensionless quantity is the ratio of the free-fall time ($\sim$ pulsation period) to the Kelvin time:

$$\frac{t_{ff}}{t_k} \sim \frac{\Pi}{t_K} \sim \frac{LR^{5/2}}{G^{3/2}M^{5/2}} \sim 10^{-12}\frac{LR^{5/2}}{M^{5/2}}, \tag{2.13}$$

if L, M, and R are in solar units. It is thus seen that for stars not differing greatly from the sun, the pulsation period is many orders of magnitude smaller than the Kelvin time.

2.4 THE "NUCLEAR" TIME

The "nuclear" time scale, t_{nuc}, is only of indirect interest in connection with pulsating stars, but knowledge of its value is useful for orientation. This time scale is, loosely speaking, the time required for the properties of a star to change appreciably as a result of nuclear evolution (changes in internal chemical composition due to nuclear transmutations). For a hydrogen-burning star we may make use of the fact that an amount of energy $\sim 0.007\, c^2 \sim 6 \times 10^{18}$ ergs (c = light velocity) is released per gram of hydrogen that is fused into helium. Assuming that $\sim 10\%$ of the mass of the star is available for this fusion, we obtain

$$t_{\text{nuc}} \sim 10^{10}\, M/L \text{ years}, \tag{2.14}$$

where again M and L are in solar units. It is seen that, normally,

$$t_{\text{nuc}} \sim 10^3\, t_K.$$

3

Some Observational Considerations

In this chapter we shall discuss briefly certain topics related to variable stars that are of primarily observational interest. Further details of the observed characteristics of the numerous types of variable stars can be found in the comprehensive and detailed discussions of, for example, Ledoux and Walraven (1958), Payne-Gaposchkin (1951, 1954), Payne-Gaposchkin and Gaposchkin (1963), Kukarkin and Parenago (1963), Hoffmeister (1971), Strohmeier and Knigge (1972), Kukarkin (1976), Pel (1978), and in certain of the references given in Chapter 1. See also the summary in J. P. Cox (1974a). Because of the vastness of the literature on this subject and the rate at which discoveries are being made, no claim is made for completeness for the most up-to-date observational results. Interested persons are advised to check the current astronomical and astrophysical literature. We have not included the *spectrum and magnetic variables* (see, e.g., Ledoux and Renson 1966), the *flare (UV Ceti)* stars (see, e.g., Lovell 1971), nor the *T Tauri* stars (see, e.g., Herbig 1962, 1978) among the pulsating variables, because it is not clear that their characteristics are necessarily directly related to pulsations. We have also not included the quasi-stellar objects, for reasons given earlier (Chapter 1); nor the pulsars, as their main observed characteristics are generally believed to be a result of *rotation* rather than pulsation (see, e.g., Hewish 1970; Cameron 1970; Ruderman 1972; Canuto 1977). We have also not included the recently observed oscillations of some of the cataclysmic variables (see, e.g., Warner and Robinson 1972; Patterson, Robinson, and Nather 1976; Warner 1976a,b; Robinson 1976; Stiening, Hildebrand, and Spillar 1979), as the nature and cause of these oscillations are unknown. Some of the material in this chapter has been borrowed from J. P. Cox (1974a).

We have summarized in Table 3.1 some of the properties of most of the recognized types of pulsating variables. In Figure 3.1 are shown the locations of some of these various types, as well as some others, on a Hertzsprung-Russell (H-R) diagram.[1]

The classical Cepheids and W Virginis stars are sometimes called

[1]This, as well as other items of astronomical nomenclature and general information, may be found in any text on introductory astrophysics, for example, Aller (1963), Swihart (1968), Unsöld (1977), Smith and Jacobs (1973), Rose (1973), or Harwit (1973).

TABLE 3.1

The Pulsating Variables*

Kind of Star	Range of Periods	Characteristic Period	Population Type	Range of Spectral Types	Absolute Magnitude (M_V)
RR Lyrae	1.5–24 h	0.5 d	II	A2–F2	0.0 to +1.0
Classical Cepheids	1–50 d	5–10 d	I	F6–K2	−0.5 to −6
W Virginis Stars	2–45 d	12–20 d	II	F2–G6(?)	0 to −3
RV Tauri Stars	20–150 d	75 d	II	G, K	~ −3
Red Semi-Regular Variables	100–200 d	100 d	I and II	(K), M, R, N, S	−1 to −3
Long Period Variables	100–700 d	270 d	I and II	M_e, R_e, N_e, S_e	+1 to −2
β Cephei Stars					
("β Canis Majoris Stars")	4–6 h	5 h	I	B1–B2	−3.5 to −4.5
Dwarf Cepheids and δ Scuti Stars	1–3 h	2 h	I	A2–F5	+2 to +3
Beat Cepheids	1–7 d	2 d	I(?)	F0–G0(?)	−1 to −3(?)
Variable White Dwarfs					
("ZZ Ceti Stars")	200–1000 s	500 s(?)	I(?)	A5–F5(?)	+10 to +15(?)

*Adapted from Table 1 of Cox (1974a), courtesy of the Institute of Physics.

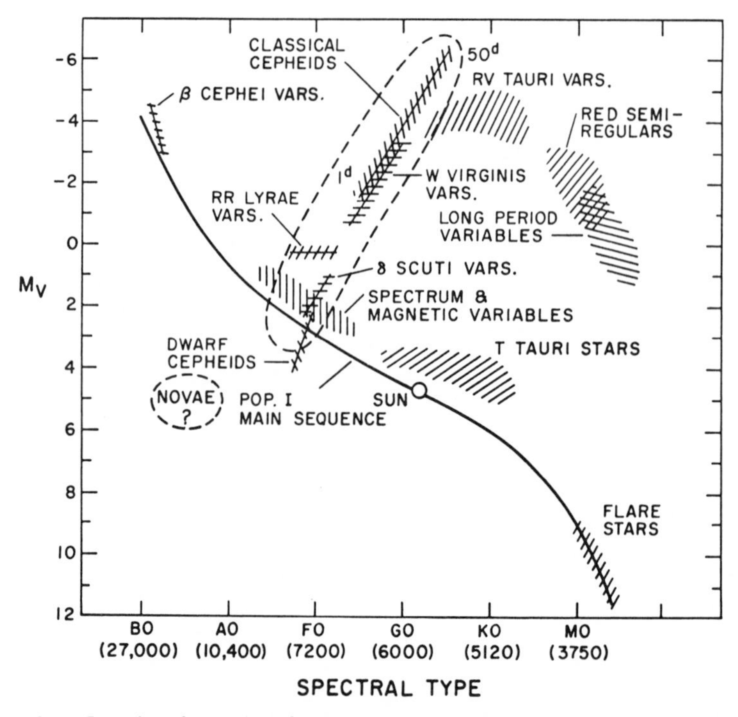

Figure 3.1. Location of a number of various types of intrinsic variables on the Hertzsprung-Russell diagram. From Figure 1 of J. P. Cox (1974a), courtesy of the Institute of Physics.

collectively "Cepheids," being regarded as counterparts, distinguished by Population type, of a single kind of star. The red semi-regular variables and the Long Period Variables are sometimes collectively referred to as the "red variables."

The group of stars in the upper part of Table 3.1 (the RR Lyrae variables, Cepheids, RV Tauri stars, and the red variables) is sometimes referred to as the *Great Sequence.* Note that, as one descends this part of the table, the characteristic periods become progressively longer and the stars become progressively redder (cooler).

Such a general correlation between period and spectral type (or color) can easily be shown to be just what would be expected for a case of radial pulsations. We may say, alternatively, that large stars have relatively long periods (long sound travel times through their diameters): such stars are, for given luminosity, relatively cool. Therefore, increasing periods and increasing coolness tend to go together.

The significance of the nearly vertical oval region shown by dashed lines in Figure 3.1 is that most pulsating stars lying in this region (the RR Lyrae variables, classical Cepheids, W Virginis variables, and dwarf Cepheids and δ Scuti variables) are thought to owe their instability to a common physical mechanism (second ionization of helium in the envelope), the details of which will be discussed in some detail in Chapter 10. This oval region is sometimes referred to loosely as the "instability strip" or "instability region." It has been suggested by Van Horn (1978), Nather (1978), and Hansen (1979) that this instability region might even include the variable white dwarfs (or "ZZ Ceti" stars, see below). However, see J. P. Cox and Hansen (1979).

For recent reviews of some of the short-period variables, see Petersen (1976), McNamara and Feltz (1978), and Breger (1979).

If one examines the frequency distribution of periods for the pulsating variables in the Galaxy, corrected for selection effects, one finds more or less well-defined peaks at the characteristic periods for the various kinds of pulsators listed in Table 3.1 (e.g., Payne-Gaposchkin 1954, p. 17). This fact suggests that the classification of pulsating stars into distinct types has some basis in reality.

The most common kind of pulsating variable, in terms of numbers per unit volume of space, is found to be, at least in the part of the Galaxy in the vicinity of the sun, the recently discovered variable white dwarfs (the "ZZ Ceti stars," McGraw 1977, Nather 1978). They appear to outnumber all other types of variables stars by a considerable factor ($>10^2$?).

3.1 CLASSICAL CEPHEIDS AND THE PERIOD-LUMINOSITY RELATION

Because of the importance of the classical Cepheids and their role in establishing the basic distance scale of the universe, through the famous period-luminosity relation, we devote here a special section to this type of variable star.

The prototype of this kind of star is δ Cephei, with a period of $5\overset{d}{.}366$ (Kukarkin *et al.* 1969, 1974, 1976). Polaris is another classical Cepheid, although the light variations are small ($<0\overset{m}{.}1$). Classical Cepheids are yellow giants and supergiants, and are therefore highly luminous (see Table 3.2 below) and visible, if not dimmed by interstellar extinction, at great distances. Classical Cepheids have been observed in about thirty external galaxies.

The periods of classical Cepheids are nearly all confined to the range 1^d–50^d, but a few Cepheids in the Large Magellanic Cloud have periods approaching 100^d; periods in the Small Magellanic Cloud extend up to about 200^d (Payne-Gaposchkin and Gaposchkin 1965). (The classical

Cepheid in the Galaxy with the longest known period is BP Her, with a period of $83^{d}.1$, according to Makarenko 1972.)

About 700 classical Cepheids are known in the Galaxy (Payne-Gaposchkin and Haramundanis 1970), and they are all closely confined to the Galactic plane and partake of the rotation of the Galaxy. They are extreme Population I objects. Because of their confinement to the Galactic plane, they are heavily obscured and reddened by interstellar dust. They are all too distant for their distances to be measured by the usual direct methods (for example, by triangulation). Hence, until recently, the only way to determine the distances of Cepheids was to use statistical methods based on the solar motion relative to the nearby stars. These methods do not always yield very accurate or reliable results. Since the mid-1950's, however, some thirteen classical Cepheids have been discovered in galactic (open) clusters (for the history of these findings, see Fernie 1969). These discoveries have made possible more accurate determinations of the distances of Cepheids (see, e.g., Kraft 1961; Sandage and Tammann 1968, 1969, 1976a,b and references therein; Geyer 1970; Schaltenbrand and Tammann 1970; Pel 1978), and hence of the zero point of the period-luminosity relation (see below).

Some properties of classical Cepheids in the Galaxy are summarized in Table 3.2 (from Cox 1974a). The masses given in this table are only estimates, based on recent stellar evolution calculations with conventional masses. These masses may be only upper limits since actual Cepheid masses may be somewhat smaller than is indicated by the evolutionary calculations (see §19.7). Unfortunately, reliable empirical masses are not available for any Cepheids, since most Cepheids are either single or members of such widely separated binaries that reliable orbital elements, and hence masses, cannot be obtained (see, e.g., Latyshev 1969; Abt 1959).

The light curves of classical Cepheids are skew symmetric and highly

TABLE 3.2

Properties of Galactic Classical Cepheids*

	Range	
Property	From	To
Period (Π)	1^{d}	50^{d}
Mean Luminosity (L)	300 $L_{\odot}$	26,000 $L_{\odot}$
Median Spectral Type	F5	G5
Mean Radius (R)	14 $R_{\odot}$	200 $R_{\odot}$
Mass (M)	$\lesssim 3.7\ M_{\odot}$	$\lesssim 14\ M_{\odot}$

*Subscript ⊙ denotes solar values.

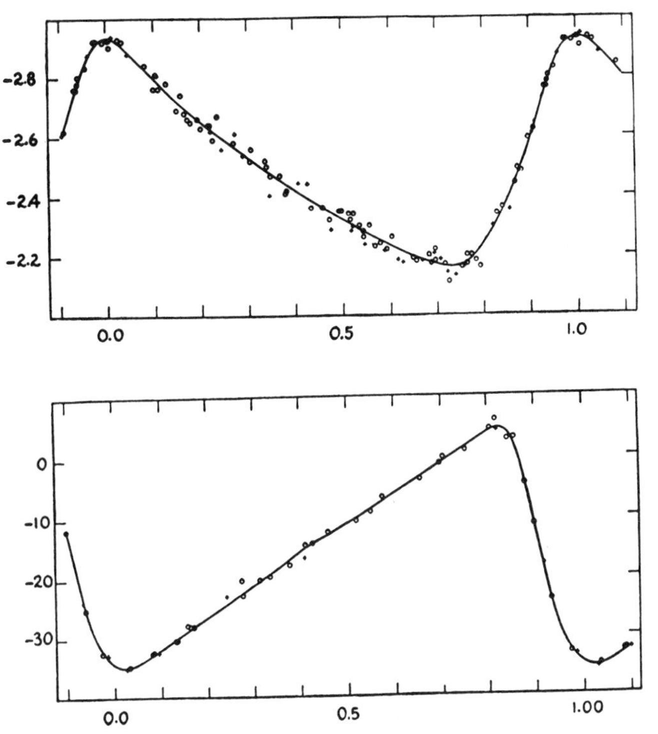

Figure 3.2. Light curve (upper figure) and radial velocity curve (lower figure) (astronomical sign convention) for δ Cephei. Abscissae are phase, and the ordinate of the light curve is apparent magnitude (arbitrary zero point). The ordinate of the velocity curve is in units of km s^{-1} and the zero-point is not corrected for the velocity of approach of the center of mass of the star, relative to the sun, of 16 km s^{-1} (from Goldberg and Aller 1943).

periodic, repeating faithfully over many periods (see Fig. 3.2). The total magnitude range (visual) is about 1^m; this increases slowly and somewhat erratically with increasing period.

The shapes of the light curves are correlated with the periods. This correlation is known as the *Hertzsprung relation,* and is illustrated, e.g., in Figure 4 of Cox (1974a) (this figure may also be found in Payne-Gaposchkin 1951). (For further discussion of the Hertzsprung relation, see, e.g., Payne-Gaposchkin 1961; Payne-Gaposchkin and Gaposchkin 1966). Note that a secondary hump often appears on the descending branch at periods between 7^d and 9^d. It should be noted, however, that the Hertzsprung relation is statistical in character, as there are many individual exceptions to it (see, e.g., Figure 5 in Cox 1974a).

The spectra and colors of Cepheids also change during the light variation. The spectra are earliest (closest to the O end of the spectral sequence) at maximum brightness, and the spectral changes are consistent with the changes in color. For δ Cephei, for example, the spectrum varies between F5 and G2 during the cycle (Kukarkin *et al.* 1969); this variation corresponds to a total change of about 1500°C in effective temperature. Most of the variation in brightness arises from the temperature variations; the radius variations are relatively small (fractional semi-amplitude around 0.05–0.10; see, e.g., Nikolov and Tsvetko 1972), and have only a minor effect on the light curves.

The radial velocity curves of classical Cepheids tend to be roughly mirror images of the light curves when the astronomical sign convention regarding radial velocities is used, as shown in Figure 3.2 (from Cox 1974a). If the velocity curve represents the motion of the stellar surface, then the phase relation between the light and velocity curves implies that the star is brightest when it is expanding through its equilibrium radius, and not when its radius is smallest, as might be expected from naive considerations. This retardation of maximum brightness behind minimum radius has been called the "phase lag discrepancy." The phase lag of maximum luminosity behind minimum radius would be about 90° if the light and velocity curves were sinusoidal. However, because of the skewness of the curves, the phase lag is actually considerably smaller than this, perhaps 0.1–0.2 periods. The physical cause of the phase lag has been clarified in recent years, and will be discussed further in Chapter 11.

The total velocity amplitude typically lies in the range 30–40 km s^{-1}, but increases slowly and erratically with increasing period Π, up to some 50–60 km s^{-1} for $\Pi \approx 30^d - 40^d$. Note that, as a result of foreshortening and limb darkening, the true velocity amplitudes are larger than the above values by a factor which is customarily taken to be 24/17 (see §3.4).

Perhaps the most important function of classical Cepheids for the astronomer is their use as powerful distance indicators; they are still the most important tool for establishing the basic distance scale of the universe (Sandage and Tammann 1971; Sandage 1972). This use is based on the well-known period-luminosity relation, which was discovered in 1912 by Leavitt of Harvard on the basis of Cepheids in the Small Magellanic Cloud (Pickering 1912). She found that the mean luminosity increases monotonically with increasing period, but she was unable to specify the zero point of the relation. The history of the determination of this zero point makes a fascinating chapter in the history of astronomy, and has been described by Baade (1956, 1963) and Fernie (1969). Suffice it to say here that the "doubling" of the size of the universe in the early 1950's was the result of the discovery by Baade, using the then newly operative 200 inch Palomar telescope, of an error in the earlier determinations of the zero point: this error had gone undetected for approximately forty years! The question of this zero point is certainly one of the most basic problems of observational astrophysics, because of its importance in the establishment of the distance scale for truly large astronomical distances.

Recent discussions of the empirical period-luminosity relation of classical Cepheids are due to Fernie (1967), Sandage and Tammann (1968, 1969, 1974, 1976a,b), Geyer (1970), van Genderin (1970), Schaltenbrand and Tammann (1970), Gaposchkin (1972), and Pel (1978). The Sandage and Tammann (1968) period-luminosity relation is shown in Figure 3.3 (from Cox 1974a). This is a composite relation, containing Galactic Cepheids as well as Cepheids found in other galaxies. These authors conclude that there is no reason to doubt that a "universal" period-luminosity relation exists for at least all the galaxies included in their study. However, the question of the universality of the period-luminosity relation is apparently not yet entirely settled (see, e.g., Fernie 1969; Gascoigne 1969).

Although the Sandage and Tammann period-luminosity relation is nonlinear, the departures from linearity are rather small. The central line

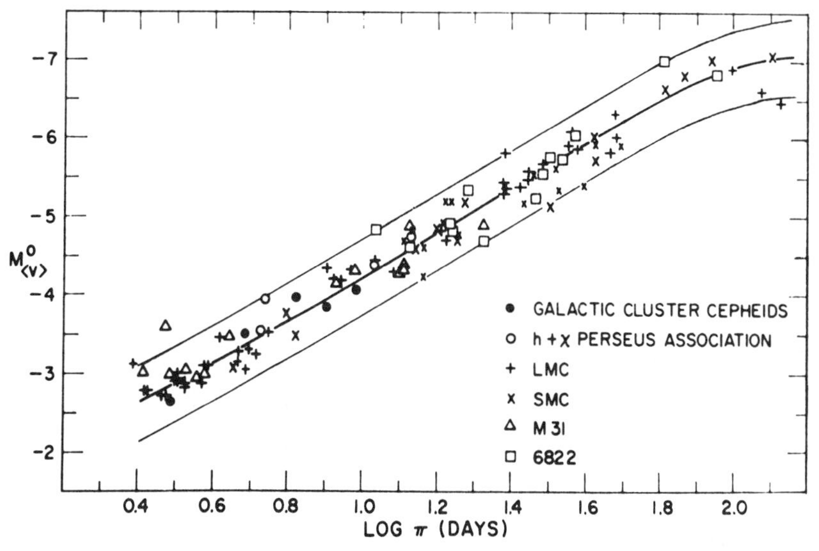

Figure 3.3. The composite period-luminosity relation of Sandage and Tammann (1968). The relation is based on Cepheids in our own Galaxy as well as in others, identified in the figure. The ordinate is absolute visual magnitude, and the superscript 0 means that the absolute magnitudes have been corrected for the effects of interstellar reddening and extinction (after Sandage and Tammann 1968). (Courtesy of *The Astrophysical Journal*, published by the University of Chicago Press, and of the authors.)

of the band shown in Figure 3.3 can be represented adequately over most of its length by the following relation:

$$M_{\langle V \rangle}{}^0 = -2.80 \log \Pi_d - 1.43 \qquad (0.4 \lesssim \log \Pi_d \lesssim 1.7), \qquad (3.1)$$

where the subscript $\langle V \rangle$ denotes an average over period, the superscript 0 means that the absolute magnitudes have been corrected for interstellar reddening and extinction, and the subscripts d mean that the periods are in days. Using the relations among $M_{\langle V \rangle}$, color $(B\text{-}V)$, T_e, and M_{bol} given by Kraft (1961), we may also write eq. (3.1) as

$$\log \left(\frac{L}{L_\odot}\right) = 1.15 \log \Pi_d + 2.47 \qquad (0.4 \lesssim \log \Pi_d \lesssim 1.7), \qquad (3.2)$$

where $L_\odot$ denotes the solar luminosity.

The scatter shown in Figure 3.3 about the central line is thought to be mostly intrinsic and a result of the finite width of the region of instability, and possibly of the presence of stars pulsating in different modes (see Chapter 10 and §19.7). The total intrinsinc width of the period-luminosity relation is approximately 1^m at a given period.

There appear to be certain differences between Cepheids in the Galaxy and in the Magellanic Clouds. Perhaps most striking are the differences in the period distributions of Cepheids in these systems. Thus, for example, in the Small Magellanic Cloud there are a great many Cepheids having $\Pi \lesssim 3^d$, whereas in the Galaxy very few Cepheids have periods as short as this. The Cepheids in the Large Magellanic Cloud are intermediate in this respect between those in the Galaxy and in the Small Magellanic Cloud (see, e.g., data summarized by Hofmeister 1967).

3.2. MORE RECENTLY RECOGNIZED TYPES OF VARIABLE STARS

Besides the types of variable stars referred to above, at least three additional types have recently received considerable attention in the astronomical and astrophysical literature. Moreover, recent discoveries have increased the membership of these types considerably. In view of these considerations, we present here brief descriptions of these types and some references to the literature.

3.2a. RAPID BLUE VARIABLES

The rapid blue variables are a somewhat loosely defined class of objects characterized by relatively blue colors and very short-period light variations. The variations are for the most part quite irregular, and sometimes described as "flickering." The periods of light variation range from some tens to some hundreds of seconds. The observational literature on these objects is summarized in Warner and Robinson (1972); Osaki and Hansen (1973a); Warner and Brickhill (1974); Brickhill (1975); Patterson, Robinson, and Nather (1977); Robinson and McGraw (1976a,b); Robinson, Nather, and McGraw (1976); Warner (1976b); Stiening, Hildebrand, and Spillar (1979); and Nather (1978). These objects are probably for the most part dwarf novae (which are thought to be close binaries; see, e.g., Kraft 1962, 1963; Warner 1976a; Patterson, Robinson, and Nather 1977) and variable white dwarfs (Richer and Ulrych 1974; McGraw and Robinson 1975, 1976; Robinson and McGraw 1976a,b; Robinson, Nather, and McGraw 1976; Van Horn 1978; Nather 1978; and Hansen 1979). In fact, it has been pointed out by McGraw (1977) that the variable white dwarfs are the most numerous of the variable stars and that they comprise a new class which he refers to as the "ZZ Ceti stars." The few rapid blue variables that do not belong to one of these two classes are of an uncertain nature (Lamb 1974; Bath, Evans, and Pringle 1974). Rapid rotation, for example, may be involved (e.g., Lamb 1974; Herbst, Hesser, and Ostriker 1974).

According to McGraw (1977) and Nather (1978), there are now known to be twelve apparently otherwise normal DA white dwarfs, with colors in the range $0.16 \leq B\text{-}V \leq 0.20$ (effective temperature $\sim 10^4$°K), which exhibit periodicities mostly in the range 200s–1000s. Further discussion of these stars can be found in McGraw (1977), Van Horn (1978), Nather (1978), Hansen (1979), J. P. Cox and Hansen (1979), and in some of the above references.

3.2b. BEAT CEPHEIDS

The beat (or "double-mode") Cepheids (they may not actually be Cepheids at all) consist of a small number of stars (according to Stobie 1977, eleven are known at present) whose light curves are not periodic. Nevertheless, these light curves can be decomposed into essentially only two (and, in one or two cases, three) periodic variations per star. The periodic light curves for each star, when added together, give back the original, observed, nonperiodic light curve. The above periodic variations are assumed to represent distinct pulsation modes, usually assumed to be the radial fundamental, first harmonic, and, when present, second harmonic. These modes are evidently for some reason simultaneously present in these stars; the modes interact with one another and produce "beats." The longest of these periods is normally between two and seven days, the next shortest period is about 70% of the longest period, and the third period, when present, is about 80% of the second period. It may be significant that the ratio of the second longest period to the longest in no case lies outside the range 0.70–0.71 (Stobie 1977, Simon 1979). These stars are located in the H-R diagram near the low-luminosity end of the Cepheid instability strip (the long, nearly vertical, oval region in Fig. 3.1). It is for this reason that they are called "Cepheids." According to Stobie (1977), nearly half the variables in the Galaxy in the appropriate period range are beat Cepheids.

Important information regarding certain aspects of stellar pulsation in general, and of these stars in particular, can be obtained from their multiple periods, largely because the period in a given mode is determined mostly by the mass and radius of the star (see, e.g., Cogan 1970; J. P. Cox, King, and Stellingwerf 1972). Hence, given two periods, both of the above quantities can in principle be determined. Discussions of these stars have been provided by Fitch (1970); Stobie (1970, 1972); Stobie and Hawardin (1972); Rodgers and Gingold (1973); Petersen (1973, 1974, 1978); Schmidt (1974); King, Hansen, Ross, and Cox (1975); Fitch and Szeidl (1976); A. N. Cox and Cox (1976); Cogan (1977, 1978a,b); Faulkner (1977a,b); Saio, Kobayashi, and Takeuti (1977); A. N. Cox, Deupree, King, and Hodson (1977); J. P. Cox (1978a); A. N. Cox, Hodson, and King (1979); see also the review papers by A. N. Cox (1978b) and J. P. Cox (1978, 1979) and the many references therein.

3.2c. LINE PROFILE VARIABLE B STARS

These stars, also called "53 Persei stars" by M. A. Smith (1979a,b), are for the most part main sequence or near main sequence stars of spectral classes mainly in the early and mid B's, say from O8 through B5. However, some of these stars are giants or supergiants, and they occupy those parts of the H-R diagram surrounding and in the general vicinity of the β Cephei stars. The line profile variable B stars are quite common, and most stars in the appropriate regions of the H–R diagram probably belong to this class (Smith 1979a).

These stars primarily exhibit temporal changes in the shapes of spectrum lines in a more or less periodic fashion, with periods ranging typically from a few hours to about two days (characteristically $\sim\frac{1}{2}$ day). These spectral line shape changes can be interpreted in terms of nonradial oscillations, in particular of g modes (M. A. Smith 1977; Smith and McCall 1978; M. Smith 1978, 1979a,b; Smith and Buta 1979; see also Chapter 17 for an explanation of the terminology). There is evidence of rather frequent changes in the character of the oscillations, with a given character persisting for, typically, about a month (Smith 1979b).

Light variations of ~0.1 magnitudes have also been detected in a few stars of this type (Buta and Smith 1979; Smith, Africano, and Worden 1979). Buta and Smith also present a rather nice discussion of the light variations accompanying nonradial stellar oscillations (see also Dziembowski 1977c).

3.3 EMPIRICAL DETERMINATION OF RADII OF PULSATING STARS

Most empirical methods of radius determinations for radially pulsating stars are based essentially on a method devised by Baade (1926) and Wesselink (1946, 1947). This method proceeds in principle as follows. If F_υ denotes the radiant flux (rate of radiation of energy per unit area) in some spectral band (normally ~700–1000 Å wide), and L_υ represents the corresponding luminosity of the star in the spectral band, then there is at each instant a simple relation between F_υ, L_υ, and the instantaneous radial distance R to the effective level in the atmosphere where the radiation in the given spectral band originates (R is approximately equal to the instantaneous stellar radius). The basic assumption underlying the Wesselink method is that F_υ is (for a given star) a function only of the *color,* measured by the color index B-V, of the star. Here B and V are apparent magnitudes, corrected for interstellar reddening, in broadband spectral regions centered, respectively, in the blue and visual (yellow-green) regions of the spectrum. If one now selects two phases during the pulsation cycle, say at times t_1 and t_2, at which the colors are equal, that is $(B\text{-}V)_1 = (B\text{-}V)_2$, then, according to the basic assumption, $F_\upsilon(t_1) = F_\upsilon(t_2)$. It then follows that

$$\frac{L_1}{L_2} = \left(\frac{R_1}{R_2}\right)^2, \tag{3.3}$$

where subscripts 1 and 2 refer to quantities at times t_1 and t_2, respectively. Hence, a measurement of the relative brightnesses of the star at two phases of equal color gives a measure of the *ratio* of the radii at these two phases.

On the other hand, if a *velocity curve* is available for the star, then the velocity, say $\dot{R}(t)$, of the stellar surface relative to the center of mass of the star can be obtained once the correction factor, say p, for converting from observed radial velocity $V(t)$ (relative to the center of mass) to $\dot{R}(t)$,

$$\dot{R}(t) = -pV(t), \tag{3.4}$$

is chosen. Assuming that the mean level in the atmosphere corresponding to the velocity curve is the same as the mean level referred to in the preceding paragraph, one than easily obtains $R_2 - R_1$ by integration of the velocity curve between t_1 and t_2.

Thus, knowing the difference $R_2 - R_1$ and the ratio R_1/R_2, both R_1 and R_2 can be obtained. By repeating the above procedure around a complete period, $R(t)$ can be found, from which a mean radius R around a cycle can be obtained.

Numerous discussions, applications, and refinements of the basic Wesselink method have been carried out (to list only a few of the more recent references): Rodgers (1957), Oke (1961a,b), Fernie (1968, and references therein), Christy (1968), Schmidt (1971), Parsons and Bouw (1971), Hill (1978), Parsons (1972), Karp (1975a), A. N. Cox and Davis (1975), Evans (1976), A. N. Cox (1979). Discussions of other methods of radius determination of pulsating stars can be found, for example, in Parsons and Bouw (1971), Parsons (1971), and Cogan (1978a).

These discussions seem to indicate that some of the assumptions underlying the usual Wesselink method are not always strictly valid (Schmidt 1971). In general, different methods of radius determination lead to different results. Evidently, radii of pulsating stars are rather poorly known, say to no better than some ten or twenty percent (e.g., Cox, King, and Stellingwerf 1972; Cogan 1978a). The anticipated direct angular diameter measurements of Cepheids by intensity interferometry (Hanbury Brown 1974) are eagerly awaited.

For an application of the Wesselink method to stars executing nonradial oscillations, see Balona and Stobie (1979).

3.4. LIMB DARKENING AND RADIAL VELOCITY

For a number of reasons, the variations in radial velocity as inferred from Doppler shifts of spectrum lines do not directly yield the actual velocity of expansion or contraction of the surface of a pulsating star. The most important correction that must be applied to infer the actual velocity of the stellar surface (relative to the center of mass of the star) from observed Doppler shifts is that which adjusts for projection effects and limb darkening. Differential motions of the layers in the regions of finite thickness where the absorption lines are formed, as well as projection effects and limb darkening, will, in general, introduce asymmetries in the observed line profiles and may thus affect the position of the "center of gravity" of the line and hence the observed radial velocity. We shall consider here only the projection effects and the effect of limb darkening, assuming that the emitting regions behave, at least from a kinematic standpoint, as a solid, spherical surface whose actual velocity, relative to the center of mass of the star, is $V \equiv \dot{R}$ (this is not the same V as the V of

§3.3). Effects of differential motions in the emitting-regions on the line profiles have been studied by, for example, Karp (1975a); Mihalas, Kunasz, and Hummer (1975); and Duval and Karp (1978).

We shall assume that the "center of gravity" of a spectrum line corresponds to a straightforward average, over the apparent stellar disk, of the line-of-sight velocity $V_r = V \cos\theta$, where V is the magnitude of the actual radial velocity of the emitting region, and θ is the angle between the vector **V** and the line of sight. In this average the *limb darkening function,* $\phi(\theta)$, must be taken into account. This function is so chosen that $\phi(\theta)$ is unity for $\theta = 0$ and decreases as θ increases. Assuming that $\phi(\theta)$ is of the form

$$\phi(\theta) = 1 - k + k\cos\theta \qquad (k \leq 1), \tag{3.5}$$

we easily obtain

$$\overline{V}_r = V \cdot (4 - k)/(6 - 2k). \tag{3.6}$$

For $k = 0$, $3/5$, and 1 we have, respectively, $\overline{V}_r/V = 16/24$, $17/24$, and $18/24$. These three values of k correspond, respectively, to *zero* limb darkening, the limb darkening predicted by the simple Eddington grey atmosphere (see, e.g., Woolley and Stibbs 1953; Mihalas 1978), and complete limb darkening. Evidently, the law of limb darkening is not very crucial to the value of $\overline{V}_r/V$, at least in the simple model we have adopted. This result is presumably why the value 24/17 is customarily used for $p = V/\overline{V}_r$ (see §3.3). Values of p obtained on the basis of more elaborate models have been reported, for example, by Parsons (1971, 1972), Karp (1975a), and A. N. Cox (1979b); but the more accurate values of p do not differ much from those given by the above simple models (for example, the value of p given by Parsons 1971 is 24/17.5).

It may be noted that, because different spectral regions generally show different amounts of limb darkening, different lines may on this account alone give conflicting radial velocities. Differential motions in the emitting layers may also cause discrepancies in the observed radial velocities as determined from different spectral lines (for example, from strong lines, formed at high levels in the atmosphere, as compared with weak lines, formed deep in the atmosphere). Generally, results from several different spectrum lines are averaged in the final radial velocity determinations.

The effects on spectral lines of *nonradial* oscillations of stars have been considered by Osaki (1971) and Dziembowski (1977c); see also M. A. Smith (1977), Smith and McCall (1978), M. A. Smith (1978), Buta and Smith (1979), and Smith, Africano and Worden (1979).

4

Some Basic Theoretical Considerations

In this chapter we shall present some of the basic equations of hydrodynamics and heat flow for a general fluid medium in fairly non-specific form and shall write them explicitly for several special cases of interest. We shall also derive an "integral" theorem that applies to a system as a whole: the energy theorem.

Detailed derivations will not, as a rule, be given, as these can usually be found in standard works on hydrodynamics and in astrophysical texts. We shall always, unless we specifically state otherwise, make certain physical simplifications. The most important of these are the following. We assume throughout that space is *Euclidean,* that is, *flat* (see, e.g., Misner, Thorne, and Wheeler 1973), so that we can always set up orthogonal Cartesian coordinate systems that span as large a region of space as we desire. We assume that mass is conserved; that is, that all material velocities are small compared with the speed of light *in vacuo*. Finally, *Newtonian* gravitation theory (as opposed to the gravitation theory of general relativity) is assumed.

As reference material for this chapter, any good texts on hydrodynamics and theoretical astrophysics should be suitable. The author has found the following especially helpful: Ledoux and Walraven (1958, §§43–54); Rosseland (1949); Landau and Lifshitz (1959); Milne-Thompson (1960); and Batchelor (1967).

4.1. EULERIAN AND LAGRANGIAN DESCRIPTIONS

We consider a general fluid medium and regard it as continuous. Thus, all elements of fluid that we deal with, even though they may be infinitesimals from the mathematical standpoint, are considered large enough to contain a large number of atoms or molecules. There are then two descriptions available for the behavior of the fluid: the *Eulerian* and the *Lagrangian*.

In the Eulerian description all physical properties of the fluid, such as *fluid velocity* $\mathbf{v}$, *total pressure* P, *mass density* ρ, *temperature* T, etc., are regarded as *field* quantities, i.e., as functions of position $\mathbf{r}$ and time t. The vector $\mathbf{r}$ may have components, for example, x, y, z (or, more generally, x_1, x_2, x_3) in a fixed Cartesian coordinate system, and $\mathbf{r}$ and t are independent variables. It is important to realize that the Eulerian position variable $\mathbf{r}$ is *not* the position of any particular fluid element, but is the position of the *point of observation*. This can clearly be varied arbitrarily, independently of the time t. (An analogy is a hole drilled anywhere in a board fence, with a fluid on the other side; $\mathbf{r}$ is the position of the *hole* in the fence.) Consequently, taking the time derivative of the Eulerian position variable $\mathbf{r}$ is in general meaningless without some qualification or explanation.

We are, however, often interested in following the motion of a particular fluid element as it moves about, and/or in observing the rate of change of some physical property, such as density, associated with the element. The derivative taken following the motion of a particular fluid element is sometimes called the *Stokes* (or *substantial* or *material*) *derivative* and will usually be denoted in this book by d/dt (sometimes the notation D/Dt is used in the literature). Noting that $\mathbf{r}$ and t are formally regarded as independent variables, we see by forming the Stokes derivative of some quantity $f(\mathbf{r},t)$ that the following operator relation is valid:

$$\frac{d}{dt} = \frac{\partial}{\partial t} + \mathbf{v} \cdot \nabla, \tag{4.1}$$

where ∇ denotes the ordinary gradient operator[1] and

$$\mathbf{v}(\mathbf{r},t) \equiv \frac{d\mathbf{r}}{dt} = \dot{\mathbf{r}} \tag{4.2}$$

is the fluid velocity. It is clear from the very concept of velocity that $\mathbf{r}$ in the last two terms in eq. (4.2) is no longer an Eulerian position variable, but rather, a *Lagrangian* position variable (see next paragraph) which gives the instantaneous position of the particular element we are following.

In the Lagrangian description the motion of a *given* fluid element is followed. Thus, the option of choosing the point of observation arbitrarily is not always available. In this description $\mathbf{r}$ denotes the position of *some particular* fluid element and consequently is no longer an independent variable. Rather, $\mathbf{r}$ is a function of time t and (in general, in three-dimensional space) of three identifying parameters, say a_1, a_2, a_3. If a_1, a_2, a_3 are the components of the position vector, say $\mathbf{a}$ (a_1,a_2,a_3), which was identical to $\mathbf{r}$ at, say, time $t = 0$, then $\mathbf{r} = \mathbf{r}(\mathbf{a},t)$, where $\mathbf{r}(\mathbf{a},0) = \mathbf{a}$. Using the Lagrangian description is similar to regarding the fluid as made up of a large number of individual mass points. The motions of these mass points are essentially what is considered, somewhat as in the "n-body" problem of classical mechanics. However, the a's do not need to be position coordinates in the Lagrangian description. In the case of one degree of freedom

[1]It can be shown that taking the gradient of some quantity is equivalent to taking the *covariant derivative* of that quantity; see, for example McConnell (1931).

(one space dimension) a might denote some physical property of the element, such as its temperature, at some prior time; or, for example, interior mass in the case of spherical symmetry (see Chap. 6 below; or any book on stellar structure, e.g., Cox and Giuli 1968, Chap. 1).

The connection between the two descriptions is provided by eq. (4.2). If $\mathbf{v}$ is a known function of $\mathbf{r}$ and t, then eq. (4.2) is a differential equation which can, at least in principle, be solved for the position vector $\mathbf{r}$ of a particular fluid element as a function of time t, with the a's serving to identify which element is being followed: $\mathbf{r} = \mathbf{r}(a_1,a_2,a_3,t)$. (The above statement can perhaps be further clarified by imagining that a snapshot is taken of the fluid at time t. In general, *some* fluid element will be located at each position $\mathbf{r}$ at that time, and $\mathbf{v}(\mathbf{r},t)$ gives the instantaneous velocity of whichever fluid element is at $\mathbf{r}$ at time t. Equation [4.2] applies to each fluid element separately; the equation of a few lines back describes the position of each of these elements as a function of time.)

Thus, in the Lagrangian description, $\mathbf{r}$ is a *dependent* variable, and the *independent* variables are the a's and t. All physical quantities must then be regarded as functions of the a's and t. The Stokes derivative in this description may be written simply as $\partial/\partial t$:

$$\left(\frac{d}{dt}\right)_{\text{Eulerian}} = \left(\frac{\partial}{\partial t}\right)_{\text{Lagrangian}}. \tag{4.3}$$

Generally, the Eulerian description is more convenient than the Lagrangian in problems with more than one degree of freedom (more than one space dimension). In problems of only one degree of freedom, however (such as spherical symmetry), the Lagrangian description is generally preferable to the Eulerian. The main reason is that the physical interpretation of the equations is clearer and more straightforward in the Lagrangian than in the Eulerian description. Also, the equations are often simpler. Therefore, we shall ordinarily use the Eulerian description when dealing with general equations and theorems in three dimensions, and the Lagrangian description when working with one-dimensional problems.

4.2. CONSERVATION EQUATIONS OF MASS, MOMENTUM, AND ENERGY

4.2a. CONSERVATION OF MASS

In this subsection we shall discuss the equations expressing mass conservation, first in the Eulerian, and then in the Lagrangian, description.

In the Eulerian description this conservation principle is usually expressed as the "continuity equation":

$$\frac{\partial \rho}{\partial t} + \nabla \cdot (\rho \mathbf{v}) = 0, \tag{4.4}$$

where ρ denotes the mass density.[2] Note that $(\rho \mathbf{v})$ is the "current density" of mass flow.

We may also expand the derivative in eq. (4.4) and make use of the basic operator relation (4.1), to obtain

$$\frac{1}{\rho}\frac{d\rho}{dt} = -\nabla \cdot \mathbf{v}, \tag{4.5}$$

which is an alternative form of the equation of mass conservation. We may, further, write $V \equiv 1/\rho$ for the specific volume (volume per unit mass of fluid), and eq. (4.5) becomes

$$\frac{1}{V}\frac{dV}{dt} = \nabla \cdot \mathbf{v}. \tag{4.6}$$

This equation shows that $\nabla \cdot \mathbf{v}$ is simply the time rate of increase of volume per unit volume (fractional rate of increase in volume) of a fixed mass of fluid as this mass moves about.

An *incompressible* fluid is one for which the density ρ of the fluid does not change during the motion of the fluid. Thus

$$\frac{d\rho}{dt} = 0, \quad \text{whence } \nabla \cdot \mathbf{v} = 0. \tag{4.7}$$

It may also be shown (see, for example, Milne-Thompson 1960) that the principle of mass conservation can be expressed in the simple form

$$\frac{d}{dt}(\rho d\tau) = \frac{d}{dt}(dm) = 0, \tag{4.8}$$

where $d\tau$ and dm are, respectively, a volume and a mass element. Consequently, when mass is conserved, the operator d/dt operating on any volume integral such as that in eq. (4.9) below, can simply be taken under the integral sign and applied to the factor(s) multiplying $(\rho d\tau)$. Thus

$$\frac{d}{dt}\int_{V(t)} (A\rho)\, d\tau = \int_V \left(\frac{dA}{dt}\right) \rho d\tau = \int_M \left(\frac{dA}{dt}\right) dm, \tag{4.9}$$

where $dm \equiv \rho d\tau$, and where the volume $V(t)$ is defined so as always to contain a fixed, but arbitrary, amount of fluid; i.e., $V(t)$ moves with the

[2]This equation is not valid if mass is not conserved. However, if *baryon* number is conserved (see, e.g., Thorne 1967a), then the equation is valid, with the proviso that ρ is to be identified formally with the baryon number density.

fluid, and so may change in size and shape. The quantity A is any *specific* physical quantity; A may be a scalar, vector, or tensor. The quantity M in the last integral in eq. (4.9) is the total amount of mass in $V(t)$. Because of this definition of V, eq. (4.9) should be used with caution in certain problems, such as in the problem of mass gain by or loss from stellar surfaces.

To obtain an expression for the principle of mass conservation in the Lagrangian description, it is convenient to regard the expression for the Lagrangian position coordinate $\mathbf{r} = \mathbf{r}(\mathbf{a},t)$ of each mass element as representing a continuous transformation of variables. This relation gives the position $\mathbf{r}$ at time t of a fluid particle which occupied the position $\mathbf{a}$ at $t = 0$. If one assumes that both $\mathbf{r}$ and $\mathbf{a}$ can be represented by components in a fixed Cartesian coordinate system, then one obtains (see, for example, Ledoux and Walraven 1958, §44, who actually consider the more general case of curvilinear coordinates) as the "integral" form of the equation of mass conservation in this description:

$$\rho(\mathbf{a},t)J(\mathbf{r}[\mathbf{a},t]) = \rho_0, \qquad (4.10)$$

where $\rho_0 \equiv \rho(\mathbf{a},0)$ is the local density of the fluid at $t = 0$. Here J is the *Jacobian* of the above transformation:

$$J(\mathbf{r}[\mathbf{a},t]) \equiv \left| \frac{\partial x_j}{\partial a_k} \right| \qquad (j,k = 1, \ldots, 3). \qquad (4.11)$$

Note that J is a function of $\mathbf{r}$, which in turn is a function of $\mathbf{a}$ and t. Hence, J varies continuously with time as the fluid particles move about, its value being unity at $t = 0$.

We can obtain the "differential" form of the mass conservation equation by forming the (Stokes) time derivative of eq. (4.10). Assuming that $J \neq 0$ (as must be true anyway in order that the a_i be soluble in terms of the x_i), we may thus obtain an equation expressing $\dot{\rho}$ in terms of $\dot{J}$, where a dot over a symbol stands for the Stokes derivative. One can readily show, using the definition of a determinant and of the fluid velocity $\mathbf{v} = \partial\mathbf{r}/\partial t = \dot{\mathbf{r}}$, that

$$\dot{J} = J \sum_i \frac{\partial v_i}{\partial x_i} = J\nabla \cdot \mathbf{v}. \qquad (4.12)$$

Using this result in the above equation, we obtain

$$\dot{\rho}/\rho = -\nabla \cdot \mathbf{v}, \qquad (4.13)$$

which is the same as eq. (4.5), obtained in terms of the Eulerian description. Although this last result has been derived under the assumption that $\mathbf{r}$ and $\mathbf{a}$ are expressed in a fixed Cartesian coordinate system, this equation is a vector relation and must accordingly be true in any coordinate system.

A simple example of eq. (4.13) will be given in Chapter 6.

4.2b. CONSERVATION OF MOMENTUM

We shall consider here the requirement of the conservation of linear momentum. This conservation principle is essentially Newton's second law, as applied to a fluid. We shall consider the resulting equation first in the Eulerian, and then in the Lagrangian, description. This resulting equation is often referred to as the equation of motion.

In the Eulerian description this equation is

$$\rho \frac{d\mathbf{v}}{dt} = -\nabla \cdot \mathsf{P} + \rho \mathbf{f}, \tag{4.14}$$

where $\mathbf{v}$ is the fluid velocity (linear momentum per unit mass), $\mathbf{f}$ is the total body or external force per unit mass, and P is the total *pressure tensor,* normally taken to be symmetric (in order that angular momentum be conserved; see, e.g., Milne-Thompson 1960). (We are here using the *dyadic* notation; see, e.g., Phillips 1933; Morse and Feshbach 1953, §1.6; or Duffey 1973.)

The derivation of eq. (4.14) assumes mass conservation. The equation of motion may also be written in a form that does not require mass conservation. The appropriate equation can be shown (see, e.g., Ledoux and Walraven 1958, §49) to be

$$\frac{\partial(\rho\mathbf{v})}{\partial t} + \nabla \cdot (\rho\mathbf{v}\mathbf{v} + \mathsf{P}) = \rho\mathbf{f}, \tag{4.15}$$

which is a conservation form of the equation. The quantity in parentheses in the second term on the left side of eq. (4.15) is sometimes called the "momentum flux density." The reason is that, in the absence of body forces ($\mathbf{f} = 0$), the rate of decrease of momentum (of volume density $\rho\mathbf{v}$) in a fixed volume of the fluid is equal to the net outward rate of flow of momentum of flux $(\rho\mathbf{v}\mathbf{v} + \mathsf{P}) \cdot \mathbf{n}$, where $\mathbf{n}$ is a unit vector pointing along the outward normal to the surface bounding the fixed volume of fluid.

If the stresses reduce to a pure hydrostatic pressure, as is the case in most applications of interest, the pressure tensor is of the form

$$\mathsf{P} = P\mathsf{I}, \tag{4.16}$$

where P is the hydrostatic pressure and I is the unit tensor. The force due to the stress acting on an element of area dS having outward normal $\mathbf{n}$ is then $-P\mathbf{n}dS$, that is, a force along the *inward* normal to dS, or a pure pressure.

In this case eqs. (4.14) and (4.15) become

$$\rho \frac{d\mathbf{v}}{dt} = -\nabla P + \rho \mathbf{f} \tag{4.17}$$

and

$$\frac{\partial(\rho \mathbf{v})}{\partial t} + \nabla \cdot (\rho \mathbf{v}\mathbf{v} + P\mathsf{I}) = \rho \mathbf{f}. \tag{4.18}$$

In the case of hydrostatic equilibrium in a static fluid ($\mathbf{v} \equiv 0$), there is no acceleration of any fluid particles, and eq. (4.17) then reduces to the equation of hydrostatic equilibrium,

$$\nabla P = \rho \mathbf{f}. \tag{4.19}$$

When turbulence, viscosity, or large-scale magnetic fields, for example, are present, their effects can usually be described in terms of a pressure tensor (e.g., Ledoux and Walraven 1958; Cox and Giuli 1968).

In the Lagrangian description the independent variables are, however, the identifying parameters a_i and the time t, rather than $\mathbf{r}$ and t. Since $P = P(a_1,a_2,a_3,t)$, we must write in a Cartesian coordinate system

$$\nabla P = \sum_i \mathbf{e}_i \frac{\partial P}{\partial x_i} = \sum_{ij} \mathbf{e}_i \frac{\partial a_j}{\partial x_i}\frac{\partial P}{\partial a_j} = \sum_j (\nabla a_j) \frac{\partial P}{\partial a_j},$$

whence eq. (4.17) becomes

$$\ddot{\mathbf{r}} = -\frac{1}{\rho} \sum_j (\nabla a_j) \frac{\partial P}{\partial a_j} + \mathbf{f}, \tag{4.20}$$

where we recall that $\mathbf{v} = \dot{\mathbf{r}}$, where the dot stands for $\partial/\partial t$. This is the appropriate equation of motion in the Lagrangian description when the stresses reduce to pure pressures. Note that this equation is fairly complicated in general.

A simple example of eq. (4.20) in the case of only one degree of freedom (one spatial dimension) will be given in Chapter 6.

4.2c. CONSERVATION OF ENERGY.

We shall find it convenient to consider the principle of conservation of energy in three forms: the *conservation of mechanical energy*, of *thermal and mechanical energy*, and of *thermal energy* alone, as do Ledoux and Walraven (1958, §§50–52).

The equation expressing conservation of mechanical energy can be obtained from the momentum equation in Eulerian form, eq. (4.14) (after dividing through by ρ), by forming the scalar product of both sides with $\mathbf{v}$:

$$\frac{d}{dt}(\tfrac{1}{2}v^2) = -\frac{1}{\rho}\mathbf{v}\cdot(\nabla\cdot\mathsf{P}) + \mathbf{f}\cdot\mathbf{v}, \tag{4.21}$$

where $v^2 \equiv \mathbf{v}\cdot\mathbf{v}$. Equation (4.21) states simply that the time rate of increase of the kinetic energy per unit mass $(1/2)v^2$ is equal to the rate at which the pressure gradient (or pressure tensor divergence) and body forces are doing work on the unit mass.

We can write eq. (4.21) in an alternative form if we make use of the identity

$$\nabla\cdot(\mathbf{v}\cdot\mathsf{P}) = \mathbf{v}\cdot(\nabla\cdot\mathsf{P}) + \mathsf{P}:(\nabla\mathbf{v}) \tag{4.22}$$

on the right side, where the operation indicated in the last term is the double dot product (see e.g., Phillips 1933) or a contraction of the two tensors shown. Thus, multiplying through by $\rho d\tau = dm$, integrating over the entire volume V of the fluid of mass M, making use of the generalized divergence theorem (see, e.g., Phillips 1933) in the first integral on the right side, assuming mass conservation, and neglecting the resulting surface integral, we obtain the conservation of mechanical energy theorem in integrated form:

$$\frac{d}{dt}\int_M \tfrac{1}{2}v^2 dm = \int_V \mathsf{P}:(\nabla\mathbf{v})d\tau + \int_M \mathbf{f}\cdot\mathbf{v}dm. \tag{4.23}$$

Another form of this theorem will be presented in §4.6 below, for the case where $\mathbf{f}$ consists only of gravitational forces.

The neglect of the surface integral referred to above is justified as long as $(\mathbf{v}\cdot\mathsf{P})\cdot d\mathbf{S}$ is zero or sufficiently small. This condition will be met if P is sufficiently small at the surface (which is usually the case) or if $(\mathbf{v}\cdot\mathsf{P})$ is nearly perpendicular to $d\mathbf{S}$ (as in, for instance, a steadily rotating star).

The meaning of eq. (4.23) can be made clear by considering the case of most interest in which the stresses reduce to pure pressures. In this case the first integral on the right side can be transformed by making use of the principle of mass conservation. We then obtain

$$\frac{d}{dt}\int_M \tfrac{1}{2}v^2 dm = \int_M \left[P\frac{d}{dt}\left(\frac{1}{\rho}\right)\right]dm + \int_M \mathbf{f}\cdot\mathbf{v}dm, \tag{4.24}$$

where the first integral on the right side is just the sum over all mass elements in the entire system, of the rate of $PdV(V = 1/\rho)$ work that the material in each such mass element is doing on its surroundings.

The second form of the energy equation, which expresses the conservation of thermal and mechanical energy, gives the rate of change of the sum of kinetic and internal (thermal) energy of a unit mass of fluid as the unit mass moves about. Here we must remember that some of the work done by the forces acting on each mass element will result in kinetic energy of mass motions rather than in internal energy. (A simple analogy is picking up a pail of water and running with it; this process will clearly impart kinetic energy to the pail and its contents, but will not necessarily heat them up.) Also, the surface forces resulting from the stresses acting on the surface of the system must be taken into account. Letting E denote the internal energy, $\mathbf{f}$ the sum of all body forces, and dq/dt the net rate of gain of heat following the motion, all per unit mass, the desired expression can be shown (e.g., Ledoux and Walraven 1958), making use of the principle of mass conservation, to be

$$\frac{d}{dt}\left(\tfrac{1}{2}v^2 + E\right) = -\frac{1}{\rho}\nabla \cdot (\mathsf{P} \cdot \mathbf{v}) + \mathbf{f} \cdot \mathbf{v} + \frac{dq}{dt}. \tag{4.25}$$

Bernoulli's theorem is a special case of eq. (4.25) (see Batchelor 1967, p. 156).

If we had not made use of the principle of mass conservation, we would have obtained the result

$$\frac{\partial(\rho E + \tfrac{1}{2}\rho v^2)}{\partial t} + \nabla \cdot \left(\rho E\mathbf{v} + \tfrac{1}{2}\rho v^2\mathbf{v} + \mathsf{P} \cdot \mathbf{v}\right) = \rho\mathbf{f} \cdot \mathbf{v} + \rho\frac{dq}{dt}, \tag{4.26}$$

which is a conservation form of the equation. The quantity in parentheses in the second term on the left side in eq. (4.26) is sometimes called the "energy flux vector," here $\mathbf{j}_E$, since in the absence of body forces ($\mathbf{f} = 0$) and of heat gains or losses ($dq/dt = 0$), the rate of decrease of the sum of the internal and kinetic energy (of volume density $\rho E + \tfrac{1}{2}\rho v^2$) in a fixed volume is equal to the total outward rate of flow of energy, $\oint_S d\mathbf{S} \cdot \mathbf{j}_E$, across the surface S bounding the fixed volume. We note that, in case $\mathsf{P} = P\mathsf{I}$, we have

$$\mathbf{j}_E = \rho\mathbf{v}\left(\tfrac{1}{2}v^2 + E + P/\rho\right), \tag{4.27}$$

where $E + P/\rho$ is the *enthalpy* per unit mass.

The principle of the conservation of internal energy is just the first law of thermodynamics in a somewhat more general form than usual. The appropriate equation can easily be obtained by combining eqs. (4.21) and (4.25), which express the conservation of, respectively, mechanical energy, and thermal and mechanical energy, and by making use of the identity in eq. (4.22). We obtain

$$\frac{dE}{dt} = -\frac{1}{\rho}\mathsf{P} : (\nabla\mathbf{v}) + \frac{dq}{dt}, \tag{4.28}$$

which is the desired expression. (This more general form of the first law of thermodynamics was also derived in Cox and Giuli 1968, §9.7.)

If the stresses reduce to pure pressures, i.e., if $\mathsf{P} = P\mathsf{I}$, then eq. (4.28) becomes

$$\frac{dq}{dt} = \frac{dE}{dt} + P\frac{d}{dt}\left(\frac{1}{\rho}\right) = \frac{dE}{dt} + P\frac{dV}{dt}, \tag{4.29}$$

where $V \equiv 1/\rho$. This equation is the form of the first law of thermodynamics as usually encountered.

For astrophysical purposes, however, three other forms of eq. (4.29) are often useful (these three forms are all equivalent to one another). These other forms are valid specifically under the following assumptions: (a) No composition changes resulting from thermonuclear reactions are occurring (this restriction, however, may be removed by adding extra terms to the equations). (b) The pressure P is a *thermodynamic* pressure, computable from a pressure equation of state, in terms of some pair, say density ρ and temperature T, of the thermodynamic variables: $P = P(\rho,T)$. (c) Similarly, the internal energy per unit mass E is computable from an energy equation of state in terms of some pair of thermodynamic variables, for example, $E = E(\rho,T)$.

These three forms are as follows (for detailed derivations, see Cox and Giuli 1968, Chap. 9 and §17.6 of the same reference):

$$\frac{d\ln P}{dt} = \Gamma_1\frac{d\ln\rho}{dt} + \frac{\rho(\Gamma_3 - 1)}{P}\frac{dq}{dt} \tag{4.30a}$$

$$\left(= \Gamma_1\frac{d\ln\rho}{dt} + \frac{\chi_T dq/dt}{c_V T}\right), \tag{4.30b}$$

$$\frac{d\ln T}{dt} = (\Gamma_3 - 1)\frac{d\ln\rho}{dt} + \frac{dq/dt}{c_V T}, \tag{4.31}$$

and

$$\frac{d\ln T}{dt} = \frac{\Gamma_2 - 1}{\Gamma_2}\frac{d\ln P}{dt} + \frac{dq/dt}{c_P T}. \tag{4.32}$$

In these equations c_V and c_P are the specific heats per unit mass at, respectively, constant volume and constant pressure, and

$$\chi_T \equiv \left(\frac{\partial\ln P}{\partial\ln T}\right)_\rho. \tag{4.33}$$

Also, the gammas are the usual adiabatic exponents:

$$\Gamma_1 \equiv \left(\frac{d\ln P}{d\ln\rho}\right)_{\text{ad}}, \qquad \Gamma_3 - 1 \equiv \left(\frac{d\ln T}{d\ln\rho}\right)_{\text{ad}},$$

$$\frac{\Gamma_2 - 1}{\Gamma_2} \equiv \left(\frac{d\ln T}{d\ln P}\right)_{\text{ad}} = \frac{\Gamma_3 - 1}{\Gamma_1}. \tag{4.34}$$

The following identities (derived in detail in the above reference) are useful:

$$\Gamma_1 = \chi_\rho + \chi_T(\Gamma_3 - 1), \tag{4.35}$$

where

$$\chi_\rho \equiv \left(\frac{\partial \ln P}{\partial \ln \rho}\right)_T; \tag{4.36}$$

$$\gamma \equiv c_P/c_V = \Gamma_1/\chi_\rho; \tag{4.37}$$

and

$$(\partial \ln E/\partial \ln \rho)_T = (P/[\rho E])(1 - \chi_T) \quad \text{(reciprocity relation).} \tag{4.38}$$

Equivalent to the last relation is:

$$\Gamma_3 - 1 = (P\chi_T)/(\rho c_V T). \tag{4.39}$$

The assumption that $\mathsf{P} = P\mathsf{I}$, where P is a thermodynamic pressure, is formally equivalent to the assumption that molecular and radiative viscosity, large-scale magnetic fields, and turbulence are not present. Hence, when these additional physical factors are present, the appropriate equations should be used with suitable caution.

4.3. HEAT GAINS AND LOSSES

We have now considered three of the four basic partial differential equations of hydrodynamics and heat flow. We must have a fourth basic equation because, before computations can be carried out involving the energy equation, we must be able to calculate dq/dt, the net rate of gain of heat per unit mass, from the prevailing physical conditions in the fluid. In general, we may say that the net heat gain by an element of matter is just the difference between the heat gained from the heat sources and the heat lost from the sinks. Let ϵ denote the total rate of gain of heat per unit mass from the sources (generally taken, in astrophysics, to be thermonuclear sources). Also, let $\mathbf{F}$ denote the total vector flux (energy flow across unit area normal to the direction of flow, per unit time) of heat due to all transport mechanisms that might be operative (radiation, conduction, convection, neutrino losses, mass loss, etc.) Then we may write

$$\frac{dq}{dt} = \epsilon - \frac{1}{\rho}\nabla \cdot \mathbf{F}. \tag{4.40}$$

The method of calculating ϵ, if it represents the rate of production of thermonuclear energy per unit mass, is described in detail in books on stellar interiors (e.g., Clayton 1968; Chiu 1967; Cox and Giuli 1968). As thermonuclear energy sources will not generally be of direct importance to us in this book, we merely note that the expression for $\epsilon = \epsilon(\rho, T,$chemical composition) can often be represented to adequate accuracy over somewhat small ranges in density ρ and temperature T by a simple power-law formula (see above references).

The most important contributions to the total heat flux **F** in astrophysics are usually the radiative and the convective fluxes. The flux due to thermal conduction, when important, can be treated formally in the same way as the radiative flux (see, e.g., Cox and Giuli 1968). The radiative flux does not ordinarily pose any severe problems, except possibly in cases of extremely rapid light variations (frequencies of the order of or larger than the reciprocal of the light travel time through the system) or in the very outermost, optically thin layers of extremely distended stellar atmospheres. We shall usually ignore all time derivatives when dealing with the radiative flux, and shall ordinarily use the approximations valid in the deep stellar interior (unless indicated otherwise). A careful and comprehensive discussion of both of these effects has been provided by Castor (1972); in fact, new developments along these lines are still emerging.

The convective contribution to **F** poses severe theoretical problems, which are compounded many times in applications to pulsating stars. In these applications one actually needs a *time-dependent* convection theory, since in some parts of a pulsating star the characteristic time scales associated with convective motions are likely to be of the same order of magnitude as the pulsation period. An adequate time-dependent theory of convection does not yet exist, although a number of attempts, ranging from relatively simple to extremely elaborate, have been made to develop prescriptions (if not real theories) for handling time-dependent convection (see, e.g., Cox 1974a, §9.8.1 for discussion of some of these attempts and references; for more recent work, see Chap. 19).

In the stellar interior it is usually adequate to compute the integrated radiative flux from a formula based on the "diffusion approximation:"

$$\mathbf{F} \approx -\frac{4\pi}{3\kappa\rho}\nabla B(T) = -\frac{4\pi}{3\kappa\rho}\left[\frac{dB(T)}{dT}\right]\nabla T, \tag{4.41}$$

where $B(T)$ is the integrated Planck function and κ is the opacity (normally the Rosseland mean opacity). A detailed discussion of the conditions under which these approximations are valid is given, for example, in Cox and Giuli (1968); see also Castor (1972). For improvements on this treatment see, for example, Unno (1965, App. A).

Although extensive tabular opacities are now available (see, e.g., A. N. Cox and Stewart 1970; A. N. Cox and Tabor 1976), simple power-law formulas for $\kappa(\rho, T,$chemical composition) are often adequate over somewhat limited ranges in density ρ and temperature T.

4.4. GENERAL DISCUSSION OF EQUATIONS

In this section we shall indicate how the basic equations considered above—the mass, momentum, and energy equation, and the equations for calculating heat gains and losses—along with the appropriate constitutive equations, are sufficient to determine completely, in principle, the future development of the fluid of interest, granting that certain information is available. By "constitutive equations" we mean, in particular, equations for calculating such things as the pressure, internal energy, opacity, and energy generation rate of thermonuclear reactions, all as functions of chemical composition, density, and temperature. The information assumed available above includes suitable initial conditions (see below), as well as a knowledge of all body forces acting on the system. For simplicity we shall assume that the fluid is not turbulent, that all stresses reduce to pure thermodynamic pressures, and that the photon mean free path is everywhere small compared to the scales of interest. More general situations do not alter the essential nature of the problem, and only complicate it.

For specificity we shall assume that all times of interest are short compared to the time required for significant changes in chemical composition to be brought about via nuclear reactions. Hence we may treat the chemical composition of the fluid as a constant. We also assume that the only body force is gravity. Hence this force can always be computed at each instant of time from the instantaneous distribution of matter (see §4.5 below).

As initial conditions, we choose the temperature T, density ρ, and fluid velocity $\mathbf{v}$, all as known functions of $\mathbf{r}$, specifying the point of observation, and the (initial) time t. From this information the temperature and velocity gradients ∇T and $(\nabla \mathbf{v})$ can be calculated; as well as the pressure P (from the constitutive equations); the pressure gradient ∇P; the heat flux divergence, $\nabla \cdot \mathbf{F}$; the thermonuclear energy generation rate, ϵ; and hence the net rate of gain of heat per unit mass, $dq/dt = \epsilon - \rho^{-1}\nabla \cdot \mathbf{F}$; the body force per unit mass, $\mathbf{f}$; and whatever other information is needed; all as functions of $\mathbf{r}$ and the (initial) time t.

It is then clear that the mass, momentum, and energy equations yield values of ρ, $\mathbf{v}$, and T at all points $\mathbf{r}$ at the somewhat later time $t + dt$. But this is precisely the same information we assumed to be known, as initial conditions, at the initial time t.

It is therefore clear that, by proceeding in the above manner step by step in time, the entire future evolution of the fluid can in principle be followed. Observe that this possibility exists *only* if *all* of the basic equations are used. It therefore appears that we have just the right number of basic equations (neither too few, nor too many) for our purposes.

4.5. THE GRAVITATIONAL FIELD

In most of this book we shall assume that the only body force is self-gravitation, which supplies the force $\mathbf{f}$ per unit mass acting on the fluid. We must accordingly calculate $\mathbf{f}$ from the gravitational potential, $\psi(\mathbf{r},t)$, which in turn is given by the solution of Poisson's equation:

$$\nabla^2\psi = 4\pi G\rho, \tag{4.42}$$

where G is the constant of gravitation. It is well known (see any book on theoretical physics or potential theory, e.g., Joos 1932) that the solution of eq. (4.42) of physical interest here is

$$\psi(\mathbf{r},t) = -G\int_V \frac{\rho(\mathbf{x},t)d\tau'}{|\mathbf{x}-\mathbf{r}|}, \tag{4.43}$$

where the integration is extended over the entire volume of the system (that is, over all regions where ρ differs from zero) (see Fig. 4.1). By taking the gradient of eq. (4.43), we see that we may write

$$\mathbf{f}(\mathbf{r},t) = -\nabla\psi(\mathbf{r},t). \tag{4.44}$$

In astrophysical contexts, $\mathbf{f}$ is usually denoted by $\mathbf{g}$, the gravitational acceleration.

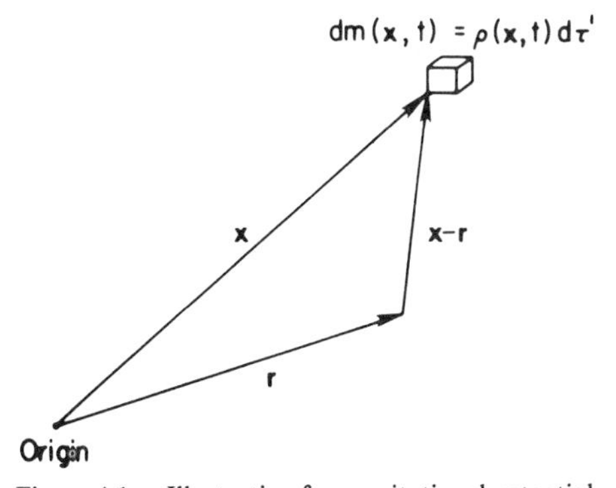

Figure 4.1. Illustration for gravitational potential.

For some purposes it is convenient to regard Poisson's equation as *two* first order differential equations, rather than as one second order differential equation. The two appropriate equations are

$$\nabla \cdot \mathbf{f} = -4\pi G\rho \tag{4.45}$$

and eq. (4.44) above.

4.6. ENERGY CONSERVATION THEOREM FOR A SELF-GRAVITATING SYSTEM

In this section we wish to apply the principle of the conservation of thermal and mechanical energy (§4.2) to a system in which the only body force is self-gravitation.

We first define the total gravitational potential energy of the system by the relations

$$\Phi \equiv -\tfrac{1}{2}G \int_V \int_V \frac{\rho(\mathbf{x},t)d\tau'\rho(\mathbf{r},t)d\tau}{|\mathbf{x}-\mathbf{r}|} \tag{4.46}$$

$$= \tfrac{1}{2}\int_V \rho(\mathbf{r},t)\psi(\mathbf{r},t)d\tau. \tag{4.47}$$

The factor 1/2 takes into account the fact that the potential energy of each pair of mass elements $\rho(\mathbf{x},t)d\tau'$ and $\rho(\mathbf{r},t)d\tau$ has been counted twice in the double integral in eq. (4.46). Both of these integrals are to be taken over the entire volume of the fluid. Equation (4.47) follows simply from the recognition that one of these volume integrals is just the expression (see eq. [4.43]) for $\psi(\mathbf{r},t)$.

We also define

$$\mathcal{T} \equiv \int_V \tfrac{1}{2}\rho v^2 d\tau, \tag{4.48}$$

and

$$U \equiv \int_V \rho E d\tau \tag{4.49}$$

as, respectively, the total kinetic energy and the total internal energy of the system. It can then be shown by integrating the appropriate equation (eq. [4.25]) over the entire volume V of the system and performing some manipulations, that

$$\frac{d\Psi}{dt} = \int_V \rho \frac{dq}{dt} d\tau - \oint_S (\mathbf{v} \cdot \mathsf{P} + \tfrac{1}{2}\rho\psi\mathbf{v}) \cdot d\mathbf{S}, \tag{4.50}$$

where

$$\Psi \equiv \mathcal{T} + U + \Phi. \tag{4.51}$$

In many cases the surface integral in eq. (4.50) will vanish. For example, P and ρ might vanish on S. If not, and if $\mathsf{P} = P\mathsf{I}$, then the normal component of the velocity $\mathbf{v}$ might vanish (as in a steadily rotating star), so that $\mathbf{v} \cdot d\mathbf{S} = 0$. If the surface integral vanishes, we have

$$\frac{d\Psi}{dt} = \int_V \rho \frac{dq}{dt}\, d\tau, \tag{4.52}$$

which states that the quantity Ψ can change only as a result of heat gains by or losses from the system. This quantity Ψ may therefore be regarded as the total energy of the system.

A star in thermal equilibrium is one in which dq/dt (see eq. [4.40]) is small everywhere in the star. For such a star, then, Ψ changes only slowly with time, according to eq. (4.52). (Because of the slow change in chemical composition that always accompanies stellar evolution, Ψ cannot be *strictly* constant: the internal structure of the star slowly changes as it evolves. Hence, the right side of eq. (4.52) is *never quite* zero, even in a star in thermal equilibrium.)

If $\epsilon = 0$, eq. (4.52) states that the energy lost through the surface is compensated for by a decrease in the total energy $\mathcal{T} + U + \Phi$ of the system. When applied to a star with no energy sources, eq. (4.53) describes the process of gravitational (or Kelvin) contraction (described in detail in books on stellar interiors).

We note a special case of eq. (4.50). We use the first law of thermodynamics, eq. (4.28), to eliminate the term in dq/dt. In case the stresses reduce to pure pressures, we have $\mathsf{P} : (\nabla\mathbf{v}) = P\rho\, d(1/\rho)/dt$ (with use of the mass equation) and, neglecting the surface integral, we have

$$\frac{d}{dt}(\mathcal{T} + \Phi) = \int_M \left[P \frac{d}{dt}\left(\frac{1}{\rho}\right)\right] dm, \tag{4.53}$$

where the integral is extended over the entire mass M of the system. This result is a special case of the conservation of mechanical energy theorem, and states that the rate of change of the sum of the kinetic and gravitational potential energies of the whole system is equal to the sum over the whole system of the rate of PdV work done by all the mass elements on their surroundings. We may note that an equivalent way of writing eq. (4.53) is

$$\left(\frac{d\Psi}{dt}\right)_{\text{ad}} = 0, \tag{4.54}$$

where

$$\left(\frac{d\Psi}{dt}\right)_{\mathrm{ad}} \equiv \frac{d}{dt}(\mathcal{T} + \Phi) - \int_M \left[P \frac{d}{dt}\left(\frac{1}{\rho}\right)\right] dm \tag{4.55}$$

and $(d\Psi/dt)_{\mathrm{ad}} = d\Psi/dt$ for the case of *adiabatic* motion, $dq/dt \equiv 0$.

Further generalization of the energy theorem (4.50) to include magnetic fields and effects of large-scale electric currents is possible, but we do not consider the matter here (see, e.g., Alfvén and Fälthammer 1963; Cowling 1957; Chandrasekhar and Fermi 1952).

5

The Linear Theory

In this chapter we shall be concerned primarily with the so-called "linearized" equations of hydrodynamics and heat flow. The general meaning of this term will be explained in §5.1, as will certain important uses of these equations. A rather detailed discussion will be presented in §5.2 of what is meant by equilibrium states in stars, and of the various types of equilibrium contemplated in stellar astrophysical discussions. The two main types of variations used in the linear theory will be defined and discussed in §5.3, and in §5.4 some of the basic linearized equations of hydrodynamics and heat flow themselves will be obtained and discussed. Finally, as a simple example of these equations, adiabatic sound waves will be considered briefly in §5.5.

5.1. INTRODUCTION

It is clear that the general equations considered in Chapter 4 form a system of nonlinear, partial differential equations, exact solutions of which are available, generally, only in certain (usually uninteresting and highly unrealistic) cases. If, however, a *particular* solution, which we shall call the "unperturbed" solution, is known, we are often interested in finding another solution (or perhaps other solutions), which we shall call the "perturbed" solution and which differs (in some sense) only slightly from the unperturbed solution. (We may think of the two solutions as representing two possible future histories of the fluid, differing from each other because of different initial conditions and both allowed by the basic equations.)

If the two solutions differ from each other only slightly, then each dependent variable for the perturbed solution can be expressed as the sum of the corresponding dependent variable for the unperturbed solution and a small correction term, that is, a small variation or perturbation. Substituting the dependent variables so expressed into the equations, noting that the unperturbed solution is also a solution of the equations, and neglecting all powers (above the first) and products of the variations, yields a system of partial differential equations whose solution gives the behavior of the *variations* of the dependent variables in time and space, provided that suitable boundary and initial conditions have been specified. The great advantage of the resulting set of equations is that they are *linear*, and well-known mathematical procedures can then be applied to obtain solutions for the variations.

An example of such a perturbation would be small oscillations of the system about some equilibrium state (see §5.2 below).

The question of the *existence* of such oscillations in nature in the above example brings us to a consideration of a more fundamental use of the linearized equations. This use is in the so-called *linear stability theory* (see, e.g., Chandrasekhar 1961, Chap. 1, for a good general discussion of this theory). This theory is concerned with the question of whether some given solution, which we may identify with our unperturbed solution, even though it may exist, is *stable* or *unstable* against a small perturbation. This solution could presumably exist in nature more than merely transiently only if it were stable against *every kind* of perturbation to which the system could be physically subjected.

If the system were a star executing small oscillations about an equilibrium configuration, one possibility is that the star was at one time unstable against such oscillations. If the star had once been unstable in this manner, then the oscillations would have arisen essentially spontaneously from the instability, because of the many small, random fluctuations in physical conditions that are always present in nature. The oscillations would then presumably have grown to their observed (generally nonlinear) condition, which itself represents a solution of the equations.

Such self-excited oscillations, which may arise from an instability against *infinitesimal* perturbations, are often called "soft" self-excited oscillations (Ledoux and Walraven 1958, p. 550). If the oscillation is a pulsation, the system is said to be *pulsationally unstable*, or "overstable" in Eddington's terminology (Eddington 1926, §135). The justification for Eddington's terminology is, apparently, that in such a system the restoring forces which accompany a perturbation from equilibrium are so strong that they lead to an ensuing perturbation on the "other side" of equilibrium that is larger than the original perturbation.

Oscillations that can grow only after a *finite* perturbation has been applied to the system are often called "hard" self-excited oscillations. An analogy is the heat of reaction of a chemical reaction: a certain ignition temperature must normally be reached before chemical burning can take place. Up to the time of this writing (1978), no examples of hard self-excited stellar oscillations have, to the best of our knowledge, been reported in the published astrophysical literature. If such oscillations are indeed not commonly found in nature, then the linear stability theory may have considerable relevance to actual pulsating stars. At least this relevance is generally taken for granted. (However, as has been pointed out by Christy 1966b, Stellingwerf 1974a, and others, the detailed behavior of a real pulsating star need not necessarily be closely governed by the predictions of the linear theory; see, for instance, J. P. Cox 1974a.)

The instability need not, of course, be of an oscillatory character. For example, it may be aperiodic. *Dynamical* and *secular instability* in stars are both examples of this kind of instability. These two kinds of instability represent increasing departures from, respectively, *hydrostatic* and *thermal* equilibrium. (However, secular instabilities sometimes have an oscillatory character; see, for example, Aizenman and Perdang 1971, 1972; Härm and Schwarzschild 1972; Hansen, Cox, and Herz 1972; Osaki and Hansen 1973b; Defouw 1973; and Hansen 1978). The relevant time scale for dynamical instability is generally of the order of the free-fall time t_{ff}; for secular instability, of the order of the Kelvin time t_K (see Chap. 2).

5.2 DISCUSSION OF EQUILIBRIUM

For the sake of discussion, we shall assume that stellar material behaves as a general fluid medium. We also assume, for simplicity, that all stresses reduce to pure thermodynamic pressures (see §4.2); that all body forces are due to self-gravitation; and that the basic energy source is provided by nuclear transmutations, that is, nuclear reactions. We then ask, is a state of complete equilibrium, characterized by *zero* fluid velocities everywhere and for all time,

$$\mathbf{v}(\mathbf{r},t) \equiv 0, \tag{5.1}$$

a solution of the equations of hydrodynamics and heat flow? (We shall always assume, unless specifically stated otherwise, that all velocities are measured relative to the center of mass of the system.)

When we allow the composition of the fluid to change as a result of nuclear transmutations, the energy equations presented in §4.2c must be generalized to take this into account. This generalization can easily be effected simply by regarding the thermodynamic quantities as functions not only of some two thermodynamic variables, but also of the chemical composition of the system. We specify this composition by giving the fractional mass abundances x_i of the various nuclear species. It can then easily be shown that an additional term should be added to one of the equations (e.g., eq. [4.30a]) following from the first law of thermodynamics. This term involves the $\dot{x}_i$, where a dot stands for the Stokes derivative d/dt.

It is clear that the nuclear energy generation rate per unit mass ϵ depends on the $\dot{x}_i$. The $\dot{x}_i$ themselves are given by (in general) n kinetic, or rate, equations:

$$\dot{x}_i = \dot{x}_i(x_1, \ldots, x_n, \rho, T) \qquad (i = 1, \ldots, n), \tag{5.2}$$

where n denotes the total number of distinct nuclear species in the system. The specific forms of these equations are not relevant to the present discussion (see, e.g., Rosseland 1949; or Clayton 1968, Chap. 7).

We shall answer the question posed a few paragraphs back by assuming that $\mathbf{v}(\mathbf{r},t) \equiv 0$ is a solution of the equations, and then examining the necessary consequences. It can then easily be shown from the basic equations presented in Chapter 4 that, without demanding highly contrived (and probably artificial) properties of the physical system, all time derivatives of the physical variables must vanish. It then follows that

$$\epsilon - \frac{1}{\rho}\nabla \cdot \mathbf{F} = 0 \tag{5.3}$$

and, moreover, that

$$\epsilon = 0, \tag{5.4}$$

and hence that

$$L = 0. \tag{5.5}$$

Thus, the system cannot lose any energy to its surroundings. In the case of a star, it would have to be at the temperature of interstellar space, at about 3 or 4°K!

We conclude that $\mathbf{v}(\mathbf{r},t) \equiv 0$ is a solution of the equations only for the uninteresting case of, say, a black dwarf star, a black neutron star, or a stationary interstellar gas cloud at the temperature of interstellar space.[1] It follows, moreover, that no system, such as a star, which loses energy to its surroundings (that is, has $L > 0$), can be in a perfectly static state with $\mathbf{v}(\mathbf{r},t) \equiv 0$. Such systems must therefore evolve through time. This same conclusion could also have been reached merely from considerations based on the general thermodynamic irreversibility of net heat losses from a system and the consequent approach to thermodynamic equilibrium. It is satisfying, however, that our equations also lead to this conclusion.

If we restrict our attention to time intervals short enough that the $\dot{x}_i$ may be neglected (in all quantities except ϵ), that is, for times considerably

[1]The "black hole" is another possibility in nature, at least in principle. However, since our treatment is inherently nonrelativistic, such an object cannot exist within the present theoretical framework.

shorter than the nuclear time (see Chapter 2), then $\mathbf{v}(\mathbf{r},t) \equiv 0$ *may* be an approximate solution of the equations. If it is, then the necessary condition (5.3) is still valid. This is just the condition of *thermal equilibrium* (energy balance throughout the system) (see, e.g., Cox and Giuli 1968, Chap. 5). The equations describing this state are the usual equations of static stellar structure, supplemented by suitable constitutive equations (see §4.4). Note that these equations of static stellar structure form a system of *ordinary* but nonlinear differential equations.

Another case of interest is one in which the system is very nearly in hydrostatic equilibrium and in which $\epsilon = 0$ (no nuclear energy generation), but $L > 0$. An example would be a gravitationally contracting, pre-main sequence star. It is easy to see that $\mathbf{v}(\mathbf{r},t) = 0$ cannot be a solution of the equations for this case, and so motions must occur. Nevertheless, these motions may be so slow that the time derivatives of $\mathbf{v}$, that is, the *accelerations* of the mass elements, may be negligibly small. In fact, the order of magnitude of these velocities must be comparable to the ratio of the stellar radius to the Kelvin time t_K (see Chapter 2):

$$|\mathbf{v}| \sim R/t_k \sim 10^{-4}\, LR^2/M \text{ cm s}^{-1}, \tag{5.6}$$

where L, R (stellar radius), and M are in solar units. For $L = R = M = 1$, $|\mathbf{v}|$ would be ~ 30 m yr^{-1}. In this case the equations would be the former equations of stellar structure, except that eq. (5.3) would have to be replaced by some form of the energy equation. Note that since all physical variables are now functions of both position and time, the above set is now a system of partial differential equations. Systems described by these latter equations are sometimes described as quasi-equilibrium states, as quasi-hydrostatic equilibrium states, or as systems in thermal imbalance.

In order to deal with systems over arbitrarily long periods of time, such that the chemical composition may change via nuclear transmutations, but with the system remaining always very nearly in hydrostatic equilibrium, it is only necessary to add the kinetic equations for the x_i (eqs. [5.2]) to the set discussed in the paragraph before last (but with the exception noted in the preceding paragraph). It is this larger set of equations that must be solved in modern stellar evolution calculations.

Unless we state otherwise, by "equilibrium" we shall always mean in this book the state described by the equations of static stellar structure.

Finally, we should point out that any "equilibrium" solution, characterized by $\mathbf{v}(\mathbf{r},t) \equiv 0$ in some approximation or other, may exist but be unstable. Any such instability may be pulsational, dynamical, or secular. The presence of such instabilities, if they exist, can often be revealed by solutions of the linearized equations, to which we turn next.

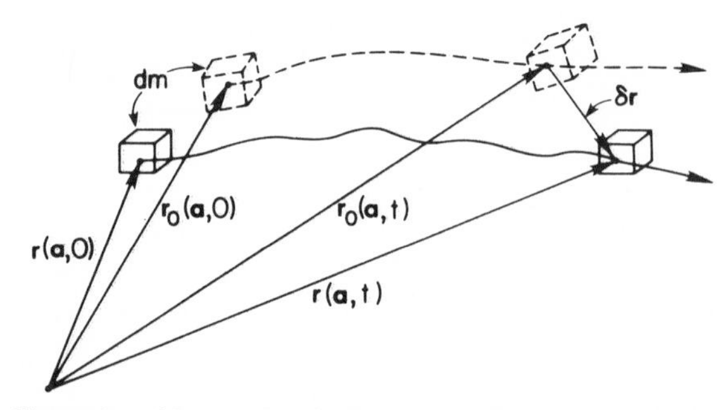

Figure 5.1. Illustration of Lagrangian displacement $\delta\mathbf{r}$. The solid wavy line and boxes show the perturbed flow, dashed lines show the unperturbed flow, both for the same fluid element of mass *dm*.

5.3. EULERIAN AND LAGRANGIAN VARIATIONS

Two types of variations, *Eulerian* and *Lagrangian,* are in general employed in the linear theory.

Consider some particular "unperturbed" solution of the equations of hydrodynamics and heat flow. We shall here denote this solution by zero subscripts. Thus, *in this solution* the position at time t of some particular fluid element characterized by the identifying Lagrangian parameters a_1, a_2, and a_3 (in three dimensions) may be represented by the vector (see Fig. 5.1)

$$\mathbf{r}_0 = \mathbf{r}_0(a_1,a_2,a_3,t) = \mathbf{r}_0(\mathbf{a},t), \tag{5.7}$$

where we are here denoting, for short, the triad of numbers a_1, a_2, a_3 by the vector $\mathbf{a}$.[2] The unperturbed solution often represents an equilibrium or quasi-equilibrium configuration, in which $\mathbf{r}_0$ may not depend on t, or may vary only slowly with t (however, the unperturbed solution need not necessarily represent such a slowly-varying configuraton at all). In the "perturbed" solution for the same physical system the position of the *same* fluid particle at time t is written as

$$\mathbf{r} = \mathbf{r}(a_1,a_2,a_3,t) = \mathbf{r}(\mathbf{a},t). \tag{5.8}$$

The *displacement* (or *Langrangian displacement*) of the fluid element at time t from its position in the unperturbed solution is then defined by the relation

$$\delta\mathbf{r} \equiv \mathbf{r}(a_1,a_2,a_3,t) - \mathbf{r}_0(a_1,a_2,a_3,t) = \delta\mathbf{r}(\mathbf{a},t). \tag{5.9}$$

[2]Of course, any triad of numbers does not necessarily constitute a *vector*. However, since a_1, a_2, and a_3 often denote the components of the position vestor $\mathbf{a}$, this notation should lead to no confusion.

It is clear that in the present Lagrangian description $\delta\mathbf{r}$ is being regarded as a function of the a's and the time. However, since for each $\mathbf{a}$ and t (that is, for each fluid particle), a corresponding position $\mathbf{r}_0$ in the unperturbed flow is determined by eqs. (5.7), we could equally well regard $\delta\mathbf{r}$ as a function of the $\mathbf{r}_0$ and t:

$$\delta\mathbf{r} = \delta\mathbf{r}(\mathbf{r}_0,t). \tag{5.10}$$

Finally, if the Lagrangian displacements $\delta\mathbf{r}$ were being described in terms of an Eulerian description, they would have to be regarded as field quantities, functions of the (arbitrary) point of observation $\mathbf{r}$ and t:

$$\delta\mathbf{r} = \delta\mathbf{r}(\mathbf{r},t). \tag{5.11}$$

Consider now any physical quantity. We denote this quantity by f in general and by f_0 in the unperturbed solution. In the Eulerian description, in which both f and f_0 are regarded as functions of $\mathbf{r}$ (point of observation) and t, it is often convenient to make use of the *Eulerian variation,* which we designate here and throughout this book with a prime. Thus, the Eulerian variation of the physical quantity f is defined as

$$f'(\mathbf{r},t) \equiv f(\mathbf{r},t) - f_0(\mathbf{r},t). \tag{5.12}$$

The Eulerian variation f' is thus the difference between the values of f in the perturbed and unperturbed solutions at a *given point of observation* $\mathbf{r}$ *and at a given time* t. Hence, in forming f' at given $\mathbf{r}$, t, we are actually, in general, comparing the properties of two *different* fluid particles in the two solutions.

We may note at this point that the operation of forming the Eulerian variation clearly commutes with both the operators ∇ and $\partial/\partial t$.

It is often desirable to compare the properties of a *given* fluid element in the two solutions at some time. The *Lagrangian variation,* denoted here and throughout this book (normally) by δf, of some physical quantity f, serves this purpose. The Lagrangian variation is defined by the relation

$$\delta f \equiv f(\mathbf{a},t) - f_0(\mathbf{a},t). \tag{5.13}$$

(In some places in the literature the Eulerian variation of a quantity f is denoted by δf, and the Lagrangian variation by Δf.)

Note that there is an alternative way of writing eq. (5.13). We may regard f on the right side as a function of $\mathbf{r}$ (point of observation) and t. The fluid element that happens to be at $\mathbf{r}$ at time t can be identified by solving eqs. (5.8) for the a's. With the a's so determined, the position $\mathbf{r}_0$ of *that* fluid element at that time in the unperturbed solution can then be found by substituting these a's into eqs. (5.7). Then, regarding f_0 on the right side of eq. (5.13) as a function of the $\mathbf{r}_0$, as found above, and t, we can be certain that we are indeed comparing the properties of the *same* fluid element in the two solutions. We then have

$$\delta f = f(\mathbf{r},t) - f_0(\mathbf{r}_0,t), \tag{5.14}$$

where $\mathbf{r}$ and $\mathbf{r}_0$ are related by eqs. (5.7) and (5.8), and $\mathbf{r} - \mathbf{r}_0 = \delta\mathbf{r}$.

The connection between the Eulerian and Lagrangian variations f' and δf, respectively, is most conveniently obtained from eq. (5.14). Subtracting and adding $f_0(\mathbf{r},t)$ from and to the right side of this equation, and making use of the definition (5.12) of f', we have

$$\delta f = f' + [f_0(\mathbf{r},t) - f_0(\mathbf{r}_0,t)]. \tag{5.15}$$

While this relation is exact, its first-order form is more convenient and more commonly recognized. This form is obtained by expanding the quantity in square brackets in a Taylor series about r_0 and retaining only linear terms in $\delta\mathbf{r}$:

$$\delta f = f' + (\delta\mathbf{r}) \cdot \nabla f_0. \tag{5.16}$$

Of course, to first order in $\delta\mathbf{r}$, ∇f_0 may be replaced, if desired, by ∇f.

Just as f' commutes with $\partial/\partial t$, so δf commutes with the Stokes derivative d/dt:

$$\frac{d(\delta f)}{dt} = \delta\left(\frac{df}{dt}\right). \tag{5.17}$$

The exact validity of this relation can most readily be seen by using the definition (5.13) of the Lagrangian variation and recalling that the very meaning of the Stokes derivative is to form the time derivative for a *given* fluid element (i.e., to keep the a's in eq. [5.13] constant during the differentiation).

However, eq. (5.17) is still valid to the first order in smallness even when an Eulerian description is being used and when, accordingly, the Stokes derivative must be written explicitly as a "convective" operator (see eq. [4.1]). This fact has been shown explicitly by two methods by Lynden-Bell and Ostriker (1967). The (first-order) validity of this equation can also be demonstrated by direct calculation, using some of the above concepts and relations.

From the above results, we may summarize by observing the following four general rules (valid at least to first order in smallness):

(1) $'$ does *not* commute with d/dt;

(2) δ does *not* commute with either $\partial/\partial t$ or ∇.

On the other hand,

(3) $'$ *does* comute with $\partial/\partial t$ and ∇;

(4) δ *does* commute with d/dt.

Let us now consider the relation between fluid velocity $\mathbf{v}$ and displacement $\delta\mathbf{r}$.

We look again at the two possible fluid flows, the perturbed and unperturbed, both solutions of the equations of hydrodynamics and heat flow (see Fig. 5.1). Let $\mathbf{r}$ and $\mathbf{r}_0$ be the position vectors of the *same* fluid element at time t in the two flows, where $\mathbf{r}$ and $\mathbf{r}_0$ are related by eqs. (5.7) and (5.8). The fluid velocity $\mathbf{v}(\mathbf{r},t)$ in the perturbed flow is then defined by the relation

$$\mathbf{v}(\mathbf{r},t) \equiv \frac{d\mathbf{r}}{dt}. \tag{5.18}$$

Similarly, the fluid velocity $\mathbf{v}_0(\mathbf{r}_0,t)$ in the unperturbed flow is

$$\mathbf{v}_0(\mathbf{r}_0,t) \equiv \frac{d\mathbf{r}_0}{dt}. \tag{5.19}$$

Thus, $\mathbf{v}$ and $\mathbf{v}_0$ describe the motions of the *same* fluid element in the two flows. Accordingly, we have

$$\mathbf{v}(\mathbf{r},t) - \mathbf{v}_0(\mathbf{r}_0,t) = \frac{d(\delta\mathbf{r})}{dt} = \delta\mathbf{v}, \tag{5.20}$$

where $\delta\mathbf{r}$ is the Lagrangian displacement, and where in the second equality we have made use of the commutativity stated above of δ and d/dt; $\delta\mathbf{v}$ denotes the *Lagrangian variation* of the velocity. We note that we may also write this second equality in the form

$$\delta\mathbf{v} = \frac{d(\delta\mathbf{r})}{dt} = \frac{\partial(\delta\mathbf{r})}{\partial t} + \mathbf{v}\cdot\nabla(\delta\mathbf{r}). \tag{5.21}$$

The *Eulerian variation* $\mathbf{v}'$ of the velocity is defined as the velocity difference between the two flows as seen *at the same point* $\mathbf{r}$ at time t:

$$\mathbf{v}' \equiv \mathbf{v}(\mathbf{r},t) - \mathbf{v}_0(\mathbf{r},t). \tag{5.22}$$

Using the (first-order) relation established above between the two types of variations (see eq. [5.16]), we have

$$\delta\mathbf{v} = \mathbf{v}' + (\delta\mathbf{r})\cdot\nabla\mathbf{v}, \tag{5.23}$$

$$\mathbf{v}' = \frac{\partial(\delta\mathbf{r})}{\partial t} + \mathbf{v}\cdot\nabla(\delta r) - (\delta\mathbf{r})\cdot\nabla\mathbf{v}. \tag{5.24}$$

Note that the first two terms on the right side of eq. (5.24) are just $d(\delta\mathbf{r})/dt$.

In the important special case in which the unperturbed solution represents an equilibrium one, with $\mathbf{v}_0 \equiv 0$, we have to first-order accuracy

$$\mathbf{v} = \delta\mathbf{v} = \mathbf{v}' = \frac{d(\delta\mathbf{r})}{dt} = \frac{\partial(\delta\mathbf{r})}{\partial t}. \tag{5.25}$$

Thus, in this case no distinction need to be made between the two types of variations of velocity or between the two types of time derivatives, and the fluid velocity $\mathbf{v}$ is itself a first-order quantity.

5.4. LINEARIZED EQUATIONS

In this section we shall write down some of the linearized equations of hydrodynamics and heat flow.

5.4a. MASS EQUATION

We start with the mass (or continuity) equation (eq. [4.4]), and write

$$\rho(\mathbf{r},t) = \rho_0(\mathbf{r},t) + \rho'(\mathbf{r},t), \tag{5.26}$$

and so forth, where zero subscripts denote quantities in the unperturbed flow and primes the Eulerian variations. Substituting eq. (5.26) into eq. (4.4), we obtain the linearized mass equation. It is

$$\frac{\partial\rho'}{\partial t} + \nabla \cdot (\rho_0\mathbf{v}' + \rho'\mathbf{v}_0) = 0. \tag{5.27}$$

We note that, because the unperturbed flow is assumed to be a solution of the equations, the linearized equations can be obtained more directly simply by treating " ′ " as an ordinary differential operator as applied to the original nonlinear equations. (However, the rules summarized near the end of §5.3 should always be adhered to.)

In the important special case where $\mathbf{v}_0 \equiv 0$, and accordingly where $\partial\rho_0/\partial t \equiv 0$ and $\mathbf{v}' = \mathbf{v}$, eq. (5.27) becomes

$$\frac{\partial\rho'}{\partial t} + \nabla \cdot (\rho_0\mathbf{v}) = 0. \tag{5.28}$$

This is a more commonly encountered form of the linearized continuity equation. (It should be remembered that, to first order in smallness, it makes no difference whether or not zero subscripts are dropped from unperturbed quantities.)

The "integrated" mass equation is also sometimes useful. It is most easily obtained by writing $\mathbf{v} = \partial(\delta\mathbf{r})/\partial t$ in eq. (5.28) and remembering that ρ_0 in that equation is not a function of time. Integrating with respect to time and choosing the constant of integration in such a way that $\rho' = 0$ when $\delta\mathbf{r} = 0$, we obtain

$$\rho' + \nabla \cdot [\rho_0(\delta \mathbf{r})] = 0 \tag{5.29a}$$

or

$$\delta\rho + \rho_0 \nabla \cdot (\delta \mathbf{r}) = 0. \tag{5.29b}$$

Even though eqs. (5.29) were derived under the assumption that $\mathbf{v}_0 \equiv 0$, they are nevertheless valid also when $\mathbf{v}_0 \neq 0$. Perhaps the simplest method of obtaining this result is to take the Lagrangian variation of the mass equation in Lagrangian form, eq. (4.10), for the case in which both position vectors $\mathbf{r}$ (position of a given fluid element at any time t) and $\mathbf{a}$ (position of the *same* fluid element at some fixed prior time such as $t = 0$) are resolved in a given Cartesian coordinate system, and then to perform a few manipulations. See also Goldstein (1950, p. 357).

Still another proof that eqs. (5.29) are valid also when $\mathbf{v}_0 \neq 0$ has been worked out by M. L. Aizenman (private communication, 1974). While space limitations prevent us from presenting Aizenman's proof in detail, we feel that at least a sketch of it will be instructive.

Aizenman substitutes eq. (5.24) for $\mathbf{v}'$ into the linearized mass equation, eq. (5.27). Then, after some manipulations and with use of some vector identities, he is able to write the equation in the form $\partial A/\partial t + \nabla \cdot (A\mathbf{v}_0) = 0$, where $A \equiv \rho' + \nabla \cdot [\rho_0(\delta \mathbf{r})]$. The only physically significant solution of this equation is the particular solution, A = constant in time and space. We then obtain $A\nabla \cdot \mathbf{v}_0 = 0$ or, since $\mathbf{v}_0$ is arbitrary, $A = 0$, which is just eq. (5.29a).

5.4b. MOMENTUM EQUATION

In this subsection we consider the momentum equation only for the case where all stresses reduce to pure pressures. Using eq. (4.1) for the Stokes derivative, and then taking the Eulerian variation of the general nonlinear momentum equation (4.17), we obtain

$$\frac{\partial \mathbf{v}'}{\partial t} + \mathbf{v}' \cdot \nabla \mathbf{v}_0 + \mathbf{v}_0 \cdot \nabla \mathbf{v}' = \frac{\rho'}{\rho_0^{\,2}} \nabla P_0 - \frac{1}{\rho_0} \nabla P' + \mathbf{f}'. \tag{5.30}$$

In the special case in which this state is an equilibrium one, with $\mathbf{v}_0 \equiv 0$, we have $\mathbf{v}' = \mathbf{v}$, and

$$\frac{\partial \mathbf{v}}{\partial t} = \frac{\rho'}{\rho^2} \nabla P - \frac{1}{\rho} \nabla P' + \mathbf{f}', \tag{5.31}$$

where we have dropped zero subscripts. This is a more commonly encountered form of the linearized momentum equation.

Sometimes it is convenient to take the Lagrangian variation of the momentum equation. Remembering that $\delta(d\mathbf{v}/dt) = d(\delta\mathbf{v})/dt = d(\delta[d\mathbf{r}/dt])/dt = d(d[\delta\mathbf{r}]/dt)/dt = d^2(\delta\mathbf{r})/dt^2$, we have

$$\frac{d^2\delta\mathbf{r}}{dt^2} = -\delta\left(\frac{1}{\rho}\nabla P\right) + \delta\mathbf{f}. \tag{5.32}$$

This equation is particularly useful in systems with spherical symmetry, and will be extensively used in Part II.

The two forms, eqs. (5.30) and (5.32), of the momentum equation are exactly equivalent, provided only that the relaton between $\mathbf{v}'$ and $\delta\mathbf{r}$ given by eq. (5.24) is used. This equivalence can be demonstrated by using the relation (eq. [4.1]) between d/dt and $\partial/\partial t$; noting that $d^2\delta\mathbf{r}/dt^2 = d\delta\mathbf{v}/dt$, that the right side of eq. (5.32) is $-(\rho^{-1}\nabla P)' + \mathbf{f}' + \delta\mathbf{r}\cdot\nabla(d\mathbf{v}_0/dt)$, and that $d\mathbf{v}_0/dt = d^2\mathbf{r}_0/dt^2$; and then realizing that $\delta\mathbf{r}\cdot\nabla(\mathbf{v}_0\cdot\nabla\mathbf{v}_0) = (\delta\mathbf{r}\cdot\nabla\mathbf{v}_0)\cdot(\nabla\mathbf{v}_0) + \delta\mathbf{r}\cdot(\mathbf{v}_0\cdot\nabla)(\nabla\mathbf{v}_0)$. See also Tolstoy (1973, Appendix V).

5.4c. ENERGY EQUATION

A very useful form of the nonlinear energy equation is eq. (4.30a), which however does not include any effects of composition changes that may be occurring as a result of thermonuclear reactions. We replace dq/dt by $(\epsilon - [\nabla\cdot\mathbf{F}]/\rho)$, where ϵ and $\mathbf{F}$ denote, respectively, the rate per unit mass of thermonuclear energy generation and the vector net heat flux. The Lagrangian variation of this equation is then

$$\frac{d}{dt}\left(\frac{\delta P}{P_0}\right) = \Gamma_{1,0}\frac{d}{dt}\left(\frac{\delta\rho}{\rho_0}\right) + (\delta\Gamma_1)\frac{d\ln\rho_0}{dt} + \frac{(\Gamma_3-1)_0\rho_0}{P_0}\delta\left(\epsilon - \frac{1}{\rho}\nabla\cdot\mathbf{F}\right)$$
$$+\left[\frac{\delta\Gamma_3}{(\Gamma_3-1)_0} + \frac{\delta\rho}{\rho_0} - \frac{\delta P}{P_0}\right]\frac{(\Gamma_3-1)_0\rho_0}{P_0}\left(\epsilon - \frac{1}{\rho}\nabla\cdot\mathbf{F}\right)_0. \tag{5.33}$$

If the system in its unperturbed state is static and in thermal equilibrium, so that $d\rho_0/dt = 0$ and $[\epsilon - (\nabla\cdot\mathbf{F})/\rho]_0 = 0$, we have

$$\frac{d}{dt}\left(\frac{\delta P}{P_0}\right) = \Gamma_{1,0}\frac{d}{dt}\left(\frac{\delta\rho}{\rho_0}\right) + \frac{(\Gamma_3-1)_0\rho_0}{P_0}\delta\left(\epsilon - \frac{1}{\rho}\nabla\cdot\mathbf{F}\right). \tag{5.34a}$$

The corresponding equation involving the temperature variations is

$$\frac{d}{dt}\left(\frac{\delta T}{T_0}\right) = (\Gamma_3-1)_0\frac{d}{dt}\left(\frac{\delta\rho}{\rho_0}\right) + \frac{1}{c_{V,0}T_0}\delta\left(\epsilon - \frac{1}{\rho}\nabla\cdot\mathbf{F}\right), \tag{5.34b}$$

where $c_{V,0}$ denotes the specific heat per unit mass at constant volume.

We may also note the general thermodynamic identities

$$\frac{\delta P}{P_0} = \Gamma_{1,0}\frac{\delta\rho}{\rho_0} + \frac{\rho_0(\Gamma_3-1)_0}{P_0}T_0\delta s, \tag{5.35a}$$

and

$$\frac{\delta T}{T_0} = (\Gamma_3 - 1)_0 \frac{\delta \rho}{\rho_0} + \frac{1}{c_{V,0}} \delta s, \tag{5.35b}$$

where δs denotes the Lagrangian variation of specific entropy s. These identities apply even if the star in its unperturbed configuration is not in complete equilibrium; they have been used by Demaret (1974).

If the linearized energy equation in terms of Eulerian variations were desired, one would proceed as follows. First, the Stokes time derivatives in eqs. (4.30a) or (4.30b) would be replaced by partial derivatives by use of eq. (4.1) (this statement of course does not apply to dq/dt). Then, the appropriate result would be obtained simply by taking the Eulerian variation of the resulting equation, making use of the commutativity rules summarized in §5.3.

In many discussions of the linear theory it is assumed that the motions are *adiabatic*. This normally means that the respective last terms in eqs. (5.34) are zero or negligible. (In the case of systems in thermal imbalance this concept must be more carefully defined; see §19.4.) Then eqs. (5.34) can be trivially integrated. Choosing the constant of integration in such a way that $\delta P = 0$ when $\delta \rho = 0$, we have

$$\frac{\delta P}{P_0} = \Gamma_{1,0} \frac{\delta \rho}{\rho_0}. \tag{5.36a}$$

The corresponding relation involving temperature variations is

$$\frac{\delta T}{T_0} = (\Gamma_3 - 1)_0 \frac{\delta \rho}{\rho_0}. \tag{5.36b}$$

5.5 APPLICATION: ADIABATIC SOUND WAVES

As an elementary application of the linearized equations, we shall briefly consider *adiabatic sound waves*. This application will emphasize the fact that, frequently, when we deal with stellar pulsations, we are in a sense dealing with a form of sound waves. We assume that the motion is *adiabatic* ($\delta[\epsilon - (\nabla \cdot \mathbf{F})/\rho] = 0$) and we neglect any changes in chemical composition. We also assume the unperturbed solution to be an "equilibrium" one, characterized by $\mathbf{v}_0 = 0$ and $[\epsilon - (\nabla \cdot \mathbf{F})/\rho]_0 = 0$; hence $\mathbf{v}' = \mathbf{v}$. Moreover, we assume that the characteristic length associated with spatial changes in P_0 and ρ_0 is long compared to the length scales of interest (short-wave acoustics), so that we can neglect ∇P_0 and $\nabla \rho_0$. Finally, we assume that no body forces are present ($\mathbf{f} = 0$).

Under these simplifying assumptions the mass, momentum, and energy equations can easily be combined into a single *wave equation* (see, for example Landau and Lifshitz 1959):

$$\frac{\partial^2 \rho'}{\partial t^2} = -{v_S}^2 \nabla^2 \rho', \tag{5.37}$$

where the *adiabatic* (or *Laplacian*) sound speed is

$$v_S \equiv (\Gamma_{1,0} P_0/\rho_0)^{1/2}. \tag{5.38}$$

It is shown by Landau and Lifshitz (1959) that the linearized momentum equation demands that $\mathbf{v}$ be parallel to $\pm\mathbf{k}$, the wave vector. Hence, the fluid velocities associated with adiabatic sound waves are parallel to the direction of propagation. In other words, such sound waves are *longitudinal,* or *compression,* waves, with pressure forces supplying all the restoring forces to the fluid particles associated with a perturbation in density. In Part III we shall encounter *gravity* waves, in which the restoring forces are supplied predominantly by gravity and not by pressure forces.

It is also shown by Landau and Lifshitz (1959) that the continuity equation demands that

$$\left|\frac{\rho'}{\rho_0}\right| = \frac{v}{v_S}. \tag{5.39}$$

This equation indicates that $|\rho'/\rho_0| \ll 1$ if $v/v_S \ll 1$. This result means that the linear theory is valid in the present application as long as the velocity fluctuations associated with the sound waves are small compared to the sound speed itself.

II

RADIAL OSCILLATIONS OF STARS

In this part we shall restrict attention specifically to purely radial, spherically symmetric oscillations of stars. Fortunately, this simple kind of motion, the most thoroughly studied of all, appears to be adequate for the theoretical treatment of most known kinds of pulsating stars observed in nature (see Chap. 3). Nonradial oscillations of stars will be considered in Part III.

In Chapter 6 we shall summarize some of the general vector equations of Chapter 4 in a form specifically adapted to the kind of motion assumed in this part, and in Chapter 7 we shall summarize some of the linearized equations of Chapter 5 in the corresponding form. The physically restricted but highly important case of linear *adiabatic* radial oscillations will be considered in Chapter 8. The physically more realistic but much more complicated case of linear nonadiabatic oscillations will be discussed in Chapter 9. Some excitation mechanisms will be reviewed, largely from a physical standpoint, in Chapter 10. The phase lag between stellar radius and luminosity will be discussed in Chapter 11. The very difficult but powerful nonlinear theory and computational techniques will be considered in Chapter 12. Finally, in Chapter 13 we shall summarize some of the simple models of stellar pulsation that have been devised to help make understandable some of the intricacies of the complex phenomenon of stellar pulsation.

6

Spherically Symmetric Radial Motion

In this chapter we shall summarize some of the relevant equations first in general spherical coordinates, and shall then specialize to the case of spherically symmetric radial motion. We shall always, except as noted otherwise, neglect turbulence and all forms of viscosity, and shall adopt the physical simplifications summarized at the beginning of Chapter 4. We shall also neglect large-scale magnetic fields and assume that the only body force is self-gravitation.

6.1. EULERIAN FORM

The system of spherical polar coordinates that we shall use is shown in Figure 6.1. Here $\mathbf{e}_r$, $\mathbf{e}_\theta$, and $\mathbf{e}_\phi$ are (dimensionless) orthogonal unit vectors pointing along the axes of the right-handed coordinate system shown. Any vector, such as velocity $\mathbf{v}$, will be resolved into components in this coordinate system:

$$\mathbf{v} = v_r \mathbf{e}_r + v_\theta \mathbf{e}_\theta + v_\phi \mathbf{e}_\theta, \tag{6.1}$$

where v_r, v_θ, and v_ϕ are in general functions of r, θ, and ϕ. The explicit form of the gradient operator ∇ in these coordinates can be found in any text on theoretical physics (e.g., Korn and Korn 1968).

Various kinds of spatial derivatives of $\mathbf{v}$ are possible ($\nabla \mathbf{v}$, $\nabla \cdot \mathbf{v}$, $\nabla^2 \mathbf{v}, \ldots$). Remembering that the unit vectors $\mathbf{e}_r$, $\mathbf{e}_\theta$, and $\mathbf{e}_\phi$ in general change direction during spatial differentiation, we may in a straightforward manner write out explicitly the nine components of the tensor $\nabla \mathbf{v}$ (which we do not give here). Since $\nabla \cdot \mathbf{v}$ is just the contracted form of this tensor, we have

$$\nabla \cdot \mathbf{v} = \frac{1}{r^2}\frac{\partial(r^2 v_r)}{\partial r} + \frac{1}{r \sin\theta}\frac{\partial(\sin\theta\, v_\theta)}{\partial\theta} + \frac{1}{r\sin\theta}\frac{\partial v_\phi}{\partial\phi}. \tag{6.2}$$

By writing $\nabla^2\Phi = \nabla \cdot (\nabla\Phi)$, where Φ is a scalar, we obtain

$$\nabla^2\Phi = \frac{1}{r^2}\frac{\partial}{\partial r}\left(r^2\frac{\partial\Phi}{\partial r}\right) + \frac{1}{r^2\sin\theta}\frac{\partial}{\partial\theta}\left(\sin\theta\frac{\partial\Phi}{\partial\theta}\right) + \frac{1}{r^2\sin^2\theta}\frac{\partial^2\Phi}{\partial\phi^2}. \tag{6.3}$$

For spherically symmetric radial motion we have $v_r \equiv v = v(r \text{ only})$, $v_\theta = v_\phi = 0$, whence $\nabla \cdot \mathbf{v}$ is given by only the first term on the right side of eq.

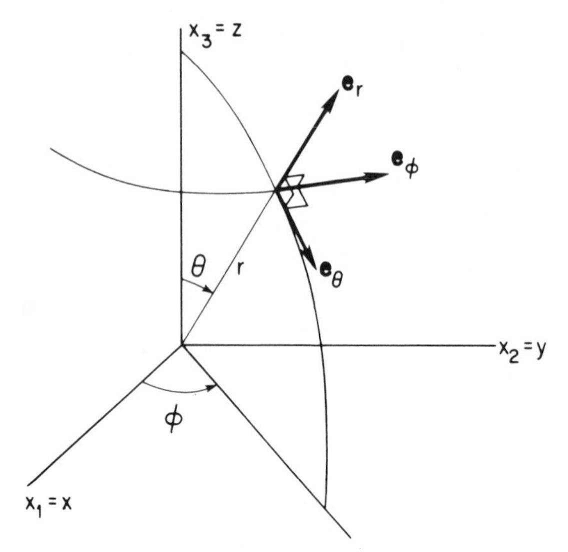

Figure 6.1. Spherical Polar Coordinates.

(6.2). Also, for spherical symmetry, $\Phi = \Phi(r \text{ only})$, and only the first term on the right side of eq. (6.3) remains. However, if $\mathbf{v} = v\mathbf{e}_r$ is a vector with only a radial component, the radial (and only) component of $\nabla^2\mathbf{v}$ consists of two terms: one similar to the first term on the right side of eq. (6.3), but with Φ replaced by v; plus an additional one (see, e.g., Morse and Feshbach 1953, p. 116; or Korn and Korn 1968).

The *mass* equation may be obtained, in the Eulerian form for purely radial motion, simply by starting with eq. (4.5) and expressing $\nabla \cdot \mathbf{v}$ therein as given by the first term on the right side of eq. (6.2).

If the only body force is self-gravitation, we must evaluate the gravitational potential $\psi(r,t)$ for our spherically symmetric system from the solution (4.43) of Poisson's equation. It is not difficult to show that, in this case, this solution takes the form

$$\begin{aligned} \psi(r) &= -\frac{Gm(r)}{r} - 4\pi G \int_r^R \rho(x) x dx \qquad (r \leq R) \\ &= -\frac{GM}{r} \qquad (r \geq R). \end{aligned} \tag{6.4}$$

Here R denotes the total stellar radius, M the stellar mass, G the constant of gravitation, and we have suppressed the explicit dependence on time t. The quantity $m(r)$ is the mass interior to a sphere of radius r (see eq. [6.10] below), and $m(R) = M$.

The gravitational force per unit mass $\mathbf{f} \equiv \mathbf{g} = -g\mathbf{e}_r$ is then obtained by taking the negative of the gradient of eq. (6.4): $\mathbf{f} = -\nabla\psi$. We obtain for the local gravitational acceleration g,

$$g = \frac{Gm(r)}{r^2}, \tag{6.5}$$

which is, intuitively, the force acting on a point unit mass located on the surface of a sphere of mass $m(r)$ and radius r.

If the stresses reduce to pure thermodynamic pressures P, the momentum equation may be written as in eq. (4.17), with the appropriate modifications for the case at hand.

The energy equation, for the case where the stresses reduce to pure thermodynamic pressures and any composition changes due to nuclear transmutations are neglected, may be written as in eqs. (4.30a) or (4.31), with dq/dt as given by eq. (4.40). The divergence of the net flux, $\nabla \cdot \mathbf{F}$, is for spherical symmetry

$$\nabla \cdot \mathbf{F} = \frac{1}{r^2}\frac{\partial(r^2F)}{\partial r} = \frac{1}{4\pi r^2}\frac{\partial L_r}{\partial r}, \tag{6.6}$$

where

$$L_r \equiv 4\pi r^2 F \tag{6.7}$$

is the "interior luminosity," or the net outward rate of flow of energy through a sphere of radius r.

In the deep stellar interior and for radiative transfer, the flux is given by eq. (4.41).

6.2. LAGRANGIAN FORM

In the Lagrangian form the mass equation for spherically symmetric radial motion is most readily obtained by starting with eq. (4.13), remembering that in this case v has only a radial component which is $\partial r/\partial t$, where r denotes radial distance and where $\partial/\partial t$ is the Stokes derivative. We regard r as a function of t and a, which will here be taken as some value of r which obtained at some prior time such as $t = 0$. Here, in keeping with the spirit of the Lagrangian description, both r and a give the radial distances out to the surface of a fixed amount of "interior mass" m at two times (see eq. [6.10] below). Using the mass equation (eq. [4.5]) for $\nabla \cdot \mathbf{v}$ for spherical symmetry, transforming from r to a as independent variables, interchanging $\partial/\partial a$ and $\partial/\partial t$ (since a and t are independent variables), integrating with respect to t and choosing the constant of integration so that $\rho = \rho_0$

when $r = a$, we obtain

$$\frac{\rho}{\rho_0} = \frac{a^2}{r^2}\frac{\partial a}{\partial r}. \tag{6.8}$$

This result is usually written more familiarly as $dm = 4\pi\rho r^2 dr = 4\pi\rho_0 a^2 da$, which just expresses the constancy of the mass dm in a thin, concentric spherical shell that moves radially with the fluid. Alternatively, the mass equation may be written in the form

$$\frac{\partial r}{\partial m} = \frac{1}{4\pi\rho r^2}, \tag{6.9}$$

where here the independent space variable is interior mass m, given explicitly by the relations

$$m = \int_0^a 4\pi\rho_0(x)x^2 dx = \int_0^r 4\pi\rho(x,t)x^2 dx. \tag{6.10}$$

The momentum equation for spherically symmetric radial motion, if all stresses reduce to pure pressures, may easily be obtained by starting with eq. (4.20) and regarding P as a function only of a (see above) and t. Combining this equation with the mass equation (6.8), we have

$$\ddot{r} = -4\pi r^2 \frac{\partial P}{\partial m} + f(m,t), \tag{6.11}$$

where $\ddot{r}$ stands for $\partial^2 r/\partial t^2$.

The energy equation in Lagrangian form is exactly the same as in the Eulerian form. However, in the Lagrangian form one generally writes

$$\frac{1}{\rho}\nabla \cdot \mathbf{F} = \frac{\partial(4\pi r^2 F)}{\partial m} = \frac{\partial L_r}{\partial m}, \tag{6.12}$$

where L_r is the interior luminosity (see eq. [6.7]). Thus, in this case $\partial L_r/\partial m$ is just the difference between the rate of energy flow *out* of the "top" of a spherical mass shell and that *into* the "bottom" of the mass shell, per unit mass of the shell.

We conclude this chapter by noting that it is possible (and sometimes useful!) to combine the mass, momentum, and energy equations into a single partial differential equation which is ordinarily of the third order in time and of the fourth order in space (see below).

We begin by taking the time derivative of the momentum equation (6.11), with $f(m,t)$ therein replaced by $-g$ (see eq. [6.5]). However, we may eliminate $\partial P/\partial m$ from the resulting equation in terms of $\ddot{r}$ and Gm/r^2 by means of the momentum equation itself. At the same time, we can express $\partial P/\partial t$ in terms of $\partial \rho/\partial t$ and $\epsilon - \partial L_r/\partial m$ by means of the energy equation (see eqs. [4.30a] and [4.40]). The quantity $\partial \rho/\partial t$ can be written in terms of $\partial(r^2 \dot{r})/\partial m$ by means of the mass equation (remembering that $\partial/\partial t$ and $\partial/\partial m$ commute because m and t may be regarded as independent variables in the Lagrangian description). We finally obtain

$$\dddot{r} - \frac{2\dot{r}\ddot{r}}{r} - \frac{4Gm}{r^3}\dot{r} - 4\pi r^2 \frac{\partial}{\partial m}\left[4\pi \Gamma_1 P \rho \frac{\partial(r^2 \dot{r})}{\partial m}\right] = -4\pi r^2 \frac{\partial}{\partial m}\left[\rho(\Gamma_3 - 1)\left(\epsilon - \frac{\partial L_r}{\partial m}\right)\right], \quad (6.13)$$

where of course Gm/r^3 could also have been expressed in terms of $\ddot{r}$ and $\partial P/\partial m$ by means of the momentum equation (6.11), with $f(m,t)$ therein replaced by $-g$. The nonradial analog of eq. (6.13) will be derived in Chapter 14.

The fact that eq. (6.13) is ordinarily of the fourth order in space may be seen from the following considerations. In the case of radiative transfer, for example, L_r involves $\partial T/\partial m$. Expressing T in terms of P and ρ by means of the equation of state of the material, we see that L_r involves (among other things) $\partial \rho/\partial m$. But it is evident from the mass equation (see eq. [6.9]) that ρ involves $\partial r/\partial m$, so that L_r actually involves $\partial^2 r/\partial m^2$. Noting that $\partial^2 L_r/\partial m^2$ appears on the right side of eq. (6.13), it is clear that this equation involves $\partial^4 r/\partial m^4$, that is, it is of the fourth order in space in this case.

7

Linearized Equations

In this chapter we shall summarize the basic linearized differential equations for purely radial, spherically symmetric motion. We adopt a Lagrangian description and shall make the simplifying assumptions stated at the beginnings of Chapters 4 and 6. Certain introductory remarks will be made in §7.1, and the linearized equations themselves will be derived in §§7.2–7.6. Finally, the linearized, nonadiabatic equation for the relative Lagrangian displacement $\delta r/r$ will be derived in §7.7.

7.1. INTRODUCTION

In this chapter we shall obtain the appropriate linearized equations for the Lagrangian variations δf of the physical variables f. The *independent* variables will be time t and (often) interior mass m, which is always the same in both the perturbed and unperturbed solutions (see Chap. 5). Hence eq. (6.10) applies, with a in the first integral here replaced by r_0, and where ρ_0 (which could also be a function of time t) and ρ are the densities in, respectively, the unperturbed and perturbed solutions. Clearly, both equalities in this equation could in principle be solved to give r and r_0 as functions of m and t.

Sometimes r_0 and t, rather than m and t, are used as independent variables. However, $\partial/\partial r_0$ commutes with the Stokes derivative $\partial/\partial t$ only if r_0 does not depend on t (but $\partial/\partial t$ *always* commutes with $\partial/\partial m$).

The *relative* displacement of the surface of a sphere containing interior mass m (or, loosely, of a mass shell) is

$$\zeta \equiv \delta r/r_0, \tag{7.1}$$

where r_0 is the appropriate radial distance in the unperturbed solution. The radial distance out to interior mass m in the perturbed solution is then

$$r = r_0(1 + \zeta). \tag{7.2}$$

We shall also often work in terms of the *relative* Lagrangian variation, $\delta f/f_0$, of some physical quantity f. Hence, in the perturbed solution we have

$$f = f_0(1 + \delta f/f_0). \tag{7.3}$$

In the linear approximation $|\zeta|$ and $|\delta f/f_0|$ are both assumed to be small compared to unity, and all powers (greater than one) and products of relative variations are neglected.

Unless we state otherwise, we shall assume throughout this chapter, and in fact throughout Part II, that the unperturbed state is one of complete equilibrium, characterized by zero velocities and both hydrostatic and thermal equilibrium. Hence, the unperturbed quantities will be assumed not to be functions of time. The case of a star in thermal imbalance is very much more complicated, and will be considered briefly in Chapter 19.

7.2. MASS EQUATION

The linearized mass equation for spherical symmetry can be obtained in a number of ways (which are all equivalent to one another). One way is to apply eq. (7.3) to both sides of the nonlinear form (6.9) of the mass equation. A simpler way is to treat the Lagrangian variation operator δ as an ordinary differential operator, and apply it directly to eq. (6.9) (suitably rewritten), interchanging δ and $\partial/\partial m$ (which is always possible) in the process. In any case the result may be written as

$$\frac{\delta\rho}{\rho_0} = -3\zeta - r_0\frac{\partial\zeta}{\partial r_0} \tag{7.4}$$

where the zero subscripts are often dropped.

7.3. MOMENTUM EQUATION

We start with the nonlinear momentum equation in the form of eq. (6.11), with $f(m,t) = -Gm/r^2$ therein. We now take the Lagrangian variation δ of this equation, remembering that here and throughout this Part (see §7.1) we are assuming that $\dot{r}_0 = 0$ and $\ddot{r}_0 = 0$, and we use the momentum equation again as applied to the unperturbed state. We obtain

$$\delta\ddot{r} = \frac{4Gm}{r_0^{\,2}}\zeta - 4\pi r_0^{\,2}\frac{\partial\delta P}{\partial m}\,. \tag{7.5}$$

The two terms on the right side of eq. (7.5) both have clear and straightforward physical interpretations. The first term is equal to the restoring force per unit mass acting on a displaced shell of matter, a force arising from purely geometrical effects. One-half of this force comes from the inverse square character of Newtonian gravitation. The other half comes from the proportionality of the total outward pressure on a sphere of radius r to r^2 ($4\pi r^2$ factor in first term on right side of eq. [6.11]). Note that the restoring force represented by this first term on the right side of eq. (7.5) is proportional to ζ. This means that this part of the total restoring force is always in the same direction as the displacement; for example, an outward displacement of a mass shell will result from this term in an outward contribution to the total restoring force.

The second term on the right side of eq. (7.5) can perhaps be more clearly interpreted if we divide and multiply δP by P_0 before carrying out the mass differentiation. After performing the differentiation, we once again make use of the momentum equation as applied to the unperturbed configuration. This term then has the approximate value $(Gm/r_0^2)\delta P/P_0$ (this approximation is valid if $\partial[\delta P/P_0]/\partial m$ is small). This term therefore gives the restoring force per unit mass resulting essentially from the pressure change δP accompanying the displacement. Note that, with the above proviso, this force has the same sign as δP, and is, for example, negative (inward) if δP decreases with an outward displacement of the shell. Thus, the *net* restoring force accompanying, say, an outward displacement of a mass shell can be negative (inward) only if the pressure decreases sufficiently with the expansion. These considerations are useful for a physical understanding of the dynamical stability of stars.

A sometimes useful form of the linearized momentum equation is obtained by writing $x \equiv r_0/R_0$, $dm = 4\pi r_0^2 \rho_0 dr_0$, and solving for $\partial(\delta P/P_0)/\partial x$. Dropping zero subscripts, we obtain

$$\frac{\partial}{\partial x}\left(\frac{\delta P}{P}\right) = \frac{R}{\lambda_P}\left(4\zeta - \frac{xR}{g}\ddot{\zeta} + \frac{\delta P}{P}\right), \tag{7.6}$$

where

$$\lambda_P \equiv -\frac{dr}{d\ln P} = \frac{P}{\rho g} \tag{7.7}$$

is the local pressure scale height in the equilibrium model, g being the local gravitational acceleration (see eq. [6.5]).

7.4. ENERGY EQUATION

For the present case of complete equilibrium in the unperturbed solution (see §7.1) and of negligible composition changes due to nuclear reactions, we may write the linearized energy equation in, for example, the form

$$\frac{\partial}{\partial t}\left(\frac{\delta P}{P}\right) = \Gamma_1 \frac{\partial}{\partial t}\left(\frac{\delta \rho}{\rho}\right) + \frac{\rho(\Gamma_3 - 1)}{P}\,\delta\left(\epsilon - \frac{\partial L_r}{\partial m}\right), \tag{7.8}$$

where L_r is the interior luminosity (see eq. [6.7]), with a somewhat similar expression for $\delta T/T$. We note that

$$\delta\left(\epsilon - \frac{\partial L_r}{\partial m}\right) = \epsilon_0 \frac{\delta\epsilon}{\epsilon_0} - \left(\frac{dL_{r,0}}{dm}\right)\frac{\delta L_r}{L_{r,0}} - L_{r,0}\frac{\partial}{\partial m}\left(\frac{\delta L_r}{L_{r,0}}\right), \tag{7.9}$$

where as usual zero subscripts denote quantities in the unperturbed state of the star. If this state is one of thermal equilibrium, then $\epsilon_0 = dL_{r,0}/dm$. Moreover, outside the energy generating regions, $\epsilon_0 = 0$ and $L_{r,0} = L$, the total stellar luminosity (constant with m). In this case only the last term in eq. (7.9) remains.

In the case of thermodynamically reversible processes, we have

$$\delta\left(\epsilon - \frac{\partial L_r}{\partial m}\right) = T_0 \frac{\partial \delta s}{\partial t}, \tag{7.10}$$

where s denotes specific entropy. This equation is valid if $\partial s_0/\partial t = 0$, which is equivalent to the assumption that $\epsilon_0 = dL_{r,0}/dm$.

The general thermodynamic identities (5.35a) and (5.35b) are also sometimes useful.

7.5. EQUATION OF RADIATIVE TRANSFER IN DIFFUSION APPROXIMATION

In the diffusion approximation (valid in the deep stellar interior, see §4.5), eq. (4.41) for the interior radiative luminosity L_r applies. Taking the logarithmic Lagrangian variation of this equation, we obtain

$$\frac{\delta L_r}{L_{r,0}} = 4\zeta - \frac{\delta\kappa}{\kappa_0} + 4\frac{\delta T}{T_0} + \frac{1}{d\ln T_0/dr_0}\frac{\partial}{\partial r_0}\left(\frac{\delta T}{T_0}\right) \tag{7.11a}$$

$$= 4\zeta - n\frac{\delta\rho}{\rho_0} + (s+4)\frac{\delta T}{T_0} + \frac{1}{d\ln T_0/dr_0}\frac{\partial}{\partial r_0}\left(\frac{\delta T}{T_0}\right) \tag{7.11b}$$

the second equality applying if the opacity $\kappa \propto \rho^n T^{-s}$.

The relatively large factor multiplying $\delta T/T_0$ in eq. (7.11b) should be noted; this factor has the value $(s + 4) \sim 7$ if $s \sim 3$ (a typical stellar interior value). Moreover, in the outer stellar layers (but below the region of hydrogen ionization), where $L_{r,0} \approx L$ and $m \approx M$ (total stellar mass), it is known (see, e.g., Schwarzschild 1958; Cox and Giuli 1968, Chap. 20) that the temperature may vary with depth in a simple manner, and we find, provided that $\partial(\delta T/T_0)/\partial r_0$ is not extremely large in these regions, that the last term in eq. (7.11b) may be rather small. Hence, the term in eq. (7.11b) containing the factor $(s + 4)$ is often the dominant term in the equation. These facts mean that under many conditions the local radiative flux can be more sensitive to the local temperature itself than to its gradient.

In the outer parts of the hydrogen ionization zone, however, the temperature exponent s typically becomes large and negative. This behavior is mainly a result of the strong dependence of the electron density on temperature in these regions. Since κ therefore increases sharply inward in these regions, $|dT_0/dr_0|$ must become very large here if the energy coming up from the deep interior is to be transported by radiation, and the above simple behavior of T_0 as a function of r_0 is then not even approximately valid (see, e.g., Cox and Giuli 1968, Chap. 20). In this case the situation is considerably more complicated than outlined above, and the last term in eq. (7.11b) cannot be neglected (see Chapter 11).

For improvements on the use of the diffusion equation in optically thin layers, see, for example, Unno (1965, App. A).

7.6. THERMONUCLEAR ENERGY GENERATION RATE

In the simplest case where we can approximate the rate per unit mass of thermonuclear energy generation by the relation $\epsilon \propto \rho^{\lambda} T^{\nu}$, we have

$$\frac{\delta\epsilon}{\epsilon_0} = \lambda \frac{\delta\rho}{\rho_0} + \nu \frac{\delta T}{T_0}. \tag{7.12}$$

This relation neglects any abundance variations and any phase delays in energy production (e.g., Rosseland and Randers 1938; Schatzman 1953; J. P. Cox 1955; and Ledoux and Walraven 1958, §66). However, for some purposes the abundance variations must be taken into account (e.g., Christensen-Dalsgaard, Dilke, and Gough 1974; Noels, Boury, Scuflaire, and Gabriel 1974; J. P. Cox 1976a).

7.7. LINEARIZED, NONADIABATIC EQUATION FOR $\delta r/r$

In this section we shall combine the linearized mass, momentum, and energy equations to obtain the third order (in time) linear, partial differential equation obeyed by $\zeta \equiv \delta r/r_0$ in nonadiabatic, radial stellar pulsations.

We start with the linearized momentum equation (7.5), and we take the Stokes derivative $\partial/\partial t$ of this equation, after writing δP as $P_0(\delta P/P_0)$. However, the quantity $\partial(\delta P/P_0)/\partial t$ in the resulting equation can be eliminated by using the energy equation in the form (7.8); and in this equation $\partial(\delta\rho/\rho_0)/\partial t$ can be eliminated by use of the mass equation (7.4). Substituting these last two expressions into the time-differentiated eq. (7.5) (after the above modifications), we obtain, after a small amount of manipulation,

$$r_0\ddot{\zeta} = 4\pi r_0^{\,2}\zeta \frac{d}{dm}\left[(3\Gamma_{1,0} - 4)P_0\right] + 12\pi r_0^{\,2}\Gamma_{1,0}P_0 \frac{\partial\dot{\zeta}}{\partial m}$$

$$+ 16\pi^2 r_0^{\,2} \frac{\partial}{\partial m}\left(r_0^{\,3}\Gamma_{1,0}P_0\rho_0 \frac{\partial\dot{\zeta}}{\partial m}\right)$$

$$- 4\pi r_0^{\,2} \frac{\partial}{\partial m}\left[\rho_0(\Gamma_3 - 1)_0\delta\left(\epsilon - \frac{\partial L_r}{\partial m}\right)\right]. \qquad (7.13)$$

However, we note the identity

$$\frac{4\pi}{r_0}\frac{\partial}{\partial m}\left(4\pi\Gamma_{1,0}P_0\rho_0 r_0^{\,6}\frac{\partial\zeta}{\partial m}\right)^{\cdot} = 12\pi\Gamma_{1,0}P_0 r_0^{\,2}\frac{\partial\dot{\zeta}}{\partial m}$$

$$+ 16\pi^2 r_0^{\,2}\frac{\partial}{\partial m}\left(\Gamma_{1,0}P_0\rho_0 r_0^{\,3}\frac{\partial\dot{\zeta}}{\partial m}\right), \qquad (7.14)$$

which can easily be proved by writing the factor $r_0^{\,6}$ as $r_0^{\,3} \cdot r_0^{\,3}$ and using the mass equation (7.4). Equation (7.13) then becomes

$$\ddot{\zeta} - 4\pi r_0\zeta\frac{d}{dm}\left[(3\Gamma_{1,0} - 4)P_0\right] - \frac{1}{r_0^{\,2}}\frac{\partial}{\partial m}\left(16\pi^2\Gamma_{1,0}P_0\rho_0 r_0^{\,6}\frac{\partial\dot{\zeta}}{\partial m}\right)$$

$$= -4\pi r_0\frac{\partial}{\partial m}\left[\rho_0(\Gamma_3 - 1)_0\delta\left(\epsilon - \frac{\partial L_r}{\partial m}\right)\right], \qquad (7.15)$$

which is the equation that was to be derived.

Note that eq. (7.15) could also have been obtained more directly by taking the Lagrangian variation of the general nonlinear partial differential equation for r (eq. [6.13]) derived in §6.2.

8

Linear Adiabatic Radial Oscillations

In this chapter we consider the case of small, purely radial, *adiabatic* oscillations of a self-gravitating gaseous sphere about its "equilibrium" configuration. We shall take this equilibrium configuration to be a static ($\dot{r}_0 \equiv 0$), spherically symmetric distribution of matter which is in both hydrostatic and thermal equilibrium ($\ddot{r}_0 \equiv 0$, $\epsilon_0 - dL_{r,0}/dm \equiv 0$, respectively), and in which the changes in chemical composition resulting from nuclear transmutations are negligible. We also make the other simplifying assumptions summarized at the beginnings of Chapters 4 and 6. An excellent, highly mathematical discussion of many topics dealt with in this chapter is provided by Ledoux and Walraven (1958, §§58–62).

This situation of small adiabatic oscillations is somewhat unrealistic, but it does give, in most cases, a good dynamical description of many features of actual pulsating stars. In particular, it gives quite accurate values for the pulsation periods and reasonably reliable results for the relative pulsation amplitude within the star. The main reason for these facts is, as will be emphasized further in Chapter 9, that the pulsations are actually nearly adiabatic throughout almost the entire stellar mass. This topic is also historically important. The mathematical theory was originated by Eddington (1918a,b). Being the simplest case to consider, it is the natural starting point for the theoretical study of pulsating stars.

Since the oscillations are assumed to be adiabatic (no net heat gains or losses by the oscillating mass elements), it is obvious that this case can yield no direct information regarding the *thermal* behavior of the star. For example, this theory cannot tell us how the luminosity of a pulsating star changes during the pulsations, as this is clearly a nonadiabatic effect (see Chapter 9). It is clear, moreover, that this theory cannot yield any direct information regarding the *existence* of the pulsations, that is, the *pulsational stability* of a star. The reason is that any small oscillations assumed to be present initially will maintain the same amplitude for all time, as the adiabatic assumption (in the absence of dissipative effects such as viscosity) implies that the system is perfectly conservative.

On the other hand, this theory can yield information regarding the *dynamical stability* of a star. This question will be considered briefly in passing in some of the following sections of this chapter.

Some general considerations, including a derivation of the linear adiabatic wave equation (LAWE), will be presented in §8.1. Standing wave solutions of this equation, and some conditions for their existence, will be considered in, respectively, §§8.2 and 8.4. Boundary conditions for these kinds of solutions will be discussed in §8.3. It will be pointed out in §8.5 that, as in most problems involving vibrations, the problem of stellar pulsations is, mathematically, an eigenvalue problem. The important period-mean density relation will be shown in §8.6 to follow from the LAWE. In §8.7 we shall consider some aspects of stellar pulsation from a largely physical standpoint. A mostly mathematical discussion of the eigenvalues will be presented in §8.8, and in §8.10 the important and useful variational property of the eigenvalues will be discussed. Some conditions for the existence of oscillatory solutions for arbitrary stellar models will be considered in §8.9. Some interesting physical interpretations of certain of the foregoing topics will be discussed in §8.11. Some practical techniques for solving the LAWE will be reviewed in §8.12, and certain results for selected stellar models will also be presented there. Epstein's now-famous results (and some more recent work) as to which parts of a star are most important in determining the period, will be discussed in §8.13. Some effects of a spatially variable adiabatic exponent Γ_1 in the stellar envelope on pulsation periods will be considered in §8.14. Finally, certain approximate formulae and results for the eigenvalues will be presented and discussed in §8.15.

8.1. GENERAL CONSIDERATIONS

In the theory of small adiabatic oscillations, the energy equations (see eqs. [5.34]) can, under the assumed conditions, be replaced by the simpler equations (5.36), where subscripts zero (often dropped) indicate the equilibrium state of the system.

The differential equation describing such oscillations can be obtained by substituting eq. (5.36a) into the linearized momentum equation and expressing $\delta\rho/\rho_0$ in terms of $\zeta(\equiv \delta r/r_0)$ by use of the linearized mass equation. The result is, after some rearrangement,

$$r_0\ddot{\zeta} = \zeta \cdot 4\pi r_0^2 \frac{d}{dm}[(3\Gamma_{1,0} - 4)P_0] + \frac{1}{r_0}\frac{\partial}{\partial m}\left(16\pi^2\Gamma_{1,0}P_0\rho_0 r_0^6 \frac{\partial\zeta}{\partial m}\right), \quad (8.1)$$

where we are using interior mass m and time t as independent variables, and a dot over a symbol denotes the Stokes derivative. This equation could also have been obtained from the linearized, nonadiabatic equation (7.15) by dropping the last term and integrating the resulting equation with respect to time. The constant of integration is set equal to zero so that $\zeta \equiv 0$ will correspond to the equilibrium (or static) state of the system.

Note that, by writing $\partial/\partial m = (4\pi r_0^2 \rho_0)^{-1}\partial/\partial r_0$ and then dropping zero subscripts, we may also write eq. (8.1) in terms of δr and its temporal and spatial derivatives. In general, the coefficients of δr and $\partial(\delta r)/\partial r$ will be found to vary appreciably in a distance comparable to R, the stellar radius. Thus, in pulsating star problems we normally deal with problems of long-wave acoustics. However, for sufficiently high "modes," the spatial wavelength of the perturbation δr will be small compared to R. In this case the resulting equation becomes essentially the equation for small, adiabatic sound waves (§5.5).

As a simple example, we might consider homologous motion, characterized by ζ constant in space throughout the star, that is, not a function of m. This example, though highly unrealistic, yields a great deal of quick insight. We also assume that Γ_1 is constant. We then have from eq. (8.1) (in which we have made use of the equation of hydrostatic equilibrium for the equilibrium model)

$$\ddot{\zeta} = -(3\Gamma_1 - 4)\frac{Gm}{r^3}\zeta. \tag{8.2}$$

We see immediately that *oscillatory* solutions ($\zeta \propto \exp[i\sigma t]$, σ real) exist only if $\Gamma_1 > 4/3$. In this case the square of the angular pulsation frequency σ is

$$\sigma^2 = \left(\frac{2\pi}{\Pi}\right)^2 = (3\Gamma_1 - 4)\frac{Gm}{r^3}, \tag{8.3}$$

where Π is the pulsation period. For $\Gamma_1 < 4/3$, we have $\sigma^2 < 0$, which corresponds to aperiodic motion on a dynamical time scale. This situation represents *dynamical instability*. In §8.9 below we shall point out that these results regarding $\Gamma_1 \gtrless 4/3$ are valid also for a *general* stellar model.

It is clear from eq. (8.3) that, because σ^2 is a constant, adiabatic and homologous oscillations in a model with $\Gamma_1 \neq 4/3$ are possible only for the *homogenous* model,

$$\rho(m) = \bar{\rho} \equiv \frac{M}{{}^4\!/_3\pi R^3} = \text{const.}, \tag{8.4}$$

where M is the stellar mass and $\bar{\rho}$ is the mean density of the star. As we shall see in Chapter 12, the above condition for the existence of homologous, oscillatory motion of the homogeneous model with $\Gamma_1 \neq 4/3$ applies also to nonlinear pulsations. However, if $\Gamma_1 = 4/3$, then $\sigma^2 = 0$ for *any* equilibrium model in which ζ is a function of t only. Hence, if $\Gamma_1 = 4/3$, then *any* stellar model can expand or contract homologously (but not pulsate) in adiabatic motion. Also as we shall see in Chapter 12, this conclusion applies to nonlinear motion as well.

8.2. STANDING WAVE SOLUTIONS OF THE LINEAR ADIABATIC WAVE EQUATION

We now assume standing wave solutions of eq. (8.1), of the form

$$\zeta(r,t) \equiv \delta r/r = \xi(r)e^{i\sigma t}, \tag{8.5}$$

where $\xi(r)$ is a function of r only and σ is a constant. (Of course, physical quantities must always be represented by real numbers. Hence the physical quantity $\delta r/r$ is actually the *real part* of the complex number appearing as the term on the right side of the last equality in eq. [8.5]. This point must always be kept in mind, but in this book we shall make this distinction explicitly only where necessary.) We shall discuss in a following section (§8.4) one condition for the existence of such standing wave solutions. Substituting eq. (8.5) into eq. (8.1), writing $\partial/\partial m = (4\pi\rho_0 r_0^2)^{-1}\partial/\partial r_0$, dropping zero subscripts, and rearranging slightly, we obtain

$$\frac{d}{dr}\left(\Gamma_1 P r^4 \frac{d\xi}{dr}\right) + \xi\left\{\sigma^2 \rho r^4 + r^3 \frac{d}{dr}[(3\Gamma_1 - 4)P]\right\} = 0. \tag{8.6}$$

This equation is sometimes referred to as the "linear adiabatic wave equation" (LAWE). The term is also used, alternatively, to refer to eq. (8.1), or even to its analogues in more general cases; see, for example, Chapter 15.

We shall point out in §8.8 that σ^2 is always real. Since all coefficients in eq. (8.6) are therefore real, it follows that purely real solutions $\xi(r)$ exist. Hence, true standing waves are possible. Therefore $Re\zeta(r,t)$ vanishes everywhere twice in every period in oscillatory motion. Since with adiabatic oscillations δP, δT, $\delta\rho$, etc., depend only on $\zeta(r,t)$ (see eqs. (5.36]), then they too must vanish everywhere, simultaneously, at these times. This state of affairs can also be expressed by saying that the star passes *precisely* through its equilibrium state twice in every period.

8.3. BOUNDARY CONDITIONS

In order to discuss the boundary conditions usually imposed on eq. (8.6), it is probably convenient first to expand out all derivatives therein, which we shall not do explicitly here.

At the center ($r = 0$), the spherical symmetry clearly demands that

$$\delta r = 0. \tag{8.7}$$

Only *regular* solutions ξ are normally of interest in physical problems. Such solutions are characterized by *finite* values of ξ and of its derivatives everywhere, in particular at the stellar center. Noting that the coefficient of $d\xi/dr$ is singular at the stellar center because of the factor $4/r$, we see that the above regularity condition also requires

$$\frac{d\xi}{dr} = 0 \tag{8.8}$$

at $r = 0$.

At the *surface* ($r = R$, stellar radius), it is usually assumed that the total pressure P vanishes (a situation in which this is not true will be considered later in this section). This assumption implies that

$$\delta P = 0 \tag{8.9}$$

at $r = R$.

The explicit expression of the surface boundary condition depends on the assumed surface boundary conditions in the equilibrium model. It is usually adequate to assume that the quantity P/ρ (proportional to temperature T for a perfect gas equation of state) becomes vanishingly small at the stellar surface. If this assumption is made, then an explicit expression for the surface boundary condition can be obtained (see eq. [8.11] below).

For a fairly general treatment of the surface boundary conditions, we may start with the form (7.6) of the linearized momentum equation, in which we have assumed that $\zeta \propto e^{i\sigma t}$. We have

$$\frac{\partial}{\partial x}\left(\frac{\delta P}{P}\right) = \frac{R}{\lambda_P}\left[\left(\frac{\sigma^2 r}{g} + 4\right)\zeta + \frac{\delta P}{P}\right], \tag{8.10}$$

where $x \equiv r/R$, λ_P is the local pressure scale height (see eq. [7.7]), and g is the local gravitational acceleration (see eq. [6.5]). Since, ordinarily, $R/\lambda_P \gg 1$ at the photospheres of real stars (and since $R/\lambda_P = \infty$ at the stellar surface if it is assumed that $P/\rho = 0$ here), then a reasonable surface boundary condition would be that the quantity in square brackets in eq. (8.10) must vanish:

$$\frac{\delta P}{P} = -\left(\frac{\sigma^2 r}{g} + 4\right)\zeta; \tag{8.11}$$

or at least that the right side of eq. (8.11) must be multiplied by the factor $(1 + O[\lambda_P/R])$, where $O(\lambda_P/R)$ may represent a power series in λ_P/R whose leading term is of order λ_P/R. We see upon substituting eq. (8.11) into eq. (8.10) that, indeed, $\delta P/P$ cannot exhibit any pathological, or singular, behavior in the surface regions.

We may note that eq. (8.11) was derived without any appeal to the assumption of adiabatic oscillations, and is therefore valid also for *non-adiabatic* (but small) oscillations.

It can be shown that eq. (8.11) will also prevent ζ from exhibiting singular behavior at the stellar surface $x = 1$.

What these last results mean is that both $\delta P/P$ and ζ vary (spatially) on a characteristic scale $\sim R$ (the stellar radius), and not on the scale of λ_P (of the order of the thickness of the stellar atmosphere; see, e.g., Cox and Giuli 1968, Chap. 20). Thus, the fractional amounts by which $|\delta P/P|$ and $|\zeta|$ can vary throughout the stellar atmosphere are of order λ_P/R, which is normally small compared to unity. These quantities are therefore nearly spatially constant throughout the thin, cool atmosphere (however, note the qualifying remarks at the end of the next section, §8.4).

We may note that the result

$$\frac{\partial}{\partial r}\left(\frac{\delta P}{P}\right) = 0 \tag{8.12}$$

applies exactly to a plane-parallel atmosphere in which the surface gravity g varies only "slowly" (see below) in time. Under these conditions instantaneous hydrostatic equilibrium implies that the pressure P at any mass level in the atmosphere is just equal to the weight per unit area of the mass lying above that level. Then, at the fixed mass level, $\delta P/P \approx \delta g/g =$ constant with respect to r, from which eq. (8.12) follows.

The term "slowly" above means that the time scale of the variations of the atmosphere associated with the pulsations (the period Π) is long compared to the natural "inertial response time," say t_a, of the atmosphere. This latter time may be taken as the time required for a unit mass to fall freely under the action of the stellar surface gravity g through a distance comparable with λ_P. Elementary considerations and use of the order-of-magnitude relation (2.8) then show that

$$t_a/\Pi \sim (\lambda_P/R)^{1/2}, \tag{8.13}$$

so that the inertial response time of the atmosphere is indeed much smaller than the pulsation period if $\lambda_P/R \ll 1$.

Hence, the boundary condition (8.11) implies a kind of "quasi-hydrostatic equilibrium," in which the atmosphere passively "floats" on the rising and sinking interior, much as a cork bobs up and down on an ocean wave.

In the case of *adiabatic* oscillations, we simply replace $\delta P/P$ in eq. (8.11) by $\Gamma_1 \delta\rho/\rho$ and use the linearized mass equation (7.4) for $\delta\rho/\rho$.

Castor (1971) has pointed out that, if radiation pressure is important at the stellar surface, the boundary condition (8.11) should be applied only to the *gas* pressure P_g; that is, P therein should be replaced by P_g. This conclusion can be understood by expanding out the derivative in eq. (8.10) and using eq. (7.7) for λ_P. Then it is clear that the correct, *general* surface boundary condition (if $\rho = 0$ at the surface) is

$$\frac{\partial(\delta P)}{\partial r} = 0 \qquad (8.14)$$

at $r = R$. Castor's conclusion then follows by integrating the general expression for $\partial(\delta P)/\partial r$, using the equation of hydrostatic equilibrium therein, making use of the near spatial constancy of r, g, and ζ in the vicinity of $r = R$, and applying this integrated expression sufficiently far out in the atmosphere that $\rho = 0$.

For adiabatic oscillations when radiation pressure is important, we may obtain a surface boundary condition still more general than the one derived from eq. (8.11). Making the approximation that $P_r \propto (1/3)T^4$ (strictly valid only in the deep interior), and using Castor's result (eq. [8.11] with P replaced by P_g), we eventually obtain for adiabatic oscillations, at $x = 1$:

$$\frac{d \ln \xi}{d \ln x} = \frac{[\sigma^2 R^3/(GM)]\beta - (3\Gamma_1 - 4\beta) + 12(\Gamma_3 - 1)(1 - \beta)}{\Gamma_1 - 4(\Gamma_3 - 1)(1 - \beta)}, \qquad (8.15)$$

where

$$\beta \equiv \frac{P_g}{P_g + P_r} \qquad (8.16)$$

is the ratio of gas to total (gas plus radiation) pressure. Equation (8.15) reduces to the correct limits for $\beta = 0$ and $\beta = 1$, as can be shown from the expressions for the Γ's for a mixture of a perfect gas and blackbody radiation (e.g., Chandrasekhar 1939, Chap. 2).

The above boundary conditions (or their equivalents) are sometimes collectively referred to as the "standing wave boundary condition" (see §8.4).

A very thorough discussion of the surface boundary conditions in the linear theory has been given by Unno (1965).

8.4. CONDITION FOR STANDING WAVES IN THE STELLAR INTERIOR

The assumption of standing wave solutions of the form $\zeta(r,t) = \xi(r)e^{i\sigma t}$ requires that the (sound) wave coming up from the stellar interior undergoes perfect reflection back into the interior at the surface. In this section we shall examine the conditions for such perfect reflection by considering the small, adiabatic oscillations of a thin, plane-parallel, isothermal atmosphere (see Baker and Kippenhahn 1965; Ledoux and Walraven 1958, §68; and Unno 1965). It is known that stellar atmospheres tend to become isothermal at small optical depths (see, for example, Cox and Giuli 1968, Chap. 20). For such isothermal atmospheres, we know that the pressure P and density ρ both go to zero exponentially with increasing height in the atmosphere: $P, \rho \propto \exp[-(r-R)/\lambda_P]$, where λ_P is the pressure scale height in the atmosphere (see eq. [7.7]), assumed constant here. We shall also assume, for simplicity, that the gammas are constant here (both in space and time).

Under these simplifying assumptions, the LAWE becomes a second-order, linear differential equation in ξ with constant coefficients; this equation is

$$\frac{d^2\xi}{dx^2} - \frac{1}{\lambda_P}\frac{d\xi}{dx} - \frac{\sigma^2}{v_S^2}\xi = 0, \qquad (8.16')$$

where x denotes height in the atmosphere above some arbitrary reference level. Such an equation has solutions of the form $\delta r \propto e^{kx}$, where k, the *wave number* for the waves we are considering, is given by the solution of a quadratic equation; and x denotes height in the atmosphere above some arbitrary reference level.

Considering the solution of this equation, two cases may be distinguished. One, which we shall refer to as case (a), is characterized by *real* values of k, and applies when $\sigma\lambda_P/v_S < 1/2$, where $v_S = (\Gamma_1 P/\rho)^{1/2}$ is the Laplacian sound speed, also assumed constant in the atmosphere. We designate the two solutions of the quadratic by k_1 and k_2, where we label these solutions in such a way that $k_1 < k_2$; writing $k = 2\pi/\lambda$, where λ is the wavelength of the wave associated with the wave number k, we have $\lambda_1 > \lambda_2$. Hence k_1 is characterized by longer sound waves than is k_2, and corresponds to a slower outward growth in amplitude than does k_2. It may be shown that, for $\sigma\lambda_P/v_S \ll 1$, we have $\lambda_1 \gg \lambda_P$; this situation therefore corresponds to "long" sound waves. Similarly, we have $\lambda_2 \approx \lambda_P$; k_2 in this instance therefore corresponds to "short" sound waves.

It should now be considered whether or not both of the above solutions are physically allowed. For this purpose let us examine the acoustic energy per unit volume. This energy is proportional to the kinetic energy per unit volume, say K. One can then easily show that $K_1 \to 0$ as $x \to \infty$. Similarly, $K_2 \to \infty$ as $x \to \infty$. Since K_2 diverges at infinity, the solution represented by k_2 is presumably not physically allowed (however, see remark at the end of this section). The only physically possible solution for case (a) would then be

$$\delta r \propto e^{i\sigma t} \cdot e^{(\sigma^2\lambda_P^2/v_S^2)(x/\lambda_P)} \quad (\sigma^2\lambda_P^2/v_S^2 \ll 1), \qquad (8.17)$$

where, in view of our simplifying assumptions, this solution is not valid if $x/\lambda_P \gg 1$. Hence, the exponential behavior in eq. (8.17) is not inconsistent with our conclusions in §8.3 that, for the standing waves of interest in the case of pulsations of the whole star, δr cannot vary very much throughout the thin atmospheric layers.

We must now consider *case (b)*: $\sigma\lambda_P/v_S > 1/2$, for which both roots of the above quadratic equation are complex. Complex values of k correspond to running waves, in general with one wave proceeding toward the center of the star and one proceeding away from the center. Since, physically, we can normally only allow *outgoing* waves, the solution representing ingoing waves must be discarded. It can also be shown that there is indeed a genuine loss of energy from the star, in the form of running acoustic waves, in case (b).

We conclude that we can have standing waves in the stellar interior only for case (a), that is, for $\sigma\lambda_P/v_S < 1/2$. Writing $v_S = (\sigma/2\pi)\lambda$, where λ is the wavelength of the sound waves associated with the angular frequency 2π/Period that we are considering, we see that the condition for the existence of standing waves in the stellar interior is

$$\lambda > 4\pi\lambda_P, \tag{8.18}$$

where λ_P is the photospheric pressure scale height.

We note that differentiation with respect to r of eq. (8.17) yields (after a little work) eq. (8.15), with β set equal to unity, except that $3\Gamma_1 - 4$ in eq. (8.15) is replaced by Γ_1. The reason for this difference is that curvature in the atmosphere has been neglected in this section. If the above analysis were to be carried through by not neglecting curvature, the correct factor in eq. (8.15) would have been obtained (see, e.g., Baker and Kippenhahn 1965, Appendix).

For the low modes of pulsating stars, however, we have $\lambda \sim R$ (stellar radius). Hence, eq. (8.18) is, roughly, $R \gg \lambda_P$. However, it can be shown fairly generally from the virial theorem that λ_P/R is of the order of the ratio of the internal energy per unit mass at the center to that at the surface. If we consider ordinary gaseous stars in which radiation pressure is not very important, the above statement implies that

$$\frac{\lambda_P}{R} \sim \frac{\text{surface temperature}}{\text{central temperature}} \sim 10^{-2} - 10^{-4} \tag{8.19}$$

for most stars. Equation (8.18) is therefore very likely satisfied, to a good approximation, for the lower pulsation modes of most stars. This equation may not be satisfied, however, for the higher modes. Also, a high-temperature corona could lead to imperfect reflection and hence to waves propagating into the corona, especially for the higher modes (see, e.g., Schatzman 1956). Moreover, the presence of a high-temperature corona, or other factors, might necessitate consideration of the solution represented by k_2 of case (a) in some circumstances (e.g., Hill 1978, and a number of papers in Hill and Dziembowski 1979). We shall not, however, consider this subject further in this part.

8.5. EIGENVALUE NATURE OF THE PROBLEM

Equation (8.6) is a second order, ordinary differential equation; therefore its solution must involve *two* constants of integration. However, this equation is also *linear* and *homogeneous* in ξ, so that one of the constants must remain arbitrary (if ξ is a solution, then so is $A\xi$, where A is an arbitrary constant). Physically, the *absolute value* of the pulsation amplitude ξ must be arbitrary. The remaining constant of integration can clearly be used to satisfy only *one* of the two boundary conditions, one at the center and one at the surface. The only other disposable parameter is σ, the angular pulsation frequency. Its value must be varied until the *other* of the above two boundary conditions is satisfied. Hence, only certain eigenfrequencies, $\sigma_0, \sigma_1, \ldots$, and the corresponding eigenfunctions, $\xi_0, \xi_1, \ldots$, exist which satisfy the boundary conditions both at the center and at the surface. In general, there are an infinite number of eigenfrequencies and corresponding eigenfunctions, and the eigenfrequencies are not in general integral multiples of the lowest, or *fundamental,* eigenfrequency σ_0. The eigenfunction ξ_0, of the fundamental mode, is the function in which there are *no zeros* (or *nodes*) of ξ in the range $0 \leq r \leq R$. In general, the eigenfunction ξ_k for the kth mode has just k nodes in the range $0 \leq r \leq R$. We shall be mostly concerned in this part with only the fundamental or lower modes.[1]

The above statements (and the proofs of them) also follow from the fact that eq. (8.6) is of the *Sturm-Liouville* type, and therefore has an extensive mathematical literature (see, e.g., Ledoux and Walraven 1958, §58; Ince 1944, Chaps. 9–11).

Some very interesting and illuminating parallels between the LAWE and the Schrödinger wave equation of quantum mechanics have been pointed out by Axel and Perkins (1971). Some earlier discussions along these lines were by Meltzer and Thorne (1966).

8.6. PERIOD-MEAN DENSITY RELATION

In this section we shall point out that the famous period-mean density relation for pulsating stars (see, e.g., Chapter 2) follows from the LAWE.

We can nondimensionalize this equation (which we shall not do expli-

[1]See footnote in §8.12c.

citly here; however, see eq. [8.58] below), and when we do so we see that there is only one parameter in the equation. This parameter is the (dimensionless) eigenvalue, and this eigenvalue is clearly the square of the dimensionless angular frequency Ω:

$$\Omega^2 \equiv \frac{\sigma^2 R^3}{GM} . \tag{8.20}$$

We may note that a simple physical interpretation may be given to Ω^2: it is equal to the ratio of the acceleration associated with the angular frequency σ of a unit mass lying on the surface of the star, to the gravitational acceleration there. This first acceleration is also equal to the centripetal acceleration of a very close satellite circling the star with angular frequency σ in a circular orbit of radius R, and for such a circular orbit this acceleration must equal the gravitational acceleration at the surface. Hence $\Omega^2 = 1$ for such a satellite orbit. We shall point out below (§8.9) that for a pulsating star with $\Gamma_1 = 5/3$, we must always have $\Omega^2 \geq 1$. Thus the pulsation period of a star must always be less than the period of such a close satellite.

When we apply the nondimensionalized form of the LAWE to a family of *homologous* stars (see, e.g., Cox and Giuli 1968, Chap. 22) all with the same value of Γ_1, we see that only *one* value of Ω applies to all members of the family. Since the mean density $\bar{\rho} \propto M/R^3$ and $\sigma = 2\pi/\Pi$ (Π = period), eq. (8.20) is a statement of the *period-mean density relation,* $\Pi\sqrt{\bar{\rho}}$ = constant, for homologous stars all having the same value of Γ_1. In terms of the pulsation constant $Q = \Pi\sqrt{\bar{\rho}/\bar{\rho}_\odot}$ ($\rho_\odot$ = mean density of the sun = 1.41 gm cm^{-3}), we have

$$Q_d = \frac{0.1159}{\Omega} , \tag{8.21}$$

where subscript d means that the units are days. Hence, Q is strictly constant (for a given mode) from one star to another only for *homologous* stars all having the same value of Γ_1.

Since real stars are not homologous to one another and since Γ_1 is in general not the same for all stars (besides not being constant), then $\Pi\sqrt{\bar{\rho}}$ is not *strictly* constant from one star to another, for a given mode. However, this quantity turns out (see §8.12c below) to be, generally, only a slowly varying function of stellar parameters. Convenient fitting formulae, giving the dependence of Q on these parameters, have been provided by J. P. Cox, King, and Stellingwerf (1972); King, Hansen, Ross, and Cox (1975); and Faulkner (1977b); see also Cogan (1970). Also, it turns out (see Chaps. 9 and 12 below) that nonadiabaticity and nonlinearity have only a small effect, normally no more than a few percent, on the value of this quantity, that is, on the periods. Consequently, most of the remarks made in this Chapter regarding periods have a much wider range of validity than for strictly linear, adiabatic oscillations.

8.7. PHYSICAL DISCUSSION

The LAWE can also be derived in a purely *physical* way, by conceptually displacing a thin mass shell adiabatically from its hydrostatic equilibrium position in a star, and then computing all the restoring forces acting on the displaced shell. In this way we see that σ^2 is proportional to the "force constant" of a simple harmonic oscillator.

The assumption that σ is constant means that the force constant is the same for all mass shells in the star: all such shells have the same "natural" oscillation frequency. In the special case of ξ (space part of ζ) independent of r_0, this constancy of the force constant can only be realized if $m/r_0^3 =$ const., that is, if the equilibrium model is the homogeneous model.

If we drop the assumption that ξ is independent of r_0, we then have an *arbitrary* equilibrium model. Again, the assumption that σ^2 is constant means that we are requiring all mass shells to have the same force constant. Since we are now dealing with an arbitrary model in which m/r_0^3 is not constant, constancy of ξ will clearly not lead to $\sigma^2 =$ constant. The function $\xi(r_0)$, in fact, is no longer arbitrary; rather, it must be such as to render σ^2 constant for the given model. Moreover, $\xi(r_0)$ must obey boundary conditions both at center and surface. For the reasons pointed out in §8.5, only certain functions $\xi_k(r_0)$, the *eigenfunctions,* exist which can satisfy all of these conditions. The corresponding *eigenvalues,* σ_k^2, are the only possible values of σ^2 for which all mass shells can have the same force constant, or the same "natural" oscillation frequency. The fact that the LAWE is a differential equation means that the restoring force on a given mass shell, and hence its oscillation frequency, depends not only on ξ but also on its derivatives. The behavior of the mass shell is therefore coupled to the behavior of other mass shells in the star, and the whole star oscillates in unison.

Thus, for example, one could imagine a situation where $\Gamma_{1,0}$ might be less than 4/3 in some region(s) of the star and greater than 4/3 elsewhere. (Such a situation must, in fact, prevail in the envelopes of essentially all stars; see, e.g., Cox and Giuli, 1968, Chap. 20.) However, the material in the region(s) where $\Gamma_{1,0} < 4/3$ would be dynamically stable if the whole star is, i.e., if the constant σ^2 is positive.

Since σ^2 may be interpreted as essentially a force constant, its value should increase as the order of the mode increases, because of the larger accelerations associated with the higher modes. Hence we can see in a general way that $|d\xi/dr|$ and $|d^2\xi/dr^2|$ ought to increase, as detailed calculations in fact show them to, as the order of the mode increases. A similar conclusion follows more rigorously from the mathematical nature of the LAWE when account is taken of its Sturm-Liouville character (see, e.g., Ince 1944, §10.3). (However, this conclusion does not necessarily follow for nonradial oscillations; see chapter 17.)

At this point it might be asked how, in nature, a star "knows" that it must oscillate with σ^2 the same for all of its parts. The immediate answer is that, in general, a star need not necessarily oscillate in this manner—in fact, there are probably many real stars (for example, various kinds of irregular variables) which do not oscillate in this way at all. A great many stars (including the classical Cepheids, RR Lyrae variables, and W Virginis variables), however, *do* appear to be pulsating in just this simple fashion, as if only one mode were present.

In an attempt to explain this phenomenon, let us suppose that σ^2 is *not* the same for all mass shells and that, moreover, $\zeta(r_0,t)$ is an essentially arbitrary function of r_0 and t (except that ζ must be such as to satisfy the relevant boundary conditions). In this case the motion would be chaotic, and steep temperature and pressure gradients would be expected. As we shall see in §8.8, arbitrary motions can be represented as a superposition of "normal modes" (i.e., eigenfunctions); and chaotic motions correspond, in general, to the presence of modes of arbitrarily high order. In discussing real stars one must abandon the ideal assumption of adiabatic oscillations, and allow for the effects of heat gains and losses, and also of friction forces, arising from viscosity. In general, these effects might both be expected to be relatively large if steep temperature and pressure gradients are present.

In an actual star in which the motion was initially chaotic, then, one would expect the highest-order modes, being accompanied by the steepest temperature and pressure gradients, to be "damped" most rapidly. Given sufficient time, only the lowest few modes, or perhaps only one of them, may be present. To say *which* mode(s) will finally be present would require a detailed analysis of both the damping mechanisms and whatever active driving mechanisms might be present in the star. If that star were pulsationally stable to *all* modes, then even the above remaining mode(s) would eventually be damped also.

What we are saying is that chaotic motions are expected to dissipate energy, and therefore to be damped, through nonadiabatic effects and friction, more strongly than orderly motions. Hence, in a general way one might expect ordered motion to be preferred in nature over chaotic motion. This expectation might be one answer to the question posed a few paragraphs back, and might provide a plausibility argument for the existence of pulsating stars possessing apparently rather simple motions.

8.8. DISCUSSION OF THE EIGENVALUES: MOSTLY MATHEMATICAL

The eigenfunctions and eigenvalues of the LAWE possess a number of important properties. For example, the eigenfunctions ξ_k are *orthogonal* to one another with respect to the element of moment of inertia $dI = r^2dm$, and all the eigenvalues σ_k^2 are *real*. These properties owe their existence to the *Hermitian* character of the LAWE.

For the kth mode, the LAWE is eq. (8.6), with $\xi(r) = \xi_k(r)$, and with the time dependence of eq. (8.5) ($\sigma \equiv \sigma_k$). We may write eq. (8.6) as

$$\mathcal{L}(\xi_k) = \sigma_k^2\xi_k, \tag{8.22}$$

where the linear operator $\mathcal{L}$ is defined by the relation

$$\mathcal{L}(y) \equiv -\frac{1}{\rho r^4}\frac{d}{dr}\left(\Gamma_1 Pr^4\frac{dy}{dr}\right) - \frac{1}{\rho r}\left\{\frac{d}{dr}[(3\Gamma_1 - 4)P]\right\}y. \tag{8.23}$$

The operator $\mathcal{L}$ has the important property that it is *Hermitian* if $P = 0$ at the stellar surface $r = R$ (equilibrium stellar radius). This means, in the present context, that if u and v are any "sufficiently regular" functions, then

$$\int_M u^*(\mathcal{L}v)r^2dm = \left[\int_M v^*(\mathcal{L}u)r^2dm\right]^* = \int_M v(\mathcal{L}u)^*r^2dm, \tag{8.24}$$

where the integrations are carried out over the entire stellar mass M, and an asterisk denotes the complex conjugate.

The Hermiticity of $\mathcal{L}$ can be proved by substituting eq. (8.23) for $\mathcal{L}$ into the left side of eq. (8.24), and using the relation $dm = 4\pi r^2\rho dr$. Performing a parts integration, making use of the vanishig of P at the stellar surface, and noting that $\mathcal{L}$ is a real operator, we see that u^* and v may be interchanged in the first integral in eq. (8.24) without affecting its value. The Hermiticity of $\mathcal{L}$ as defined in eq. (8.23) is therefore proved.

The reality of all the eigenvalues σ_k^2 of $\mathcal{L}$ now follows readily from the Hermiticity of $\mathcal{L}$, as is well known to anyone who has studied quantum mechanics (see also §15.2). Positive values of σ_k^2 (σ_k real) correspond to oscillations of constant (in time) amplitude, whereas negative values of σ_k^2 (σ_k imaginary) correspond to dynamical instability (it has, in fact, been shown by Cox 1967 that either dynamical collapse or dynamical expansion are inevitable in the case of dynamical instability).

Solutions of the LAWE corresponding to positively or negatively damped oscillations (σ_k complex) therefore do not exist in this case (however, solutions of the LAWE characterized by σ_k complex *do* exist if motions in the unperturbed model are present; see Chapter 15 and §19.1).

This statement is just another way of saying that the theory of purely conservative, adiabatic oscillations can ordinarily yield no direct information regarding the question of pulsational instability in a star.

The orthogonality of the eigenfunctions with respect to r^2dm can also easily be demonstrated, but we will not do so here (see §15.2). If ξ_k and ξ_l are any two eigenfunctions, the orthogonality condition may be expressed in the form

$$\int_M \xi_k^* \xi_l r^2 dm = J_k \delta_{kl}, \tag{8.25}$$

where

$$J_k \equiv \int_M |\xi_k|^2 r^2 dm = \int_I |\xi_k|^2 dI \tag{8.26}$$

is the *oscillatory moment of inertia* for the kth mode, δ_{kl} is the Kronecker delta ($\delta_{kl} = 0$ if $k \neq l$, 1 otherwise), and

$$I \equiv \int_M r^2 dm \tag{8.27}$$

is the moment of inertia of the star about its center.

More general solutions of the linear adiabatic (or nonlinear, nonadiabatic) wave equation can always be expressed in terms of the eigenfunctions ξ_k of $\mathcal{L}$, since these eigenfunctions are known to be *complete* (see e.g., Ince 1944, Chap. 11). Thus, if $\zeta(r,t)$ is any function for which the integral $\int_M |\zeta(r,t)|^2 r^2 dm$ exists, we can write

$$\zeta(r,t) = \sum_k q_k(t) \xi_k(r), \tag{8.28}$$

where the $q_k(t)$ are functions only of the time t, and are given for all k by virtue of the orthogonality of the ξ_k by the relation

$$q_k(t) = \frac{1}{J_k} \int_M \xi_k^*(r) \zeta(r,t) r^2 dm. \tag{8.29}$$

8.9. CONDITIONS FOR EXISTENCE OF OSCILLATORY SOLUTIONS

In this section we shall examine some conditions for the existence of oscillatory solutions of the LAWE for an *arbitrary* spherical stellar model, under the conditions of our basic assumptions in this part. A discussion of much of the contents of this section may also be found in Ledoux and Walraven (1958, §60).

We first write the expression for the eigenvalue σ_k^2 as an integral over the entire star, as follows:

$$\sigma_k^2 = \frac{1}{J_k} \int_M \xi_k^* (\mathcal{L}\xi_k) r^2 dm. \tag{8.30}$$

We consider in detail only the case where Γ_1 is constant throughout the star. Then eq. (8.30) may be written (after inserting the specific form of $\mathcal{L}$ as given in eq. [8.23] and performing a few manipulations) as the sum of two integrals one of which is always greater than or equal to zero, the other of which is positive for $\Gamma_1 > 4/3$, for a star in hydrostatic equilibrium. Thus, if $\Gamma_1 > 4/3$, then $\sigma_k^2 > 0$ for all k. Oscillatory solutions therefore exist for *all* modes if $\Gamma_1 > 4/3$, regardless of the nature of the equilibrium stellar model.

If $\Gamma_1 = 4/3$, one of the above two integrals vanishes, and it then follows that, in general, $\sigma_k^2 \geq 0$, all k. For the *fundamental* mode, $k = 0$, however, it can be shown from the variational property of the LAWE (see §8.10) that $\sigma_0^2 = 0$ for this case (we shall not explicitly show this here). Thus, any star with $\Gamma_1 = 4/3$ is *neutrally* stable in the fundamental mode ($\sigma_0^2 = 0$). It also follows from this variational property that $d\xi_0/dr = 0$, or that ξ_0 is independent of r for this case. The displacement $(\delta r)_0$ of a given interior mass level in this mode is therefore proportional to r for this mass level, and we have a case of *homologous* expansion or contraction on an arbitrary time scale.

Because of the Sturm-Liouville nature of the LAWE, the eigenvalues can be ordered in such a way that $\sigma_k^2 > \sigma_0^2$, $k > 0$ (see, e.g., Ince 1944, Chap. 10). It then follows that, for $\Gamma_1 \geq 4/3$, $\sigma_k^2 > 0$, $k > 0$. This result means that the above neutral stability exists *only* for the fundamental mode for $\Gamma_1 = 4/3$; all higher modes are oscillatory. These higher modes, however, are not normally of much interest if the model is neutrally stable for the fundamental mode.

If $\Gamma_1 < 4/3$ and constant, it can be shown, again from the variational property referred to above, that, at least for the fundamental mode, $\sigma_0^2 < 0$. The star is thus dynamically unstable at least in the fundamental mode and possibly also in a few of the next higher modes. However, since, to order of magnitude, $|d\xi_k/dr|^2 \sim |\xi_k|^2/\lambda_k^2$, where λ_k is the "wavelength" of the kth mode, and since one of the above two integrals contains $1/\lambda_k^2$, then this integral increases in value, and hence causes σ_k^2 to increase, as k increases. On the other hand, the other of these two integrals is not very sensitive to the order of the mode. Hence, even if $\Gamma_1 < 4/3$, oscillatory solutions must exist for sufficiently high modes.

The conclusion in the preceding paragraph can also be understood intuitively. Note that one of the above two integrals contains the factor $(3\Gamma_1 - 4)$, which arises essentially from the gravitational forces and the spherical geometry; whereas the other integral expresses essentially the compression forces in the gas (cf. §8.1). It is clearly the former integral that gives rise to dynamical instability. Obviously, the compression forces become more important than both the gravitational forces and the effect of the spherical symmetry, as λ_k becomes small compared to the stellar radius R. In this limit we deal essentially with plane sound waves where only the compression forces are important. Again, oscillatory solutions for the higher modes may not be of much interest if the star is dynamically unstable in the fundamental mode.

If Γ_1 is not constant, the discussion is more involved, and such clearcut conclusions as those above regarding dynamical stability cannot be drawn (see, for example, Ledoux and Walraven 1958, §58). Nevertheless, it is useful at this point to re-write the expression for $\sigma_k{}^2$ given by eq. (8.30) in terms of certain averages.

First, we define the following average:

$$\left\langle \Gamma_1 r^2 \left| \frac{d\xi_k}{dr} \right|^2 \right\rangle \equiv \frac{\int_V \Gamma_1 P r^2 |d\xi_k/dr|^2 dV}{\int_V P dV}, \tag{8.31}$$

where $dV = 4\pi r^2 dr$ is a volume element, and the integrals are extended over the entire volume V of the star. We also define an average value of $|\xi_k|^2$:

$$\langle |\xi_k|^2 \rangle \equiv \frac{1}{I} \int_M |\xi_k|^2 r^2 dm = \frac{1}{I} \int_I |\xi_k|^2 dI = J_k/I, \tag{8.32}$$

where J_k is the oscillatory moment of inertia of the star (see eq. [8.26]) and I is the moment of inertia of the star about its center (see eq. [8.27]).

We now define an average value of $3\Gamma_1 - 4$ by the relation

$$\int_0^R (3\Gamma_1 - 4) P \frac{d}{dr} (4\pi r^3 |\xi_k|^2) dr$$
$$= (\overline{3\Gamma_1 - 4}) \int_0^R P \frac{d}{dr} (4\pi r^3 |\xi_k|^2) dr. \tag{8.33}$$

We define, finally, another kind of average of $|\xi_k|^2$:

$$\int_\Omega |\xi_k|^2 \, d(-\Omega) = (\overline{|\xi_k|^2})(-\Omega), \tag{8.34}$$

where Ω is the *gravitational potential energy* of the star in its equilibrium state (see eq. [2.11]), not to be confused with the dimensionless angular oscillation frequency defined in eq. (8.20). The quantity $d(-\Omega)$ is just the integrand in eq. (2.11).

We now assemble the various terms, make use of the virial theorem in the form (2.10), and perform a few parts integrations and other manipulations. We obtain, finally,

$$\sigma_k^2 = \left\{ \frac{1}{\langle |\xi_k|^2 \rangle} \left\langle \frac{\Gamma_1}{3} r^2 \left| \frac{d\xi_k}{dr} \right|^2 \right\rangle + \frac{(\overline{3\Gamma_1 - 4})(\overline{|\xi_k|^2})}{\langle |\xi_k|^2 \rangle} \right\} \left(\frac{-\Omega}{I} \right) \tag{8.35}$$

$$\equiv \alpha(-\Omega/I), \tag{8.36}$$

where the dimensionless quantity α is defined by eqs. (8.35) and (8.36).

We note that, except for rather high modes, α is generally of order unity, since the two kinds of average of $|\xi_k|^2$ in eq. (8.35) are not likely to be greatly different. Also, since $|d\xi_k/dr|^2 \sim |\xi_k|^2/\lambda_k^2$, where λ_k is the "wavelength" of the kth mode, the first term in curly brackets in eq. (8.35) is expected to be of order R^2/λ_k^2 (R = stellar radius), roughly the square of the number of "wavelengths" in a stellar diameter. Hence, for rather high modes, we expect that $\sigma_k^2 \sim (R^2/\lambda_k^2)\sigma_0^2$. We also note that the second factor in the above equation, $(-\Omega/I)$, depends only on the equilibrium stellar model and not on the pulsation properties thereof.

The fact that, according to eq. (8.36), $\sigma_k^2 \sim (-\Omega/I)$ (at least for the lower modes) illustrates the general similarity of (radial) stellar pulsations to more familiar oscillating systems.

Finally, we may note two other forms in which eq. (8.35) can be written; these forms are instructive and useful. First, we note that, in the expression for the gravitational potential energy Ω, we may express the interior mass m in terms of $\bar{\rho}(r)$, the *mean density interior to r*:

$$\bar{\rho}(r) \equiv m/(\tfrac{4}{3}\pi r^3). \tag{8.37}$$

We then obtain

$$\sigma_k^2 = \alpha \cdot \tfrac{4}{3}\pi G\bar{\rho} \cdot \left\langle \frac{\bar{\rho}(r)}{\bar{\rho}} \right\rangle, \tag{8.38}$$

where $\bar{\rho}$ denotes the mean density of the entire star, and angular brackets denote an appropriate average of the indicated quantity over the whole star.

This equation shows that, at least for k fairly small, $\sigma_k^2 \sim G\bar{\rho}$, which is just the period-mean density relation referred to above (see Chap. 2 and §8.6). Moreover, eq. (8.38) shows that σ_k^2 tends to increase with increasing mass concentration, i.e., with increasing values of the density average in this equation. Physically, a star with large mass concentration is more "tightly bound" than a star with small mass concentration, and we intuitively expect a higher pulsation frequency in the former case than in the latter. However, the dependence of σ_k^2 on mass concentration is not monotonic (see §8.12c); this fact is accounted for, in terms of eq. (8.38), by the factor α.

Second, we again make use of the virial theorem in the form (2.10) to express $-\Omega$ in terms of stellar mass M and of suitably defined averages, over the whole star, of Γ_1 and of the Laplacian sound speed v_S. We also write $I = M\langle R^2\rangle$, where $\langle R^2\rangle^{1/2}$ is a sort of radius of gyration, of order R, of the star. We obtain

$$\sigma_k^2 = \alpha \cdot 3 \left\langle \frac{1}{\Gamma_1} \right\rangle \frac{\langle v_S^2\rangle}{\langle R^2\rangle}. \tag{8.39}$$

This relation shows that, at least for the lower modes (smaller k), $\sigma_k^2 \sim \langle v_S^2\rangle / \langle R^2\rangle$, or that the corresponding pulsation period is $\Pi_k \sim (\langle R^2\rangle / \langle v_S^2\rangle)^{1/2}$. Thus, as was stated in Chapter 2, the pulsation period in one of the lower modes of a pulsating star is indeed of the order of the sound travel time through a stellar diameter.

We shall conclude this section by noting that eq. (8.35) can be used to derive a *lower limit* to the value of σ_k^2 for a star with a given, constant value of Γ_1 and a mean density $\bar{\rho}$. If we assume that the mean density interior to r, $\bar{\rho}(r)$, does not increase outward, we conclude that

$$\sigma_k^2 \geq (3\Gamma_1 - 4) \cdot {}^4\!/\!_3 \pi G \bar{\rho} \qquad (\text{all } k), \tag{8.40}$$

where the value of this lower limit is equal to the value of σ^2 for the fundamental mode of the *homogeneous* model of the same Γ_1 and $\bar{\rho}$ as the star being considered.

8.10. VARIATIONAL PROPERTY OF THE EIGENVALUES

Aside from facilitating physical interpretation, the expressions for σ_k^2 derived in §8.9 would appear to be of only academic interest. This is so because, in order to be able to evaluate the integrals in these expressions, it is first necessary to know the eigenfunctions $\xi_k(r)$. Finding these may involve simultaneously finding the eigenvalues σ_k^2, or at any rate usually depends on prior knowledge of the σ_k^2. The question then arises as to the practical utility of these expressions.

We shall show in this section that the eigenvalues obey a variational principle. This principle makes these expressions (actually, expressions which *look* very much like those derived in §8.9) very useful for obtaining estimates of the values of the σ_k^2 without actually knowing the ξ_k, that is, without having to solve the complete eigenvalue problem. The existence of a variational principle derives ultimately from the Hermiticity of the operator $\mathcal{L}$ (see §8.8), as is well known. Our proof of the variational principle somewhat follows a proof given by Lynden-Bell and Ostriker (1967).

To explain the variational principle, we let $u(r)$ be any sufficiently regular function of the equilibrium radial distance r. We now define the real number Σ^2:

$$\Sigma^2 \equiv \frac{1}{J}\int_M u^* \mathcal{L}(u) r^2 dm, \tag{8.41}$$

where J is defined exactly as in eq. (8.26), except with ξ_k replaced by u; the linear operator $\mathcal{L}$ is defined in eq. (8.23); and the integrations are all carried out over the entire stellar mass M.

We now contemplate subjecting $u(r)$ to an arbitrary, small *variation*, $\Delta u(r)$, at every radial distance r. When $u(r)$ is so varied, we expect that the value of Σ as defined by eq. (8.41) will change by the amount $\Delta\Sigma$. This change $\Delta\Sigma$ can be calculated by taking the variation of eq. (8.41), treating the Δ operator as an ordinary differential operator. We note that Δ can be taken inside the integral signs since this variation does not apply to r or to the associated value of interior mass m; that is, r remains fixed during the variation.

Multiplying eq. (8.41) by J and taking the above variation, we obtain an expression relating $\Delta\Sigma$, u, u^*, Δu, Δu^*, $\mathcal{L}(u)$, and $\Delta\mathcal{L}(u)$. However, since $\mathcal{L}$ is linear, $\Delta\mathcal{L}(u) = \mathcal{L}(\Delta u)$. We also note that, because of the Hermiticity of $\mathcal{L}$, Δu and u^* can be interchanged in this expression. Finally, we make use of the reality of Σ^2 and $\mathcal{L}$. We obtain

$$2J\Sigma\Delta\Sigma + 2Re\left\{\int_M (\Delta u^*)[\Sigma^2 u - \mathcal{L}(u)]r^2 dm\right\} = 0, \tag{8.42}$$

where $Re\{\ldots\}$ means the real part of the indicated quantity.

Now, we see that, if $u(r)$ is a solution of the equation

$$\Sigma^2 u - \mathcal{L}(u) = 0, \tag{8.43}$$

then $\Delta\Sigma = 0$ even if $\Delta u^* \neq 0$. Conversely, if $\Delta\Sigma = 0$ and $\Delta u^*(r)$ is an arbitrary (but small) function of r, then $u(r)$ must be a solution of eq. (8.43). But eq. (8.43) is just the LAWE, whence $u(r) \equiv \xi_k(r)$ and $\Sigma^2 = \sigma_k^{\,2}$, the eigenfunction and eigenvalue, respectively, of the LAWE for the kth mode. We have therefore proved the following: Those solutions of eq. (8.43) for which Σ is an *extremum* ($\Delta\Sigma = 0$) with respect to arbitrary small variations of $u(r)$, are the *eigenfunctions* $\xi_k(r)$, and the associated eigenvalues are the corresponding *adiabatic eigenvalues* $\sigma_k^{\,2}$, both for the kth mode.

The practical utility of the variational principle lies in the fact that Σ^2 goes through a true extremum as the function $u(r)$ is varied continuously, the extremal value itself being $\Sigma^2 = \sigma_k^{\,2}$ and occurring when $u(r) \equiv \xi_k(r)$. Hence, the error in Σ^2 for a given trial function $u(r) \approx \xi_k(r)$ is generally much smaller than the error in $u(r)$. A remarkably accurate approximation to the value of σ_k^2 is therefore often provided by only a very rough trial function $u(r)$ (see §8.15). Alternatively, we may say that, if $u(r)$ differs everywhere from $\xi_k(r)$ by a small quantity of order ϵ, then Σ^2 differs from σ_k^2 by a small quantity of order ϵ^2.

For some practical applications of this variational principle, see, for example, Goertzel and Tralli (1960, Chap. 15); see also Ledoux and Walraven (1958, §58).

For the *fundamental* mode ($k = 0$) of radial oscillations this variational principle is particularly useful, for in this case Σ^2 obeys a *minimal* (special case of an extremal) principle. This fact derives from the Sturm-Liouville nature of the LAWE, in particular from the fact that the eigenvalues can in this case be *ordered* (see, e.g., Ince 1944, Chap. 10).

We can now state the minimal principle which applies to the fundamental mode of radial oscillations: The function $u(r)$ which leads to the lowest possible minimum of Σ^2 is the *fundamental eigenfunction,* and this lowest possible minimum value of Σ^2 is σ_0^2, the *fundamental eigenvalue.*

This minimal principle can be used to prove some of the assertions which were made in §8.9 (but we shall not demonstrate these proofs explicitly here). First, one may prove that $\sigma_0^2 = 0$ and $\xi_0(r) =$ const. for $\Gamma_1 = 4/3$; and that $\sigma_0^2 < 0$ for $\Gamma_1 < 4/3$; both regardless of the stellar model. (However, according to the general theory of relativity, dynamical instability actually sets in at a value of Γ_1 slightly larger than 4/3, as was first shown by Chandrasekhar 1964a,b,c,; also see §19.5.) Second, eq. (8.3) for the angular frequency in the fundamental mode of the *homogeneous* model having constant density $\rho(r) = \bar{\rho}$ (mean density) throughout and constant Γ_1, can be derived from this principle; also, the fact can be established that $\xi_0(r) =$ const. for this model.

Finally, the above minimal principle leads to an *upper bound* (though now a *lowest* upper bound) on σ_0^2, as was pointed out by Ledoux and Walraven (1958, §60). We consider the case $\Gamma_1 \neq$ constant. In this case we can write Σ^2 as in eq. (8.36). If we take the simple trial function $u(r) =$ constant and use the above minimal principle, we have

$$\sigma_0^2 \leq \overline{(3\Gamma_1 - 4)}_{u\,\text{const.}}(-\Omega/I) \equiv \Sigma^2_{u\,\text{const.}} \tag{8.44}$$

where a simple expression for $\overline{(3\Gamma_1 - 4)}_{u\,\text{const.}}$ may easily be obtained (but we do not do so explicitly here).

We therefore conclude that, if $\Gamma_1 =$ constant, σ_0^2 must lie in the range

$$(3\Gamma_1 - 4)\,{}^4\!/_3\pi G\bar{\rho} \leq \sigma_0^2 \leq \Sigma^2_{u\,\text{const.}}, \tag{8.45}$$

where the lower limit applies to the homogeneous model (see above), and the upper limit is defined in eq. (8.44).

8.11. SOME PHYSICAL INTERPRETATIONS

In this section we shall consider, for the most part, the physical interpretation of the integral expressions, exemplified by eq. (8.30), for the square of the eigenfrequency σ_k^2 in the kth mode of radial, adiabatic, linear pulsations. Ledoux and Walraven (1958, §59) have shown that the variational property of the eigenvalues of the LAWE (see §8.10) corresponds, physically, to the fact that a star, being a mechanical system, must obey Hamilton's principle of classical mechanics.

Since squares of small quantities are involved in these expressions, one must be rather careful when complex notation is being used. For example, $Re(\zeta^2) \neq [Re(\zeta)]^2$.

We may write $\zeta \equiv \delta r/r$ in the form

$$\zeta(r,t) = \xi(r)e^{i\sigma t} = |\xi(r)|e^{i\phi(r)}e^{i\sigma t}, \tag{8.46}$$

where $|\xi|$ is the absolute value of the *amplitude* of ζ and where ϕ is the *phase* of ζ, in general a function of r, but independent of r in the case of pure standing waves. What we really mean, for example, by ζ^2 is then $(Re[\zeta])^2$.

Let us now calculate the total kinetic energy $\mathcal{T}$ in the star residing in the radial pulsations, still assuming purely oscillatory solutions (σ real). This is

$$\mathcal{T} = \int_M \tfrac{1}{2}\{\partial[Re(\zeta)]/\partial t\}^2 r^2 dm, \tag{8.47}$$

where the integral is extended over the entire mass of the star. One may show from eq. (8.47) that $\mathcal{T}$ will go through zero only in the case of standing waves, and that in this case $\mathcal{T}$ will vanish twice in every period. In general, however, the minimum value of $\mathcal{T}$ will be positive, and the total kinetic energy never quite vanishes at any time in a period, as is also intuitively evident when the motion is not that of a standing wave. In any case, the *average* value $\overline{\mathcal{T}}$ of the total kinetic energy over a period of duration $\Pi = 2\pi/\sigma$ is

$$\overline{\mathcal{T}} = \tfrac{1}{4}\sigma^2 \int_M |\xi|^2 r^2 dm. \tag{8.48}$$

By eq. (8.26) we see that

$$\sigma^2 J = 4\overline{\mathcal{T}}. \tag{8.49}$$

Hence, the left side of eq. (8.41) (after multiplying through by J_k) is just $4\overline{\mathcal{T}}_k$, or 4 times the total average kinetic energy in the kth mode. (Henceforth, we shall omit the subscript k, as everything we say is valid for all values of k.)

We consider now the right side of eq. (8.41) (after multiplying through by J_k). For its interpretation, let us calculate the total work done by the pressure gradient and gravity forces acting on the star when it is displaced by a small amount from (hydrostatic) equilibrium.[2]

Suppose that a given mass shell whose equilibrium radius is r_0 has already been displaced by an amount $\delta r' \equiv \eta$ from its equilibrium position. We may then calculate, to first order in small quantities, the net force acting on the shell as a result of such a displacement, say per unit mass. The work done per unit mass on the shell when it is subjected to a further displacement $d\eta$ is then obtained by multiplying the above force by $d\eta$. After performing a few manipulations and making use of the linearized mass equation, we obtain a result which contains (among other things) $d(\delta P/P_0)$. We eliminate this last quantity by using the energy equation (7.8) with eq. (7.10) used therein (we are thus assuming thermodynamically reversible processes). Finally, we integrate the result over the total displacement $\delta r(r_0,t) = \zeta r_0$ and over the entire stellar mass M. The total work done on the star during the displacement by the forces accompanying the displacement is then given by the sum of several integrals, over the entire star, involving ζ, $\delta\rho/\rho_0$, the entropy variation, and its Stokes derivative.

We note that only that part of the above result involving ζ and $\delta\rho/\rho_0$ is truly conservative, that is, a function only of the final displacement itself and independent of the path of the displacement, or of the manner in which the final displacement was reached. This part of the result thus has the properties of a true potential energy. The remaining terms depend for their value on what heat exchanges with the surroundings have occurred during the displacement, and hence on the path taken to reach the final displacement. These terms are therefore not conservative in the above sense. They represent the work done on the star by the nonadiabatic, nonconservative forces acting during the displacement.

We accordingly write the above work W in the form

$$W = -\delta\Phi + \text{nonadiabatic terms,} \tag{8.50}$$

where it turns out that

$$\delta\Phi \equiv \int_M \left[\frac{1}{2}\frac{\Gamma_{1,0}P_0}{\rho_0}\left(\frac{\delta\rho}{\rho_0}\right)^2 - \frac{2GM}{r_0}\zeta^2\right] dm \tag{8.51}$$

(accurate to second order in small quantities) is the effective potential

[2]Note that we do not include in this the work done as a result of expansion or compression of the mass shells; thus we consider here only the work which can appear as *kinetic energy* (see §4.2).

energy of the star in its perturbed condition, with $\delta\Phi = 0$ corresponding to a *zero* displacement from hydrostatic equilibrium. The negative sign has been used in eq. (8.50) since the potential energy of a system is always taken (aside from the arbitrary zero point) as the work done *against* the forces acting on the system.

To make more apparent the *sign* of $\delta\Phi$ as given by eq. (8.51), we may express $\delta\Phi$ in terms only of $\zeta = \delta r/r_0$ and $\partial\zeta/\partial r_0$. We use the linearized mass equation, expand out the square term, and perform an integration by parts, making use of the fact that $P = 0$ at the stellar surface. Dropping zero subscripts and rearranging slightly, we finally obtain

$$\delta\Phi = \tfrac{1}{2}\int_0^R \left\{4\pi\Gamma_1 P r^4 \left(\frac{\partial\zeta}{\partial r}\right)^2 - 4\pi r^3\zeta^2 \frac{d}{dr}\left[(3\Gamma_1 - 4)P\right]\right\} dr, \quad (8.52)$$

where we recall that $\zeta(r,t)$ here is an essentially *arbitrary*, but *real*, function of r and t.

We note, first, that if Γ_1 = constant $> 4/3$, then $\delta\Phi > 0$ for *any* displacement $\zeta \neq 0$ (assuming the star in its unperturbed state to be in hydrostatic equilibrium). Moreover, the fact that $\delta\Phi$ contains no first-order terms means that the *slope* of $\delta\Phi$ versus "displacement" is *zero* to first order at $\zeta = 0$ (again making the above assumption). Hence $\delta\Phi$ possesses a true minimum (with minimum value zero) with respect to every conceivable displacement from hydrostatic equilibrium.

Second, and finally, for Γ_1 = constant $\leq 4/3$, we see that $\delta\Phi \leq 0$ if $\zeta(r,t)$ is constant with respect to r, and that, again, the slope of $\delta\Phi$ with respect to "displacement" is zero at $\zeta = 0$. Thus, if $\zeta(r,t)$ is either constant with respect to r or sufficiently slowly varying, then either neutral stability on an arbitrary time scale or dynamical instability are possibilities. However, it may be that $\delta\Phi > 0$, even if $\Gamma_1 \leq 4/3$, if the gradients of $\zeta(r,t)$ are steep enough. This conclusion is not in contradiction with our earlier results, since $\zeta(r,t)$ here is not necessarily an eigenfunction of the LAWE. In fact, the above conclusion is reasonable, since we know, for example, that sufficiently high modes of the LAWE are dynamically stable even if $\Gamma_1 < 4/3$.

In order to complete our interpretation of eq. (8.41), we take the time average, $\overline{\delta\Phi}$, of $\delta\Phi$ as given by eq. (8.52) over a complete period of duration $\Pi = 2\pi/\sigma$, treating the motion as purely periodic with angular frequency σ. Bearing in mind the above precautionary remarks about complex notation, we obtain

$$\overline{\delta\Phi} = \tfrac{1}{4}\int_M \xi^* \mathcal{L}(\xi) r^2 dm, \quad (8.53)$$

where $\mathcal{L}$ is the linear operator defined in eq. (8.23).

On the basis of the above discussion, we see that the physical interpretation of eq. (8.41) is just that

$$4\overline{\mathcal{T}} = 4\overline{\delta\Phi}, \tag{8.54}$$

or that in oscillatory motion the average kinetic energy is equal to the average potential energy. Note that, even though eq. (8.41) applies specifically to the eigenvalues $\sigma_k{}^2$ and eigenfunctions $\xi_k(r)$ of the LAWE, our interpretation of this equation as given in eq. (8.54) does not require that σ^2 and $\xi(r)$ be, respectively, an eigenvalue and an eigenfunction of the LAWE.

The result (8.54) is actually quite general, and is sometimes known as *Rayleigh's Principle.* It says that in any oscillating system the frequency of oscillation adjusts itself so that the average kinetic and potential energies are equal. For any given form $\xi(r)$ of the disturbance, this equation provides a simple and powerful method for estimating the vibrational frequency of an oscillating system (e.g., Temple and Bickley 1956).

The total pulsation energy of a star is

$$\Psi = \mathcal{T} + \delta\Phi, \tag{8.55}$$

where $\Psi = 0$ in the static state of the star. The average of Ψ over a period is

$$\overline{\Psi} = 2\overline{\mathcal{T}} = 2\overline{\delta\Phi}. \tag{8.56}$$

For pure standing waves, Ψ is strictly constant, and the pulsation energy is alternately purely kinetic and purely potential, as in a simple one-dimensional pendulum. If the pulsations have a running wave component ($\phi[r] \neq$ const. in eq. [8.46]), then Ψ is not strictly constant, but oscillates about $\overline{\Psi}$ with a small amplitude and with angular frequency 2σ (recall that Ψ involves the *squares* of trigonometric functions; see, for example, eq. [8.52]).

The validity of the statements in the above paragraph becomes evident when one realizes that the quantity $\delta\Phi$, given to second-order accuracy in small quantities by eq. (8.52), can be expressed *generally,* that is, *nonlinearly,* by noting its relation to the sum of the gravitational and internal energies E_g and U respectively (the gravitational energy was denoted by Φ in §4.6), of the star. This relation is, for a spherically symmetric distribution of matter,

$$\delta\Phi = \delta_{\text{ad}}(E_g + U) = \delta_{\text{ad}} \int_M \left(-\frac{Gm}{r} + E \right) dm, \tag{8.57}$$

where E is the internal energy per unit mass and the symbol δ_{ad} means that the indicated variation is to be carried out *adiabatically*. It can be shown explicitly that the negative of $\delta\Phi$ as given by eq. (8.57) is indeed equal to the work done, through a whole period, by the pressure gradient and gravitational forces during an adiabatic displacement over the whole star.

The above remarks make it clear that the total pulsation energy Ψ, as given to second order accuracy by eq. (8.55), is a special case of the more general total energy Ψ defined in §4.6. Hence, the general energy theorem discussed there applies exactly to the Ψ of eq. (8.55).

Equation (8.52) for $\delta\Phi$ can also be obtained directly from eq. (8.57) by carrying out the indicated variation. This variation must, however, be carried at least to second order in small quantities, since the first-order variation vanishes identically if the unperturbed system is in hydrostatic equilibrium.

The expansion of $\delta\Phi$ to third order has been carried out by Stothers and Frogel (1967). The third-order variation can give information regarding the direction (expansion or contraction) in which a perturbation will grow in the case of a dynamically unstable system. This third-order variation shows, for example, that for Γ_1 = constant $< 4/3$ and for homologous motion (ζ spatially constant), the direction of the instability is, perhaps not surprisingly, in the direction of the initial perturbation.

8.12. SOLUTION OF THE LAWE

In this section we shall consider practical methods for obtaining numerical solutions of the linear adiabatic wave equation (LAWE) for arbitrary equilibrium stellar models, and shall also summarize the results of actual solutions of this equation for certain selected stellar models. Two general procedures for the solution of this equation will be considered. One of these procedures, to be discussed in §8.12a, involves fitting techniques. This procedure was used almost exclusively until only the past few years, and is still used by many workers in the field. The other of these procedures, to be discussed in §8.12b, is based on matrix algebra and on properties of Sturm sequences. Finally, results for selected stellar models will be discussed in §8.12c.

8.12a. FITTING METHODS

We write the space part of the LAWE, eq. (8.6), in the nondimensional form

$$\xi'' = \left(\frac{R}{x\lambda_P}\right)\left\{\left(x - \frac{4\lambda_P}{R}\right)\xi' - \left[\frac{\Omega^2 x^3}{\Gamma_1 q} - \frac{(3\Gamma_1 - 4)}{\Gamma_1}\right]\xi\right\} + \left(\frac{R}{x\lambda_P}\right)(x\xi' + 3\xi)\frac{d\ln\Gamma_1}{d\ln P}, \quad (8.58)$$

where ξ is the space part of $\zeta \equiv \delta r/r_0$, primes denote differentiation with respect to $x \equiv r_0/R_0$, zero subscripts have been dropped, Ω is the dimensionless angular frequency defined in eq. (8.20), and $q \equiv m/M$, were m is the mass interior to r_0 and M is the total stellar mass. Finally, λ_P is the local pressure scale height. It is interesting to note that $(R/x\lambda_P) = V/x^2$, where $V \equiv -d \ln P/d \ln x$ is the V of stellar structure theory (e.g., Schwarzschild 1958).

We first consider integrations of eq. (8.58) that begin at the surface ($x = 1$) and proceed inward. We normalize ξ to unity at the surface: $\xi_{\rm surf} = 1$. We also assume throughout this subsection that Γ_1 is constant throughout the star, so that the last term in eq. (8.58) may be neglected. Effects of a variable Γ_1 will be discussed in §8.14.

In the surface regions of a star (where $x \approx 1$) we know that λ_P/R is normally small; hence we may neglect, for purposes of discussion, λ_P/R in the first term in curly brackets in eq. (8.85). The standing wave boundary condition discussed in §8.3 requires that the quantity in curly brackets in eq. (8.58) be very small in the surface regions. We therefore have, at $x = 1$,

$$\xi'_{\rm surf} = \Gamma_1{}^{-1}[\Omega^2 - (3\Gamma_1 - 4)]. \qquad (8.59)$$

We recall that the value of Ω^2 is not known *a priori*, so that Ω^2 must be regarded as a trial parameter in outside-in integrations. It should be noted that, if radiation pressure is important at the stellar surface, eq. (8.59) should be replaced by formulas such as eq. (8.15).

We now note that, for the homogeneous model with constant Γ_1, we have that $\Omega^2 = 3\Gamma_1 - 4$ for the fundamental mode. Moreover, we pointed out in §8.9 that this model has the smallest possible value of Ω^2 of all stellar models for this mode for given values of Γ_1 and mean density $\bar{\rho}$. It therefore follows that $\xi'_{\rm surf} \geq 0$ (if $\Gamma_1 \geq 4/3$) for the kinds of models under consideration (Γ_1 = constant) and for the adopted surface boundary condition (8.59).

From the above results we may conclude that for the fundamental mode of models of the above kind (Γ_1 = constant $\geq 4/3$), ξ either decreases inward in the surface regions, or remains constant in the case of the homogeneous model or if $\Gamma_1 = 4/3$. Also ξ' decreases inward (if $\Gamma_1 > 4/3$), and even more rapidly than does ξ, since x^3/q decreases inward in the surface regions. (Note that, since Ω^2 for realistic models is normally several times larger than unity [see §8.12c below], then, roughly, $\xi' \propto \Omega^2$ for $x \approx 1$.)

Hence, as the solution is followed inward with a given trial value of Ω^2, ξ normally decreases steadily, at least initially. Numerical calculations show that, if the trial value of Ω^2 is too small for the given model, then ξ will eventually (proceeding inward) diverge positively, that is, become large and positive. If this trial value is too large, ξ will eventually diverge negatively, that is, become large and negative. The correct value of Ω^2 is that for which ξ decreases more-or-less monotonically all the way to the stellar center.

Consider now the inside-out solution, the one which starts at $x = 0$, satisfies the central boundary condition, and proceeds outward. Note that $x^3/q \rightarrow \bar{\rho}/\rho_c$ as $x \rightarrow 0$, where ρ_c denotes the central density of the model. Note also that $R/\lambda_P \rightarrow 0$ as $x \rightarrow 0$. Hence, from eq. (8.58) we see that the solution which obeys the central boundary condition (see eq. [8.8]) will not exhibit singular behavior in the central regions.

We note that $\bar{\rho}/\rho_c \ll 1$ in most realistic stellar models, particularly for highly centrally concentrated models. Hence, the term in eq. (8.58) containing Ω^2 is usually small or negligible in the central regions of such models. (This statement, however, is not true for the homogeneous model or in models with only a small central mass concentration, or for very high modes in centrally concentrated models.) In such cases, then, the LAWE is practically independent of Ω^2 in the central regions. Physically, this near lack of dependence on Ω^2 just reflects the fact that the local gravitational acceleration may be large compared with the acceleration resulting from the pulsations in the deep stellar interior: the interior regions in such cases move outward and inward so slowly that at each instant of time they may be very nearly in hydrostatic equilibrium.

We see, then, that outside-in integrations leave Ω^2 undetermined, whereas inside-out integrations leave ξ_c (central value of ξ) undetermined. The correct solution may be obtained by matching the inside-out and the outside-in integrations at some intermediate point, that is, by requiring that $d \ln \xi/dx$ be the same at the fitting point for the two solutions. In this way both the eigenvalue Ω^2 and the eigenfunction $\xi(x)$ for the mode and model under consideration may be found.

Analytic solutions of the LAWE can be obtained for certain sufficiently simple stellar models. Some of these solutions are worked out and discussed in Rosseland (1949, Chap. 3; however, see Vaughan 1972).

8.12b. MATRIX METHODS

Here we shall describe a method, based on matrix algebra, for numerically solving the LAWE for an arbitrary equilibrium stellar model. Analogous methods have been used, for example, by Castor (1971) and Davey (1970). These methods are ideally suited to the use of electronic digital computers. Some of the material of this subsection has been based on unpublished notes of J. Castor, and used with his kind permission.

The immediate goal is to write the linearized equation of motion, eq.

(8.6), in a finite difference form, of which there are many possibilities. We assume the time dependence of eq. (8.5), in which, in this section, we replace σ by ω. The form we shall adopt here is consistent with a widely used scheme for numerically carrying out nonlinear calculations (see, e.g., A. N. Cox, Brownlee, and Eilers 1966). This form differs in some details from those referenced above, but these details are irrelevant in the present discussion, which will illustrate the principles involved.

In this scheme the model is conceptually divided into J discrete, spherical, concentric mass zones, the mass in each zone remaining constant in time. This zoning may comprise the entire star, or only the outer envelope surrounding a rigid, spherical core. The zones are numbered 1, 2, . . . , J, from the inside out, and there are $J + 1$ interfaces, numbered 0, 1, 2, . . . , J. The ith zone has mass $M_{i-1/2}$, and is bounded by interfaces of radii r_{i-1} and $r_i (i = 1, \ldots, J)$. Intrinsic variables, including $M_{i-1/2}$, pressure $P_{i-1/2}$, specific volume $V_{i-1/2}$, temperature $T_{i-1/2}$, and so on, are regarded as evaluated at the "midpoints" of zones and to have constant values within each zone. The extrinsic variables, for example, r_i, L_i (interior luminosity), are regarded as evaluated at the zone interfaces (see Fig. 8.1). Thus, the finite difference form of eq. (8.6) is, in this scheme,

$$\omega^2 \delta r_i = -\frac{4Gm_i}{r_i^3}\delta r_i + 4\pi r_i^2 \frac{(\delta P)_{i+1/2} - (\delta P)_{i-1/2}}{\mathcal{D}M_i}, \tag{8.60}$$

where $m_i \equiv M_{r,i}$ is the mass interior to interface i, and

$$\mathcal{D}M_i \equiv \tfrac{1}{2}(M_{i-1/2} + M_{i+1/2}) \tag{8.61}$$

is the "effective" mass which determines the inertia associated with the ith interface.

However, for adiabatic oscillations δP and δV are simply related (see, e.g., eq. [5.36a]). We write this relation in finite-difference form and

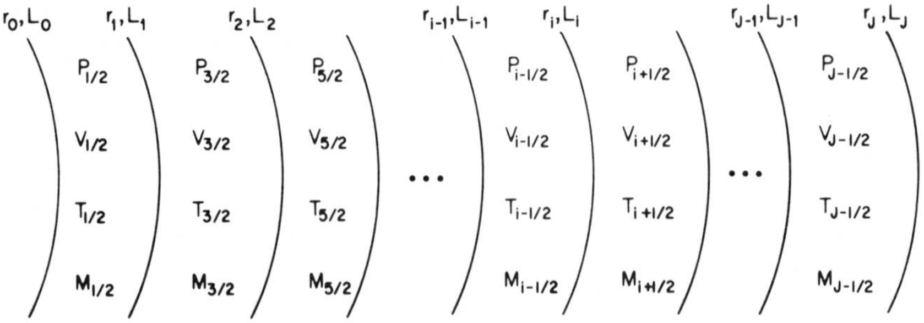

Figure 8.1. Schematic illustration of zoning adopted in the matrix method of solving the LAWE described in the text.

define new dependent variables X_i:

$$X_i \equiv (\mathcal{D}M_i)^{1/2}\delta r_i \qquad (i = 0, 1, \ldots, J). \tag{8.62}$$

The reason for introducing these new variables will become clear presently. Equation (8.60) may then be written in the form

$$-A_{i+1}X_{i+1} + (B_i - \omega^2)X_i - C_{i-1}X_{i-1} = 0, \tag{8.63}$$

where the coefficients depend only on the equilibrium model, are defined only for $i = 1, 2, \ldots, J - 1$, and are given by rather complicated expressions (their exact forms are not relevant for the present discussion).

Clearly, now, we cannot apply eq. (8.63) to $i = 0$, since there is no X_{-1} (that is, δr_{-1}). Instead, we use the central boundary condition:

$$X_0 = 0 \text{ (i.e., } \delta r_0 = 0). \tag{8.64}$$

Also, we cannot apply eq. (8.63) to $i = J$, because A_{J+1} and B_J involve $P_{J+1/2}$, $M_{J+1/2}$, etc., which are not defined. Moreover, there is no X_{J+1} (that is, δr_{J+1}). Thus, we must replace eq. (8.63) for $i = J$ by a surface boundary condition of some kind, such as the "standing wave" condition

$$\left(\frac{\delta P}{P}\right)_{J-1/2} = -\left(\frac{\omega^2 r_J^{\,3}}{GM} + 4\right)\frac{\delta r_J}{r_J}, \tag{8.65}$$

where the quantity on the left side depends on δr_J and δr_{J-1}, and G and M denote, respectively, the constant of gravitation and the total stellar mass.[3]

Hence, dropping X_{-1} and X_{J+1}, assuming that $X_0 = 0$, and using a surface boundary condition, imply that the X_i $(i = 1, \ldots, J)$ can be regarded as a *column matrix* X being operated on by a matrix whose elements are A_i, B_i, C_i, etc. (but special values must be assigned to A_J, B_J, and C_{J-1}). Note that the introduction of the new dependent variables has rendered the matrix operating on X *symmetric;* this property will be useful in the following.

We now define the symmetric, tridiagonal matrix $\mathcal{M}$ as a matrix whose principal diagonal is made up of the b_i, whose superior (upper) contiguous "diagonal" is made up of the a_i, and whose inferior (lower) contiguous "diagonal" is made up of the c_i, where

$$\begin{aligned} b_i &\equiv B_i, & i &= 1, \ldots, J; \\ a_i &\equiv A_{i+1}, & i &= 1, \ldots, J - 1; \\ c_i &\equiv C_{i-1}, & i &= 2, \ldots, J. \end{aligned} \tag{8.66}$$

[3]Strictly speaking, eq. (8.65) is not precisely "centered." However, this probably does not matter very much, particularly if $\delta P/P$ does not vary greatly in the immediate surface regions (probably a reasonably good approximation in most cases).

Thus, the problem of solving the LAWE with the present methods is equivalent to solving the eigenvalue matrix problem

$$(\mathcal{M} - \omega^2 I)X = 0, \tag{8.67}$$

where I denotes the unit matrix, for the eigenvalues and eigenvectors.

Since $\mathcal{M}$ is real and symmetric, its eigenvalues are all real (see, for example, Bellman 1970, p. 35). (This property is consistent with what we already know about the LAWE; see §8.8.) The eigenvalues are the J real roots of the characteristic polynomial:

$$|\mathcal{M} - \omega^2 I| = 0, \tag{8.68}$$

where vertical bars denote the determinant of the indicated quantity.

We may conveniently evaluate this determinant and find the eigenvalues by use of the properties of *Sturm sequences* (see below). First, however, we define the mth-order determinant $D_m(\lambda)$ as the determinant of the symmetric, tridiagonal matrix whose first m rows and columns are the same as those of $\mathcal{M} - \lambda I$. We now write this determinant as the sum of two $(m-1)^{\text{th}}$ order determinants by expanding $D_m(\lambda)$ in terms of the elements of the last column. We then in turn expand the determinant multiplying a_{m-1} in terms of the elements of the last row. We obtain

$$D_m(\lambda) = (b_m - \lambda)D_{m-1}(\lambda) - a_{m-1}c_m D_{m-2}(\lambda). \tag{8.69}$$

This is a recursion relation useful for evaluating $D_m(\lambda)$, which is clearly an mth order polynomial in λ. Differentiating eq. (8.69) with respect to λ, we obtain another useful relation:

$$D'_m(\lambda) = -D_{m-1}(\lambda) + (b_m - \lambda)D'_{m-1}(\lambda) - a_{m-1}c_m D'_{m-2}(\lambda), \tag{8.70}$$

where the prime here denotes differentiation with respect to λ.

We now define the sequence of polynomials:

$$\begin{aligned} &D_0(\lambda) = 1, \\ &D_1(\lambda) = b_1 - \lambda, \\ &D_2(\lambda), D_3(\lambda), \ldots, \end{aligned} \tag{8.71}$$

where $D_2(\lambda), D_3(\lambda), \ldots$ are obtained successively from eq. (8.69). We note that $D_J(\lambda) = |\mathcal{M} - \lambda I|$ is the characteristic polynomial of our problem.

It is not difficult to show that the sequence of polynomials defined by eq. (8.71) forms a *Sturm sequence* (Ralston 1965, p. 351). Because of this, we may make use of *Sturm's theorem* (Ralston 1965, p. 352). Let $q_m(\lambda)$ be the number of *sign changes* in the sequence of polynomials $\{D_m(\lambda)\}$, that is, the number of times $D_k(\lambda)$ changes sign as k takes on, successively, the values $0, 1, 2, \ldots, m$. We then have the result: $q_m(\lambda)$ is the number of roots (eigenvalues) of $D_m(\lambda)$ that are less than λ. In particular, if $q_J(\lambda)$ is the number of sign changes in $D_0(\lambda), D_1(\lambda), \ldots, D_J(\lambda)$, then $q_J(\lambda)$ is the number of zeros of $D_J(\lambda)$ which are less than λ.

The procedure for finding the eigenvalues of the LAWE by means of the present matrix methods may then be summarized in the following steps:

1. Evaluate $D_0(0), D_1(0), \ldots, D_J(0)$ successively, using the recursion relation (8.69). If there are *any* sign changes, then D_J has negative roots, and dynamical instability is implied.
2. If $q_J(0) = 0$, then begin using a Newton-Raphson method of iteration, with the first iterate being $\lambda^{(0)} = 0$. Setting $D_J(\lambda^{[n+1]}) = 0$; writing the value of λ in the $(n+1)^{\text{st}}$ iteration, $\lambda^{(n+1)}$, as the value in the nth iteration, $\lambda^{(n)}$, plus a correction; and expanding $D_J(\lambda^{[n+1]})$ to first order in the correction in a Taylor series about λ^n; we obtain the approximate formula
$$\lambda^{(n+1)} = \lambda^{(n)} - D_J(\lambda^{[n]})/D'_J(\lambda^{[n]}). \tag{8.72}$$
The $D'_J(\lambda)$ are obtained from the recursion relation (8.70). This procedure quickly converges to $\lambda_0 = \omega_0{}^2$, the fundamental eigenvalue.
3. Given λ_0, try values of $\lambda > \lambda_0$ until a value has been found such that $q_J(\lambda) = 2$. Use of the Newton-Raphson scheme as described above will quickly yield $\lambda_1 = \omega_1{}^2$.
4. Continue as in 3, for as many eigenvalues as one wishes to find.

Let us assume that we have found the eigenvalues $\lambda_0, \lambda_1, \ldots$. The final question that we must answer, now, is, how do we obtain the eigenvector $\delta r_0, \delta r_1, \ldots, \delta r_J$, corresponding to each eigenvalue λ_i?

We recall that the X_i (proportional to δr_i, see eq [8.62]) are given by eq. (8.63). If we set $i = 1$ in this equation and remember that we have set $X_0 = 0$, the resulting equation relates X_1 and X_2 linearly. Adopting $i = 2$ would bring in X_3, which could then in turn be expressed in terms of X_2, and so on. We therefore assume that for any i,
$$X_{i-1} = d_{i-1}X_i, \tag{8.73}$$
where the d_{i-1} are as yet undetermined. If we apply eq. (8.73) to $i = 1$, we can show that, since $X_1 \neq 0$ in general, $d_0 = 0$. To find the values of the remaining d_i, we substitute eq. (8.73) into eq. (8.63) and solve for X_i. The result is an equation similar to eq. (8.73), only with i replaced by $i + 1$. In this way we obtain the recursion relation
$$d_i = \frac{A_{i+1}}{B_i^2 - \omega^2 - C_{i-1}D_{i-1}} \qquad (i = 1, \ldots, J-1), \tag{8.74}$$
from which $d_0, d_1, \ldots, d_{J-1}$ can all be evaluated successively.

TABLE 8.1

Solution of Adiabatic Wave Equation for Selected Stellar Models* ($\Gamma_1 = 5/3$).†

Model	$\rho_c/\bar{\rho}$	$\sigma_0{}^2 R^3/GM$	$Q_0 = \Pi_0 (\bar{\rho}/\bar{\rho}_\odot)^{1/2}$ (days)	Π_1/Π_0	$(\xi_R/\xi_c)_0$	$(\xi_R/\xi_c)_1$	Node of ξ_1**
Homogeneous (polytrope $n = 0$)‡	1	1	0.1158	0.28	1	-0.4	$x = 0.845$ $q = 0.603$
Polytrope $n = 1$	3.30	1.892	0.0842	0.396	1.24	—	—
Linear model, $\rho = \rho_c(1 - x)$ (Stothers and Frogel 1967)	4.00	1.836	0.0846	0.389	—	—	$x = 0.76$ —
Convective model (polytrope $n = 1.5$)	5.99	2.712	0.0703	0.465	1.42	-3.39	$x = 0.72$ $q = 0.83$
Polytrope $n = 2$	11.4	4.00	0.0579	0.548	2.37	—	—
Epstein's model 4 with external convective zone§	1.9×10^5	4.30	0.0558	0.543	7.89×10^3	-1.462×10^4	$x = 0.76$ —
Standard model (polytrope $n = 3$)	54.2	9.261	0.0383	0.738	22.41	-59.12	$x = 0.67$ $q = 0.993$
Polytrope $n = 3.5$	1.53×10^2	12.69	0.0325	0.772	2.55×10^2	—	—
Polytrope $n = 4$	6.22×10^2	15.38	0.0295	0.779	5.95×10^3	—	—
Epstein's model 4 (Epstein 1950)	1.9×10^6	14.082	0.0309	0.765	6.124×10^5	-1.05×10^6	$x = 0.76$ $q = 0.9999$

*This is Table 3 of Cox (1974a), reproduced by permission of the Institute of Physics.

†The information for this table was obtained (except as noted otherwise) from Ledoux and Walraven (1958, Table 12), Hurley, Roberts, and Wright (1966) and Pollack and Hansen (1970, unpublished calculations).

‡*Polytropic* stellar models are models characterized by the relation P = constant $\rho^{(n+1)/n}$, where n is the polytropic index. Detailed discussions of the properties of polytropic models may be found in, e.g., Chandrasekhar (1939, Chap. 4) and Cox and Giuli (1968, §23.1).

§In this model (Ledoux 1955), the fractional thickness of the convection zone is 0.94; the fractional mass of the convection zone is 0.70.

**Here x denotes fractional radius, q fractional interior mass.

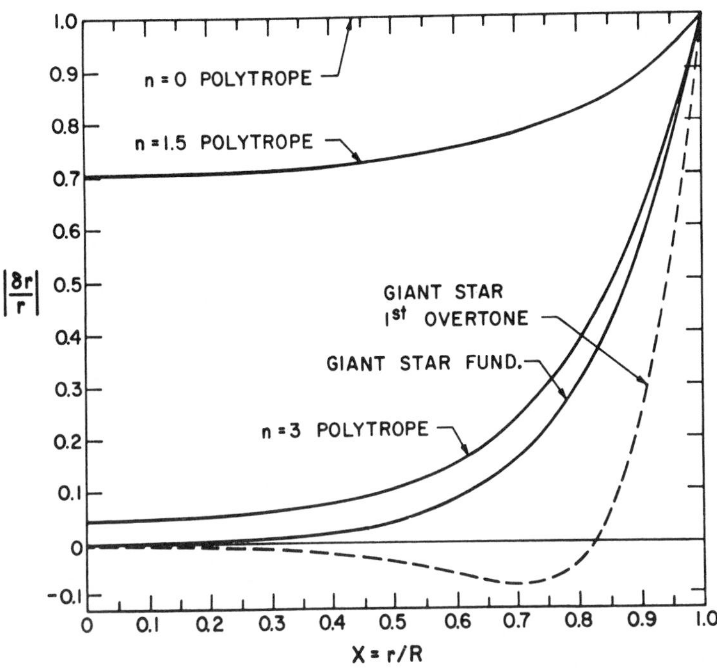

Figure 8.2. Eigenfunctions for the fundamental (solid lines) and first harmonic (broken line) modes for selected stellar models. Abscissa is fractional distance from the stellar center; ordinate is relative radius semi-amplitude. This is Fig. 14 of Cox (1974a), reproduced by permission of The Institute of Physics.

To get the X_i, now, we adopt the normalization (say) $\delta r_J = 1$, that is, $X_J = (\mathcal{D}M_J)^{1/2}$. Then the X_i, $i = J - 1, \ldots, 1$, are obtained successively from eq. (8.73), thus completing the solution for all the X_i.

8.12c. SUMMARY OF RESULTS FOR SELECTED STELLAR MODELS

Table 8.1 (which is essentially Table 3 of J. P. Cox 1974a, reproduced by permission of the Institute of Physics) summarizes some properties of solutions of the LAWE, for both the fundamental and first harmonic modes,[4] for certain selected stellar models all having $\Gamma_1 = 5/3$ (see also Table 12 of Ledoux and Walraven 1958). Some representative eigenfunctions are shown in Fig. 8.2 (this is Fig. 14 of Cox 1974a, reproduced by permission of The Institute of Physics).

[4]The term "first harmonic" is actually a misnomer, as this term implies a commensurability between the lowest and next-lowest eigenfrequencies. The correct term is "first overtone." However, to comply with conventional usage in the subject, we shall use these terms interchangeably throughout this book.

Two of the noteworthy features of Table 8.1 and Fig. 8.3 are the following. First, the dimensionless eigenfrequency $\Omega_0^2 = \sigma_0^2 R^3/GM$ in the fundamental mode generally increases, though not monotonically, with increasing mass concentration as measured by $\rho_c/\bar{\rho}$. This behavior is what we would expect (see §8.9). Second, the ratio $|\xi_R/\xi_c|$ of surface to central values of ξ is strongly correlated positively with $\rho_c/\bar{\rho}$, for both fundamental and first harmonic. Hence, stars having $\bar{\Gamma}_1$ not close to 4/3, and with great central mass concentration, have much larger pulsation amplitudes in the surface regions than in the central regions. The pulsations effectively do not penetrate down to the relatively massive cores of such stars, where most of the mass is located.

This behavior of the relative pulsation amplitude has far-reaching implications regarding the cause of the pulsations in common types of pulsating stars, as most such stars are giants or supergiants (see, for example, Cox 1974a, §9.1), and hence must be, according to current ideas of stellar evolution, highly centrally condensed. These implications will be discussed in Chapter 10. The above behavior of the relative pulsation amplitude with depth in a star also has important implications as to which regions in a star contribute to the determination of the period (see §8.13).

The above results suggest that radial stellar pulsation (at least in giant- and supergiant-like stars) is a fairly superficial phenomenon affecting, for all practical purposes, only the outer stellar layers. This suggestion is strengthened by a consideration of the amount of pulsation energy possessed by such a pulsating star (Cox 1967). Here it is shown that the total pulsation energy (kinetic plus potential) of a typical Cepheid, averaged over a period, is some eight orders of magnitude smaller than the gravitational energy. It is basically the relatively great central mass concentration of real stars (particularly giant- and supergiant-like stars) that makes radial stellar pulsations normally a rather superficial phenomenon energetically.

8.13. WEIGHT FUNCTIONS

We consider in this section a now-classical paper by Epstein (1950). He made use of the integral expression (8.30) for σ^2, for any mode k. After numerically solving the LAWE for $\xi_k(r)$, he then evaluated the integrand in the numerator of eq. (8.30). This integrand may be regarded as a sort of "weighting function," whose relative magnitude indicates which parts of the star give the greatest contribution to σ_k^2. Epstein found that for the fundamental mode of highly centrally concentrated models, the weighting function had a single strong maximum in the vicinity of $x(\equiv r/R) \approx 0.75$, and was very small in the central regions of the star. Hence, we may say that the period in the fundamental mode is determined (for centrally concentrated models) primarily by conditions in the *envelope,* say around $x \approx 0.75$, and that the period is almost independent of conditions in the *central* regions where most of the mass is located.

For the first harmonic, the weighting function had a single sharp maximum at $x \approx 0.85$. Hence, the above conclusions apply also to this mode.

A similar study has been carried out more recently by Petersen (1975), using more realistic stellar models for the evaluation of the weighting functions. This work resulted in a clarification of certain points left a bit equivocal in Epstein's (1950) study.

The above conclusions have an immediate application, at least to centrally concentrated stellar models, that is often employed in modern pulsation calculations. If the central regions of a model have only a small effect on the periods, these regions may often be treated only very crudely in calculations of the periods. For example, in calculations of the kind described by King, Cox, Eilers, and Davey (1973), the central regions of the model (containing most of the complicated details of the past evolution of the star) are usually replaced by a rigid core, which of course does not participate in the pulsations at all. Practical tests are usually made in this kind of calculation to insure that the radius of the core is small enough not to affect appreciably the computed periods.

An application of the idea involved in weight functions to nonradial stellar oscillations is considered briefly in §17.13.

8.14. EFFECT OF VARIABLE Γ_1 IN THE ENVELOPE

The effect of a spatially variable adiabatic exponent $\Gamma_1(\equiv[d \ln P/d \ln \rho]_{ad})$ on the periods of radial, adiabatic pulsations of a star can be discussed either on the basis of the LAWE itself or of the integral expressions derived in §8.9. Such a spatially variable Γ_1 would obtain, for example, in regions of a star in which some abundant element, such as hydrogen or helium in one of its two ionization states (neutral or singly ionized), is in the midstages of ionization. In such regions Γ_1 drops to a value near unity, reflecting the fact that here most of the work of adiabatic compression, for example, goes into ionization energy, and only a relatively small amount is left over for increasing the kinetic energy of thermal motions, that is, the temperature. This effect applies, qualitatively, to all modes of radial pulsations. However, for the reasons given in §8.7, the effect is largest for the fundamental mode, and it is this mode that we specifically have in mind.

In either case the conclusion is the same: the presence of an ionization zone in the outer parts of a star would, through the effect of the ionization on Γ_1, *decrease* the eigenfrequency, or *increase* the pulsation period. This conclusion is indirectly confirmed by detailed calculations.

8.15. APPROXIMATE FORMULAE AND RESULTS FOR THE EIGENVALUES

In §8.10 we found that the real number Σ^2 could be expressed by a formula very similar to eq. (8.36), where the dimensionless quantity $\alpha(u)$ is defined in eq. (8.35). We know that $\Sigma^2 = \sigma_k^2$, the eigenvalue in the kth mode, when $u(r) = \xi_k(r)$, the eigenfunction in this mode. Moreover, the eigenvalues of the LAWE are known to obey an extremal principal, which becomes a minimal principle in the case of the fundamental mode. Accordingly, if we take as a trial function $u(r)$ = constant, we obtain an upper limit to σ_0^2, given by eq. (8.44). An approximate expression for the square of the fundamental angular frequency is then

$$\sigma_0^2 \approx \Sigma^2_{u\,\text{const}}, \tag{8.75}$$

where $\Sigma^2_{u\,\text{const}}$ is defined in eq. (8.44). The approximate expression (8.75) is expected to yield fairly accurate values of the periods of radial pulsations of stars, except possibly for highly centrally concentrated stars (see below).

Some results are shown in Table 8.2, constructed from information given by Ledoux and Pekeris (1941) and by Rosseland (1949) (see also Hurley, Roberts, and Wright 1966). In this table $\rho_c/\bar{\rho}$ denotes the ratio of central to mean density, Γ_1 is assumed constant throughout the star, and $\Pi_0/\Pi_{0,\,\text{exact}}$ denotes the ratio of the fundamental pulsation period as computed from eq. (8.75), to the period as obtained by actually solving the eigenvalue problem. We see that the above expectation is generally

TABLE 8.2

Comparison of Exact and Approximate Fundamental Periods of Selected Stellar Models

Model	$\rho_c/\bar{\rho}$	Γ_1	$\Pi_0/\Pi_{0,\,\text{exact}}$
Polytrope $n = 3/2$ (completely convective star)	5.991	5/3	0.996
Polytrope $n = 2$	11.40	1.428	0.964
Polytrope $n = 3$ (standard model)	54.18	5/3(?)	0.957
Cowling Model (opacity $\propto \rho T^{-3.5}$)	37.0	5/3	0.907
Polytrope $n = 4$	622.4	1.428	0.682

confirmed. We may conclude that the approximate expression (8.75) yields fundamental periods to within about 10% accuracy or better for stars having mass concentrations less than or equal to those for main sequence stars (i.e., for $\rho_c/\bar{\rho} \lesssim 60$). Even for highly centrally concentrated stars such as yellow or red giants or supergiants (most common types of pulsating stars are thought to be of this kind, see, e.g., J. P. Cox 1974a, §9.1), this approximate expression may nevertheless yield periods accurate to within a factor of, say, 2 or 3. A better approximation for highly centrally concentrated models is indicated below.

We note that eq. (8.75) would probably be a good approximation for *any* equilibrium model in which Γ_1 is constant and has a value near 4/3. The reason is that, as was pointed out in §8.10, the eigenfunction approaches constancy as $\Gamma_1 \rightarrow 4/3$, and such constancy was assumed for the trial function used to obtain eq. (8.75).

Ledoux, Simon, and Bierlaire (1955) attempted to improve on eq. (8.75), and adopted $u(r) = \bar{\rho}/\bar{\rho}(r)$ as a trial function. Here $\bar{\rho}(r)$ is the mean density interior to radial distance r, and $\bar{\rho}$ is the mean density of the whole star. These authors then derived an approximate formula for σ_0^2, for Γ_1 = constant, based on this trial function. This formula is given by Ledoux and Walraven (1958) as eq. (60.14). According to the authors, this equation gives good results for a wide range of stellar models.

9

Linear, Nonadiabatic Radial Oscillations

In this chapter we restore the nonadiabatic terms which were dropped in Chapter 8. However, we still assume small, purely radial oscillations about a configuration which is assumed to be in complete—hydrostatic and thermal—equilibrium. We also neglect any changes in chemical composition resulting from nuclear reactions, and make the other simplifying assumptions summarized at the beginnings of Chapters 4 and 6. (Specifically, we neglect, unless stated otherwise, viscous stresses and other irreversible effects.)

Since the system is now no longer conservative, we can expect to find solutions with complex frequencies, that is, solutions characterized by securlarly increasing or decreasing pulsation amplitudes. It is therefore possible to investigate the important question of pulsational stability, or, in other words, the causes of pulsations in stars.

An overview of the problem, from a largely physical standpoint, will be summarized in §9.1, and a discussion of some mathematical aspects of the problem will be given in §9.2. Since these topics have already been discussed at some length in, for example, Cox (1974a) and Cox and Giuli (1968, Chap. 27), our remarks here will be kept relatively brief. In one subsection of §9.2, a practical method of solution of the mathematical problem will be outlined. Integral expressions for the eigenvalues will be discussed in §9.3. In §9.4 we shall consider, and slightly generalize, Eddington's evaluation of the stability coefficient.

9.1. OVERVIEW: PHYSICAL DISCUSSION

Since these matters have already been discussed in some detail in the above-mentioned references, we shall here merely summarize some of the main ideas. The interested reader is advised to consult the above references for further details or information.

It is interesting to note that, for the star as a whole, these nonadiabatic effects are normally expected to be quite small, of order Π/t_K (Π = pulsation period, t_K = Kelvin time), which was shown in Chapter 2 to be, generally, many orders of magnitude smaller than unity. This circumstance is one reason why the simple adiabatic theory gives such a good description of many features, such as the periods, of pulsating stars. The assumption of adiabatic pulsations in stars may often be a much better approximation than the assumption of adiabatic sound waves in terrestrial applications.

Because of the relatively low temperatures in the outer layers of stars, the luminosity variations, δL_r, are usually effectively "frozen in," that is, δL_r = function (time only), in these regions, and the variations in physical conditions here are significantly nonadiabatic.

The above nonadiabatic effects, though very minute throughout most of a pulsating star, do, however, lead to a slow change in the pulsation amplitude, as is well known. This slow change in amplitude is brought about essentially by small phase shifts between, say, maximum pressure P and maximum density ρ. These phase shifts, in turn, arise from the time derivatives in the energy equation (4.30a) or (5.34a).

The *direction* of the effect of these phase shifts can easily be seen from the energy equation. If, for example, $dq/dt > 0$ (local heat *gains*) at maximum density, then P will still be increasing when ρ has reached maximum. Thus, maximum pressure will in this case come *after* maximum density, and maximum δP will *lag* behind maximum $\delta\rho$.

The effects of these phase shifts on the pulsation amplitude can readily be seen from the conservation of mechanical energy theorem in the form (4.24). If we integrate this equation over a complete period of duration Π, assuming that the system returns precisely to its initial state at the end of the period, we observe that the conservative gravitational energy term vanishes. We thereby obtain an expression for the change $\Delta\mathcal{T}$ in kinetic energy of the star during the period; clearly, we have $\Delta\mathcal{T} = W$, where W is the total work done on the star by the sum of the gravity and pressure gradient forces acting over a period by all the mass elements in the star. This work is just equal to the PdV work done by all these mass elements on their surroundings throughout a period.

The above discussion leads to a simplified picture of pulsational stability in stars. We can picture the star as if it were made up of a large (infinite) number of elementary, independent Carnot-type heat engines, one such engine corresponding to each shell of mass dm. If the sum over all mass elements in the star of the above work integrals is *positive*, then $\Delta\mathcal{T} > 0$ and we have pulsational *instability;* if the above sum is *negative*, then $\Delta\mathcal{T} < 0$ and we have pulsational *stability*.

The above picture enables us easily to identify which portions of a star are "driving" regions and which are "damping" regions. The criterion for identifying the above two types of region is as follows: *a region that is gaining heat at maximum compression is a driving region*. And, conversely, *a region that is losing heat at maximum compression is a damping region.* Note that this criterion is quite general and does not depend on any assumptions about the mechanism of heat transfer.

We may now define the *e*-folding time τ_d for decay of the pulsation amplitude:

$$\frac{1}{\tau_d} \equiv -\frac{1}{2}\frac{\langle dW/dt \rangle}{\langle \delta\psi \rangle}, \tag{9.1}$$

where $\delta\psi$ is the total (kinetic plus potential) pulsation energy of the star (this total pulsation energy will be defined more precisely below), and angular brackets denote an average of the indicated quantity over a period of duration Π. The factor $1/2$ takes into account the fact that the pulsation *amplitude* grows or decays only one-half as fast as does the pulsation *energy*. In the limit of vanishingly small amplitudes (linear theory), the right side of eq. (9.1) becomes the *stability coefficient* κ:

$$\kappa = 1/\tau_d. \tag{9.2}$$

Hence, *stability* (damped oscillations) implies that $\kappa > 0$, whereas *instability* (growing oscillations) implies that $\kappa < 0$.

Perhaps the simplest mechanical analogy to nonadiabatic stellar pulsations is a one-dimensional simple harmonic oscillator in which the restoring force is not exactly in phase with the displacement. An everyday example is pushing a child in a swing.

It should be noted that restoring forces which are not in phase with the displacement will in general lead to phase shifts in the displacements of mass shells, relative to one another, in different parts of the star. Hence, nonadiabatic oscillations are not in general expected to be strictly standing waves, as detailed calculations have confirmed: there must always be a (possibly small) running wave component. The system therefore never at any time in the cycle ever passes precisely through its equilibrium state. A simple mechanical analogy is a pendulum which is not constrained to oscillate in a plane. In real pulsating stars, however, according to detailed calculations, the running wave component is very small throughout the main body of the star. The wave is almost a pure standing wave, except possibly in the outermost stellar layers.

There are several ways of getting the explicit form of $\langle dW/dt \rangle = W/\Pi$ in eq. (9.1) for the linear theory (the above considerations actually apply to nearly sinusoidal pulsations of *any* amplitude). One way is to linearize all quantities entering into W directly (see the first term on the right side of eq. [4.24]). However, because of the assumption of perfectly periodic motion in the above considerations, all quantities entering into W must be expanded at least to *second* order in small quantities. Making use of the linearized energy equation (7.8) and carrying out the relevant cyclic integrations, we obtain

$$\left\langle\frac{dW}{dt}\right\rangle = \left\langle\frac{d\mathcal{T}}{dt}\right\rangle = \frac{1}{\Pi}\int_M dm \int_0^{\Pi} (\Gamma_3 - 1)_0 \frac{\delta\rho}{\rho_0}\,\delta\left(\epsilon - \frac{\partial L_r}{\partial m}\right) dt. \tag{9.3}$$

An alternative (but essentially equivalent and more straightforward) method of obtaining eq. (9.3), which avoids the necessity of carrying all expansions to second order in small quantities, is the following. The linearized momentum equation (7.5) is multiplied through by $\delta\dot{r} = r_0\dot{\zeta}\,(\delta\dot{r} \equiv \partial[\delta r]/\partial t,\ \zeta \equiv \delta r/r_0)$, and then integrated over the entire stellar mass. The general thermodynamic identity (5.35a) is also used, and, after a few manipulations and slight rearrangement, the following is obtained:

$$\frac{d\delta\psi}{dt} = -\int_M (\Gamma_3 - 1)_0 T_0(\delta s)\,\frac{\partial}{\partial t}\left(\frac{\delta\rho}{\rho_0}\right) dm, \tag{9.4}$$

where

$$\delta\psi = \delta\mathcal{T} + \delta\Phi, \tag{9.5}$$

$$\delta\mathcal{T} \equiv \int_M \tfrac{1}{2}\,(\delta\dot{r})^2\, dm, \tag{9.6}$$

$$\delta\Phi \equiv \int_M \left[\frac{1}{2}\frac{\Gamma_{1,0}P_0}{\rho_0}\left(\frac{\delta\rho}{\rho_0}\right)^2 - \frac{2Gm}{r_0}\zeta^2\right] dm. \tag{9.7}$$

We note that this $\delta\Phi$ is identical to a similar quantity defined in the linear adiabatic theory, see §8.11. As was shown there, $\delta\Phi > 0$ for *any* small perturbation, provided that $\Gamma_{1,0} > 4/3$. Also see §8.11 for the relation of $\delta\psi$ to the ψ of §4.6.

Now, we integrate eq. (9.4) over a complete period of duration Π, assuming that the system returns precisely to its initial state at the end of the period. Integrating the right side of eq. (9.4) by parts, and assuming thermodynamically reversible processes (see eq. [7.10]), we obtain an equation which is identical to eq. (9.3).

We now assume a time dependence of the form

$$\frac{\delta\rho}{\rho} = \left(\frac{\delta\rho}{\rho}\right)_{\rm sp} e^{i\omega t} = \left|\left(\frac{\delta\rho}{\rho}\right)_{\rm sp}\right| \exp\,[i(\omega t + \phi_\rho)], \tag{9.8}$$

where ω is the angular oscillation frequency, zero subscripts have been dropped, subscripts "sp" mean the "space part" (generally complex) of the indicated quantity, vertical bars mean the absolute value (≥ 0) of the indicated quantity, and ϕ_ρ denotes the phase, generally a function of m (mass interior to radial distance r), of $(\delta\rho/\rho)_{\rm sp}$. Similarly, we write

$$\delta\left(\epsilon - \frac{\partial L_r}{\partial m}\right) = \delta\left(\epsilon - \frac{\partial L_r}{\partial m}\right)_{\rm sp} e^{i\omega t} = \left|\delta\left(\epsilon - \frac{\partial L_r}{\partial m}\right)_{\rm sp}\right| \exp\left[i(\omega t + \phi_q)\right], \qquad (9.9)$$

where ϕ_q is the phase of $\delta(\epsilon - \partial L_r/\partial m)_{\rm sp}$. We now substitute eqs. (9.8) and (9.9) into eq. (9.3) and assume that $\omega = \sigma$ is purely real in the evaluation of the time integral. We obtain

$$\left\langle \frac{d\delta\psi}{dt} \right\rangle = \tfrac{1}{2}\, C_r, \qquad (9.10)$$

where the complex quantity C is defined by the relation

$$C \equiv \int_M (\Gamma_3 - 1)\left(\frac{\delta\rho}{\rho}\right)^*_{\rm sp} \delta\left(\epsilon - \frac{\partial L_r}{\partial m}\right)_{\rm sp} dm \qquad (9.11a)$$

$$= C_r + iC_i, \qquad (9.11b)$$

in which subscripts r and i denote, respectively, real and imaginary parts. The quantity C is often referred to as the "work integral." The total pulsation energy (kinetic plus potential) in the case of nearly sinusoidal oscillations is then (see §8.11)

$$\delta\psi = 2\overline{\delta\mathcal{T}} = \frac{1}{2}\sigma^2 J, \qquad (9.12)$$

where $\sigma = 2\pi/\Pi$ and J is the oscillatory moment of inertia of the star in the mode considered (see §8.9). The expression for the stability coefficient is then

$$\kappa = -\frac{C_r}{2\sigma^2 J}, \qquad (9.13)$$

and this can be shown (see §9.3 below) to be an accurate expression if $|\kappa/\sigma| \ll 1$. The motion of the mass elements in the case of nonadiabatic oscillations is therefore nearly periodic, with a slowly increasing or decreasing amplitude:

$$\zeta(m,t) = \xi(m) e^{i\sigma t} \cdot e^{-\kappa t}, \qquad (9.14)$$

where κ is given to good approximation by eq. (9.13). Hence, the oscillations are *excited* if $\kappa < 0$, or if $C_r > 0$; and *damped* if $\kappa > 0$, or if $C_r < 0$.

The main effect of the imaginary part C_i of C is to affect the pulsation *period* (see §9.3 below). We have

$$\sigma \approx \Sigma + \frac{C_i}{2\Sigma^2 J}, \qquad (9.15)$$

where Σ^2 is the real number defined in §8.10, and the last term is usually small compared to the first term on the right side.

It should be noted that the quantities in the integrand of C should, strictly speaking, be evaluated from the correct *nonadiabatic* eigenfunctions for the problem of interest. These eigenfunctions are, however, often not known in advance. In this case eq. (9.11) may be regarded as an approximate expression for C, in which the correct eigenfunctions are replaced by suitable trial functions. Such a situation is not entirely unsatisfactory, and the use of trial functions for the approximate computation of C has been extensively used in the theory of stellar pulsation.

A common procedure for the approximate evaluation of C, widely used both in the past and present, is making the *quasi-adiabatic* approximation. This approximation consists in evaluating the integrand of C on the basis of the solutions of the LAWE (see Chap. 8). Specifically, the solution $\xi(m)$ of the LAWE for the model and mode of interest is obtained, and then this function is used to compute $\delta T/T$, $d(\delta T/T)/dr$, and so on, from the adiabatic relations, eqs. (5.36). From these can be computed $\delta(\epsilon - \partial L_r/\partial m)$, which is part of the integrand of C. This procedure may provide a reasonable approximation to the correct value of C if the outer, nonadiabatic regions of the star are not very extensive. However, this procedure should always be used with considerable caution. One possibly serious defect of the method is that δL_r, computed in this manner, does not exhibit the freezing-in behavior described earlier, demanded of physically significant solutions. This behavior of the quasi-adiabatically computed δL_r causes its contribution to C in these outer, nonadiabatic regions to be spuriously large. One simple approximate procedure to circumvent this difficulty is to cut off the integration in C at the transition region (see §10.1).

Crude estimates of the values to be expected of the damping times for pulsation amplitude, τ_d, may be found in Cox and Giuli (1968, Chap. 27).

Table 9.1 summarizes the order-of-magnitude results of detailed calculations for the damping times, in units of the fundamental pulsation periods, of several common types of pulsating stars (from Cox 1974a). We

TABLE 9.1

Damping Times for Typical Pulsating Stars

Type of star	$\lvert \tau_d/\Pi_0 \rvert$*
Classical Cepheid RR Lyrae Variable	10^2–10^3
Delta Scuti Variable Dwarf Cepheid	10^4–10^6
W Virginis Variable	10–20
Long Period Variable (Mira Variable)	1–10

*Π_0 = fundamental pulsation period.

see that in essentially all cases (with the possible exception of the Mira variables) the damping times are considerably larger than the fundamental pulsation periods.

9.2. MATHEMATICAL ASPECTS

In this section we shall investigate some of the mathematical aspects of the problem of linear, radial, nonadiabatic pulsations of stars. Certain physical aspects of this problem were discussed in §9.1. The problem will be discussed from a rather general standpoint in §9.2a, and modern methods of obtaining the numerical solution of this problem for models of stars or stellar envelopes will be outlined in §9.2b. Finally, we shall summarize in §9.2c certain key features of such solutions as applied to stellar models.

9.2a. GENERAL

The appropriate equations for the present problem are the linearized mass, momentum, energy, radiative transfer, and energy generation rate equations. (In principle, the qualifying adjective "radiative" should be replaced by "energy" in the term "radiative transfer." However, since quantitative and reliable theories of convective heat transfer do not now exist, and are not likely to exist for some time to come, the above qualifying adjective is justified on practical grounds.) The Lagrangian variations of all variables are assumed to contain the time dependence $e^{i\omega t}$, where $\omega = \pm\sigma + i\kappa$, and σ are κ both real. The pulsation *period* is determined by the value of σ, and κ is the stability coefficient (see §9.1). The two possible signs of σ have no physical significance, and merely correspond to $(i\omega)$ and its complex conjugate $(i\omega)^*$ (see §9.3). The dependent variables are the space parts of, say, the radius, density, temperature, and luminosity variations of a mass shell. These variables will be denoted, in accordance with the notation in the rest of this chapter, by, respectively, ξ, $\delta\rho/\rho$, $\delta T/T$, and $\delta L_r/L_r$ when regarded as relative variations, and the subscript "sp" will be dropped. Also, zero subscripts will usually be dropped from equilibrium quantities. In general, the above variables must all be considered complex, since their phases will not in general be constant throughout the star.

With the above time dependence, the basic, partial differential equations become *ordinary* differential equations, with radial distance as the independent variable. These equations are *linear, homogeneous,* and of the *fourth* order in complex variables, or of the *eighth* order in real variables. There are accordingly *four* complex constants of integration, plus the complex angular frequency ω.

Given an equilibrium model of a star or stellar envelope, there are four boundary conditions, two at the center or base of the envelope, and two at the surface. These boundary conditions serve to determine four of the above five complex constants. Because of the homogeneity of the problem, one of these five complex constants must remain arbitrary, and may be used for normalization. For example, one might require (as is usually done in such calculations) that ξ_R (surface value of ξ) $= 1$. The complex eigenvalue is ω.

Reasonable general boundary conditions usually require that all relative pulsation variables be *regular* both at center and surface. At the center it clearly must be that

$$\delta r = 0 \tag{9.16}$$

and

$$\delta L_r = 0. \tag{9.17}$$

These are also the conditions usually assumed at the base of the envelope in the case of an envelope-only model (however, see §9.2.b). Since both r and L_r vanish at the center (L_r varies as r^3 near $r = 0$), the above regularity condition then requires that $\delta r/r = \xi$ and $\delta L_r/L_r$ both remain finite at the center (this requirement of course does not apply to $\delta L_r/L_r$ if L_r should be identically zero throughout the core of the star).

The requirements (9.16) and (9.17) yield *two* complex relations among ω and the central (or base-of-the-envelope) values of the pulsation variables. The two relations derive from the mass and energy equations. In the case of a complete stellar model, the mass equation yields the relation

$$\left(\frac{\delta\rho}{\rho}\right)_c = -3\xi_c, \tag{9.18}$$

where subscript c denotes central values. The energy equation then yields the result

$$\left(\frac{\delta T}{T}\right)_c = (\Gamma_3 - 1)_c\left(\frac{\delta\rho}{\rho}\right)_c - \frac{i}{\omega}\left(\frac{\epsilon}{c_V T}\right)_c \left[\lambda_c\left(\frac{\delta\rho}{\rho}\right)_c + \nu_c\left(\frac{\delta T}{T}\right)_c - \left(\frac{\delta L_r}{L_r}\right)_c\right] \tag{9.19}$$

(see eq. [7.12]). We note that eqs. (9.18) and (9.19) contain the unknown parameters ω, ξ_c, and $(\delta L_r/L_r)_c$. In the case of an envelope-only model these two equations would be replaced by two different, but equivalent, equations, and the unknown parameters would be ω, $d\xi/dr$, and $d(\delta L_r/L_r)/dr$, the last two quantities representing the values of these derivatives at the base of the envelope.

The specific surface boundary conditions depend on the assumptions made regarding the surface regions of the equilibrium model, and can involve certain subtleties (see, e.g., J.P. Cox 1963; Baker and Kippenhahn 1962; Unno 1965; Castor 1971; Iben 1971b; and Langer 1971). Unless the stellar atmosphere is very extended, one can usually assume with sufficient accuracy the standing wave boundary condition (say eq. [8.10]). This boundary condition may also be written in the form (see §8.3)

$$\left(\frac{d\delta P}{dr}\right)_R = 0. \tag{9.20}$$

This equation is valid if the density vanishes at the stellar surface.

The other surface boundary condition is obtained from the luminosity equation. Pure radiative transfer may normally be assumed in the surface regions, but various degrees of approximation in the description of the radiative transfer may be employed. The simplest case, and historically one of the earliest considered (see for instance Cox 1963), is one in which the surface temperature T_R is assumed to be zero. In this case the above regularity condition requires that

$$\delta T_R = 0. \tag{9.21}$$

If it is assumed that the diffusion form of the radiative transfer equation is valid at the surface, then eq. (9.21), together with the assumption that $T_R = 0$ and the above regularity condition, lead to the relation (see eq. [7.11])

$$\left(\frac{\delta L_r}{L_r}\right)_R = 4\xi_R - n_R\left(\frac{\delta \rho}{\rho}\right)_R + (s + 4)_R\left(\frac{\delta T}{T}\right)_R, \tag{9.22}$$

where n_R and $-s_R$ denote the surface values of, respectively, the density and temperature exponents in the opacity law.

In the more realistic case of photospheric boundary conditions, the appropriate boundary condition for δT may be obtained as follows. If the stellar atmosphere is assumed to be thin compared to the stellar radius and if the atmosphere is not changing "too rapidly" (see §8.3), then the temperature T as a function of the normal optical depth τ is given approximately by the grey-atmosphere relation (see, e.g., Woolley and Stibbs 1953; Mihalas 1978). Now, the variation ΔL in luminosity L associated with the variations Δr_{ph} and ΔT_e in, respectively, the photospheric radius and effective temperature of the star is given by taking the variation of the relation $L \propto r_{ph}{}^2 T_e{}^4$. If a variation δT in the temperature of a mass element is assumed to be a result of both variations in T_e and in τ (the normal optical depth down to the mass element), the temperature-optical depth relation referred to above shows that, at small normal optical depths ($\tau \ll 1$), $\delta T/T \approx \Delta T_e/T_e$. If this relation is used in the above equation involving ΔL, Δr_{ph}, and ΔT_e, and if we assume that $\delta r \approx \Delta r_{ph}$ and $\delta L_r \approx \Delta L$ for a given mass level in the atmospheric regions, then we obtain

$$\left(\frac{\delta L_r}{L_r}\right)_R = 2\xi_R + 4\left(\frac{\delta T}{T}\right)_R. \tag{9.23}$$

This last relation is normally used as the surface boundary condition for δT in this case.

Note that, whichever of eqs. (9.22) or (9.23) is used as the surface boundary condition for δT, *two* parameters are left undetermined by these surface boundary conditions. These are ω and *one* of the pair $(\delta T/T)_R$, $(\delta L_r/L_r)_R$. The reason is that the boundary condition (9.20) determines $(\delta P/P)_R$ (assuming ξ_R to have been chosen), which in turn depends on $(\delta\rho/\rho)_R$ and $(\delta T/T)_R$ through the equation of state. Recall that the central boundary conditions left ω, ξ_c (or $[\delta\rho/\rho]_c$), and $(\delta L_r/L_r)_c$ undetermined. The condition that the four variables ξ, $\delta P/P$, $\delta T/T$, and $\delta L_r/L_r$ be continuous throughout the star is sufficient to determine values for all four of the above unknown (complex) quantities, for each mode of interest. It can therefore be seen that, in principle at least, satisfaction of the boundary conditions is needed to determine the detailed characteristics of nonadiabatic stellar pulsations.

It is useful to note from the energy equation, say eq. (7.8), that

$$\left[\frac{d}{dr}\left(\frac{\delta L_r}{L_r}\right)\right]_R = 0 \tag{9.24}$$

if the pressure P vanishes at the surface (usually an excellent approximation). This result is consistent with the concept of freezing-in of the luminosity variations in the outer stellar layers, as discussed in §9.1.

It can also be shown (see Ledoux and Whitney 1961) that, if the density ρ vanishes at the surface,

$$\left[\frac{d^2}{dr^2}\left(\frac{\delta L_r}{L_r}\right)\right]_R = 0. \tag{9.25}$$

It should be noted that eqs. (9.24) and (9.25) are not boundary conditions, merely conclusions which are valid under the stated assumptions.

9.2b. METHODS OF SOLUTION

The numerical solution of the above eigenvalue problem involves certain computational difficulties (see, for example, Baker and Kippenhahn 1962; J.P. Cox 1963) and it is only in recent years that sufficiently powerful methods have been developed to overcome these difficulties (Castor 1971;

Iben 1971b). (Baker and Kippenhahn 1965 employed techniques somewhat similar to those of the above authors, but did not actually obtain ω as an eigenvalue of the full nonadiabatic problem.)

The difficulties owe their origin basically to the fact that the pulsations are very nearly adiabatic throughout the star. Consequently, conventional fitting techniques, based on the general considerations of §9.2.a, encounter severe practical problems unless special techniques are employed. For example, outside-in integrations that obey the surface boundary conditions always contain a small admixture of solutions which diverge rapidly as the numerical solution is followed inward. These unwanted divergent solutions eventually show up and, in practice, invalidate the outside-in integrations.

The above difficulty was avoided by Baker and Kippenhahn (1962) and Cox (1963) by replacing the interior boundary conditions by the above interior adiabatic condition itself, that is, by requiring that the outside-in integrations approach adiabaticity as the solution is followed inward. In this way the solutions of the equations were accurately determined, for given ω, only by those outer regions cooler than some 10^5°K (usually amounting to a depth of some 5 to 50% of the stellar radius). In these early investigations the angular pulsation frequency σ was either assumed known from observations or theory (J.P. Cox 1963) or obtained from the solution of the *adiabatic* wave equation (Baker and Kippenhahn 1962). The stability coefficient κ was set equal to zero in obtaining the *solution* of the differential equations; κ was then evaluated by use of integral expressions (see §9.3), in which the above solutions were used. This method of obtaining κ is quite accurate if $|\kappa/\sigma| \ll 1$, as is normally the case (see §9.1). References to earlier approximate methods for dealing with nonadiabatic effects may be found in, for example, J.P. Cox (1967, 1963); Ledoux and Walraven (1958); and Rosseland (1949).

The more modern computational methods do not actually include the core of the star in general. However, these methods may be extended in a straightforward (and almost trivially easy) way to include the stellar core if necessary or desirable (see, e.g., Ziebarth 1970a,b; Davey 1970, 1973).

We shall now briefly outline the method proposed and used by Castor (1971) for the solution of the linear, nonadiabatic wave equation (LNAWE). Some of the following is based on unpublished lecture notes of John Castor and used with his kind permission.

We shall first show that the appropriate linearized differential equations may be written in a special way, on which the method is based. In this method the *momentum* and *energy* equations are regarded as basic, for only they normally contain explicit time derivatives. The momentum equation may be written as in eq. (7.5), and the energy equation, for thermodynamically reversible processes (the only kind we consider), as in eq. (7.10). Note that the momentum equation contains $\delta\ddot{r} \equiv \partial^2(\delta r)/\partial t^2$ on the left side, and δr and $\partial(\delta P)/\delta m$ on the right side. The energy equation contains $T\delta\dot{s}$ on the left side, and $\delta\epsilon$ and $\partial(\delta L_r)/\partial m$ on the right side. However, $\delta\epsilon$ may be regarded as a function of $\delta\rho$ and δT (see eq. [7.12]); similarly, δL_r may be regarded, at least for radiative transfer in the diffusion approximation, as a function of δr, $\delta\rho$, δT, and $\partial(\delta T)/\delta m$ (see, e.g., eqs. [7.11]). Now δT and δP may be regarded as functions of $\delta\rho$ and $T\delta s$, by virtue of the thermodynamic identities (5.35). In turn, $\delta\rho$ may be regarded as a function of δr and $\partial(\delta r)/\partial m$, through the mass equation (e.g., eqs. [5.24b] or [7.4]). Thus, the right side of the momentum and energy equations may be regarded, ultimately, as functions of δr and $T\delta s$ and their derivatives with respect to m. Adopting the time dependence $\exp(i\omega t)$, the left sides of these equations are seen to be functions only of δr and $T\delta s$. We may therefore adopt δr and $T\delta s$ as the basic dependent variables, and the momentum and energy equations may be written in operator form as

$$\omega^2\delta r = G_1(\delta r) + G_2(T\delta s), \tag{9.26a}$$

$$i\omega T\delta s = K_1(\delta r) + K_2(T\delta s), \tag{9.26b}$$

where the G's and the K's are linear operators. It may be shown in a straightforward but tedious way that

$$G_1(y) \equiv -\frac{4Gm}{r^3}y - 4\pi r^2\frac{\partial}{\partial m}\left[\frac{\Gamma_1 P}{V}\frac{\partial}{\partial m}(4\pi r^2 y)\right], \tag{9.27}$$

$$G_2(y) \equiv 4\pi r^2\frac{\partial}{\partial m}\left(\frac{\Gamma_3 - 1}{V}y\right), \tag{9.28}$$

and, for radiative transfer in the diffusion approximation,

$$\begin{aligned} K_1(y) \equiv & \left[\left(\frac{\partial\epsilon}{\partial V}\right)_T - \left(\frac{\partial\epsilon}{\partial T}\right)_V\frac{(\Gamma_3 - 1)T}{V}\right]\frac{\partial}{\partial m}(4\pi r^2 y) \\ & - \frac{\partial}{\partial m}\left(\frac{L_r}{\pi r^3}4\pi r^2 y\right) \\ & + \frac{\partial}{\partial m}\left\{\frac{L_r}{\kappa}\left[\left(\frac{\partial\kappa}{\partial V}\right)_T - \left(\frac{\partial\kappa}{\partial T}\right)_V\frac{(\Gamma_3 - 1)T}{V}\right]\frac{\partial}{\partial m}(4\pi r^2 y)\right\} \\ & - \frac{\partial}{\partial m}\left\{(4\pi r^2)^2\frac{ac}{3\kappa}\frac{\partial}{\partial m}\left[\frac{4(\Gamma_3 - 1)T^4}{V}\frac{\partial}{\partial m}(4\pi r^2 y)\right]\right\}, \end{aligned} \tag{9.29}$$

$$K_2(y) \equiv \left(\frac{\partial\epsilon}{\partial T}\right)_V\frac{1}{c_V}y + \frac{\partial}{\partial m}\left[\frac{L_r}{\kappa}\frac{1}{c_V}\left(\frac{\partial\kappa}{\partial T}\right)_V y\right] +$$

$$\frac{\partial}{\partial m}\left[(4\pi r^2)^2 \frac{ac}{3\kappa}\frac{\partial}{\partial m}\left(\frac{4T^3}{c_V} y\right)\right]. \tag{9.30}$$

In these equations $V \equiv 1/\rho$ is the specific volume, ϵ is the energy generation rate per unit mass, κ is the opacity, and the remaining symbols have their usual meanings; also, we have dropped zero subscripts from all equilibrium quantities. We note that G_1 is a differential operator of the second order; G_2 is of the first order; K_1 is of the third order; and K_2 is of the second order. Also, we note that $G_1(\delta r)/r$ is the same as $\mathcal{L}(\delta r/r)$, where $\mathcal{L}$ is the linear operator defined by eq. (8.23).

It is interesting to examine the G's and K's for the case where all coefficients can be treated as constants. This case gives one a feeling for the order-of-magnitude effects of these operators. We also retain, following Castor, only the highest-order derivative terms in each case. One can then derive relatively simple order-of-magnitude expressions for $|G_1|$, $|G_2|$, $|K_1|$, and $|K_2|$, on the basis of which physical interpretations can be made. For example, if we replace $\partial/\partial r$ by l^{-1}, where l is a characteristic length, we may show that

$$|G_1| \sim (\text{sound travel time across } l)^{-2} \tag{9.31}$$

and

$$|K_2| \sim (\text{cooling time by radiation of region of characteristic dimension } l)^{-1}. \tag{9.32}$$

It is also interesting to note that

$$\left|\frac{G_1}{G_2}\right| \sim \left|\frac{K_1}{K_2}\right| \sim v_S^2 \frac{\partial}{\partial r}, \tag{9.33}$$

where v_S is the Laplacian sound speed. It therefore follows from eq. (9.33) that

$$\left|\frac{G_1(\delta r)}{G_2(T\delta s)}\right| \sim \left|\frac{K_1(\delta r)}{K_2(T\delta s)}\right| \sim \left|\frac{\delta\rho/\rho}{[(T\delta s)/(c_V T)]}\right|, \tag{9.34}$$

where we have used the approximation $|\partial(\delta r)/\partial r| \sim |\delta\rho/\rho|$. Also, $v_S^2 = \Gamma_1 P/\rho \sim c_V T$.

Equation (9.34) says that in both the basic equations (9.26a) and (9.26b), the ratio of the first to the second terms on the right side is of the general order of the ratio of the relative density fluctuations $\delta\rho/\rho$ during a period, to the magnitude of the corresponding heat exchange $\delta q = T\delta s$ per unit mass, expressed as a fraction of the internal energy, approximately $c_V T$, of that mass. In the deep interior of a star where $c_V T$ is relatively large, $|G_1(\delta r)| \gg |G_2(T\delta s)|$ and $|K_1(\delta r)| \gg |K_2(T\delta s)|$. It is only in the outer stellar layers that the two terms of each pair become of comparable magnitude.

We may say that the term $G_2(T\delta s)$ contains the effects of *nonadiabaticity* on stellar pulsations. Also, the term $K_1(\delta r)$ gives the heat flows resulting from the *adiabatic* temperature fluctuations, which in turn are produced by the dynamical motions. The approximation $-i\omega T\delta s = K_1(\delta r)$ is sometimes called the quasi-adiabatic approximation (see §9.1).

In case $G_2 = 0$ or $K_1 = 0$, we have, respectively, either pure adiabatic pulsations (Chapter 8) or pure heat flow.

Using a differencing scheme such as that used in connection with our earlier discussion of the LAWE (see §8.12.b), we see that eqs. (9.26) can be replaced by the matrix equations

$$\omega^2 \mathbf{X} = G1\mathbf{X} + G2\mathbf{Y}, \tag{9.35a}$$

$$i\omega \mathbf{Y} = K1\mathbf{X} + K2\mathbf{Y}, \tag{9.35b}$$

where $\mathbf{X}$ and $\mathbf{Y}$ are column matrices; $\mathbf{X}$ consists of $J + 1$ elements (J = number of zones in the model), and $\mathbf{Y}$ consists of J elements. Also,

$$\left.\begin{aligned} X_i &= (DM_i)^{1/2}\delta r_i, \\ Y_{i-1/2} &= (T\delta s)_{i-1/2}, \end{aligned}\right\} \quad (i = 1, \ldots, J) \tag{9.36}$$

where $\mathcal{D}M_i$ was defined in §8.12.b, and $G1$, $G2$, $K1$, and $K2$ are each $J \times J$ matrices many of whose elements are zero.

With the differencing scheme adopted in §8.12.b, we see that the momentum equation (9.35a) involves, for each X_i (but not close to either boundary), X_{i-1}, X_{i+1}, $Y_{i-1/2}$, $Y_{i+1/2}$; three elements of $G1$; and two elements of $G2$. Similarly, for each $Y_{i-1/2}$ (again, not close to either boundary), the energy equation (9.35b) involves $Y_{i-3/2}$, $Y_{i+1/2}$, X_{i-2}, X_{i-1}, X_i, X_{i+1}; four elements of $K1$; and three elements of $K2$. Thus, with this differencing scheme, eqs. (9.35) relate seven successive "points" in the model, where we regard a "point" as either an interface or the midpoint of a mass zone.

The explicit expressions for the matrix elements here are in general fairly complicated, and some specific examples are given by Castor (1971). At any rate, their values depend on the equilibrium model, boundary conditions, and so on, and are assumed known.

The boundary conditions might be something like the following. At the *center* or at the base of an envelope:

$$\begin{aligned} \delta r_0 &= 0 \qquad (\text{i.e., } X_0 = 0), \\ \delta L_0 &= 0 \end{aligned} \tag{9.37}$$

At the *surface* there are some options (see §9.2a), but reasonable choices might be:

$$\left(\frac{\delta P}{P}\right)_{J-1/2} = -\left(\frac{\omega^2 r_J^{\,3}}{GM} + 4\right)\frac{\delta r_j}{r_J},$$

$$\frac{\delta L_J}{L_J} = 2\,\frac{\delta r_j}{r_J} + 4\,\frac{\delta T_{J-1/2}}{T_{J-1/2}}. \qquad (9.38)$$

The specific boundary conditions adopted show up in the values of the matrix elements in eqs. (9.35), especially near one or the other (or both) of the two boundaries. Near these boundaries, eqs. (9.35) take on simplified forms, because quantities are not defined inside the center or outside the surface.

More specifically, as pointed out by Castor (1971), to apply the boundary conditions, it is necessary to discard the innermost and outermost interfaces (0 and J), and also the innermost and outermost zones (1 and J). It can be shown that the boundary conditions (9.37) and (9.38), as applied to these zones and interfaces (plus whatever other zones and interfaces are involved), in addition to the remaining equations, supply just the correct number of equations for the $2J + 1$ unknowns of the problem. These considerations are equivalent to dropping quantities defined only inside the center (such as X_{-1}) and outside the surface (such as X_{J+1}), and giving special values to the appropriate matrix elements.

Equations (9.35) may be written as a single matrix equation (Castor 1971) if one interlaces the variables by regarding $\mathbf{Z}$ as a column matrix possessing $2J$ elements, which are (setting $X_0 = 0$) $Y_{1/2}, X_1, Y_{3/2}, X_2, \ldots, Y_{J-1/2}, X_J$. Thus, eqs. (9.35) may be written in the form

$$\mathcal{M}\mathbf{Z} = 0, \qquad (9.39)$$

where $\mathcal{M}$ is a complex matrix possessing $(2J) \times (2J)$ elements (because its elements are complex); $\mathcal{M}$ is a band matrix whose "band" is at most seven elements wide. Note that the first row of $\mathcal{M}$ involves only the energy equation (9.35b); the last row involves only the momentum equation (9.35a).

We consider now the solution of the matrix equation (9.39) for the eigenvalues (ω^2) and eigenvectors. Because this equation is *homogeneous,* it cannot be solved for the eigenvectors until the eigenvalues are known, and they are of course not known *a priori.*

Castor (1971) assumes that the corresponding *adiabatic* problem has been solved, so that the adiabatic eigenvectors are known. He then uses these to obtain *first guesses* at the nonadiabatic eigenfrequencies, essentially by the use of integral expressions of the kind discussed in §9.3. But such first guesses are still not correct for eq. (9.39), so the following procedure is adopted.

Castor (1971) drops the last equation of eqs. (9.39) from the set; this last equation is just the momentum equation as applied to the outermost interface. The matrix equation (9.39) is then replaced by the matrix equation

$$\mathcal{M}'\mathbf{Z}' = \mathbf{W}, \tag{9.40}$$

where the matrix $\mathcal{M}'$ is the same as $\mathcal{M}$, but without the last row and column; $\mathcal{M}'$ contains $(2J - 1) \times (2J - 1)$ elements; and Z' is the same as $\mathbf{Z}$ except minus one element (this is X_J, which is eventually set equal to unity for normalization). Here $\mathbf{W}$ is a column matrix all of whose elements are zero except for the last few, which are, in general, nonvanishing. Since eq. (9.40) is now no longer homogeneous, the ω therein does not have to be the eigenfrequency. Thus, *given* a value for ω, eq. (9.40) may be solved for the $2J - 1$ unknowns $Y_{1/2}, X_1, Y_{3/2}, \ldots, X_{J-1}, Y_{J-1/2}$ in terms of X_J. Since this value of ω is not necessarily the eigenvalue, then the last equation of the set, which was dropped, is now not necessarily satisfied. This last equation may be written as

$$f(\omega) \equiv G1(J,1)X_{J-1} + G2(J,1)Y_{J-1/2} + [G1(J,2) - \omega^2]X_J, \tag{9.41}$$

where $f(\omega) = 0$ when $\omega = \omega_0$, the eigenfrequency, and where the G's are matrix elements.

Castor (1971) then notes that X_{J-1}, $Y_{J-1/2}$, and X_J are each rational functions (equal to the ratio of two polynomials), since each of them is a solution of a system of simultaneous, linear, algebraic equations. Consequently, $f(\omega)$ is regular except at its poles (see any text on advanced calculus, for example, Franklin 1940). Thus, any reasonable root-finding scheme for regular functions should serve to find the eigenfrequency, that is, that value of $\omega(=\omega_0)$ for which $f(\omega) = 0$.

The method adopted by Castor (1971) is the secant method, which is based on using two previous iterates, ω_{k-1} and ω_k, to estimate a new iterate, ω_{k+1}, assuming that $f(\omega)$ is linear in the vicinity of ω_0.

Using this method, one then iterates the procedure described above until the equation $f(\omega) = 0$ is satisfied to some predetermined accuracy criterion. (Of course, all of these manipulations must be carried out with complex arithmetic.) The *second* trial value of ω needed for this method may be obtained, for example, by arbitrarily changing the first trial value somewhat. In this way one eventually obtains ω_0, the eigenfrequency for the mode of interest, as well as the corresponding eigenvectors $\mathbf{X}$ and $\mathbf{Y}$.

A difficulty with the above procedure is noted by Castor (1971). He points out that $f(\omega)$ has a *pole*, $\omega = \omega_p$, near each zero, $\omega = \omega_0$. This pole corresponds to a value of ω such that $\delta r_J = 0$. Since all the X_i and $Y_{i-1/2}$ are usually normalized by setting $X_J = 1$, it follows that, when $\omega = \omega_p$, $Y_{i-1/2}/X_J = \infty$, $X_i/X_J = \infty$ $(i = 1, \ldots, J-1)$. If $|\omega - \omega_0| > |\omega - \omega_p|$, then the above secant method will either fail or converge only slowly.

Castor's remedy for this trouble is to divide $f(\omega)$ by any function which has a pole at the pole of $f(\omega)$. He chooses $X_1(\omega)$ for this function, and defines

$$g(\omega) \equiv f(\omega)/X_1(\omega). \tag{9.42}$$

The advantage of choosing X_1 for this arbitrary function is that it has no zeros for any of the low pulsation modes. Castor then suggests using $g(\omega)$ in place of $f(\omega)$ in the above discussion regarding the secant method. This procedure improves convergence and at the same time gets around the above pole difficulty. For further information and details, see Castor's (1971) paper.

Finally, we shall attempt to specify precisely what is meant by the term "mode" when applied to linear but nonadiabatic pulsations. Since such pulsations never carry the system precisely through its equilibrium state (see §9.1), *nodes*, characterized by the vanishing of ξ, do not actually exist in nonadiabatic pulsations. Hence, the presence of nodes cannot be used to characterize the modes (a somewhat analogous situation obtains in the case of nonradial oscillations; see Chapter 17). By "mode" we shall mean simply one of the eigensolutions of the linear, nonadiabatic equations. When nonadiabatic effects are small or, roughly equivalently, when $|\kappa/\sigma| \ll 1$ (these conditions are well satisfied in most types of pulsating stars, see §9.1), a well-defined one-to-one correspondence seems to exist between the modes in the nonadiabatic and adiabatic problems, and the ordering of the modes is similar in the two classes of problems. A "node" in the case of nonadiabatic pulsations is normally a spherical surface on which $|\xi|$ attains a sharp minimum, becoming quite small, and across which the phase of ξ changes by very nearly π radians. Calculations show that nonadiabatic effects, unless enormous, have very little effect on the location of nodes.

9.2c. GENERAL FEATURES OF LINEAR NONADIABATIC RADIAL OSCILLATIONS

In this subsection we shall summarize some of the general features of solutions of the linear nonadiabatic wave equation (LNAWE) for "realistic" stellar models lying in the general "Cepheid region" of the Hertzsprung-Russell diagram. The comments here are based on models which are assumed to be radiative throughout, even though their parameters may, according to current ideas, in some cases be such that convection might actually be present in the envelopes. The primary justification for this restriction to radiative models is the generally unsatisfactory state of our knowledge of convection in real stars.

1. The nonadiabatic periods, at least in the fundamental mode, usually seem to be shorter than the corresponding adiabatic periods, but normally by only a small amount. The difference is typically only a few tenths of a percent, but it can amount, in some cases where the equilibrium luminosity/mass ratio is relatively large, to some 2 or 3%. Wood (1976), in dealing with much larger nonadiabaticities, found that the nonadiabatic periods were sometimes larger than the adiabatic periods, and that the percentage difference between the two was often considerably larger than the above values. Other interesting effects regarding the eigenfunctions were reported. Analogous results have been obtained by King, Wheeler, Cox, and Hodson (1978) and by King, Wheeler, Cox, Cox, and Hodson (1979).

2. The run of the nonadiabatic relative pulsation amplitude, $\xi(r)$, does not differ much, at least below the hydrogen (H) ionization zone, from $\xi_{ad}(r)$, the corresponding adiabatic relative pulsation amplitude. Hence $\xi(r)$ (normalized to unity at the surface) is nearly real, so that the pulsations are nearly standing waves. (However, $\delta L_r/L_r$ is nearly in phase with $\delta\rho/\rho$ only in the interior. Near the surface, $\delta L_r/L_r$ may differ appreciably in phase from $\delta\rho/\rho$, as is shown by the existence of the phase lag in the surface luminosity variations; see Chapter 11.)

In radiative models in the general "Cepheid region" of the Hertzsprung-Russell diagram, $\xi(r)$ generally remains close to unity, and may even exceed unity (proceeding inward), in the regions of the star in and above the H ionization zone. This behavior is strictly a nonadiabatic effect, and is due, ultimately, to the fact that $|dT_0/dr|$ is generally very large in the H ionization zone of radiative models (see Chapter 11). Below this zone, $\xi(r)$ rapidly approaches $\xi_{ad}(r)$ (proceeding inward).

3. The pulsations rapidly approach adiabaticity as the interior is approached. This means that the adiabatic relations, eqs. (5.36), become increasingly better approximations as the interior is approached.

9.3. INTEGRAL EXPRESSIONS FOR THE EIGENVALUES

We have seen that the problem of small, nonadiabatic, radial stellar pulsations can be solved accurately only by obtaining numerical solutions of a complicated, high-order eigenvalue problem. These solutions clearly depend on the transfer mechanism assumed for the heat flow. However, considerable insight into the overall time behavior of these solutions, and of some of the various possibilities, can be gained by adopting an approach based on formal integral expressions for the eigenvalues. These integral expressions are general in the sense that they do not involve any explicit assumptions regarding the physical nature of the transfer mechanism. For these integral expressions to yield accurate numerical results for the eigenvalues, however, the functions appearing in them must be the *eigenfunctions,* which are not known before the specific eigenvalue problem has been solved. However, as was the case with the integral expressions derived in §8.8 for the eigenvalues of the *adiabatic* pulsation problem, reasonable approximations to the nonadiabatic eigenvalues can often be obtained by using approximate trial functions in place of the eigenfunctions. These trial functions are essentially arbitrary, aside from being sufficiently regular. Because, however, the nonadiabatic problem does not have the Hermitian property, these integral expressions do not possess the extremal character as was the case with adiabatic pulsations (see §8.10). Hence, the trial functions should be as accurate as possible if the integral expressions are to provide good approximations to the eigenvalues. The use of integral expressions has played an important role in the historical development of the theory of stellar pulsation.

Detailed derivations and discussions of these integral expressions have been provided by, for example, Ledoux (1963); Cox and Giuli (1968, Chap. 27); J. P. Cox, Hansen, and Davey (1973); J. P. Cox, Davey, and Aizenman (1974); Davey (1974); and Aizenman and Cox (1974, 1975a). (The last five references concern the use of integral expressions in the thermal imbalance problem; see §19.4.) Our discussion here will therefore amount only to a summary.

These integral expressions are based on the linear nonadiabatic wave equation, one form of which is eq. (7.15). A time dependence of all variations is assumed, as in eqs. (9.8) and (9.9); the LNAWE is multiplied through by $\xi^* r^2 dm$ (an asterisk denotes the complex conjugate); and then an integration over the entire stellar mass is performed. In this way one obtains the cubic equation

$$i\omega(\omega^2 - \Sigma^2) = C/J, \tag{9.43}$$

whose solutions provide the desired integral expressions for ω. The quantity J is the oscillatory moment of inertia of the star in the mode considered (see eq. [8.26]). Also, Σ^2 is given by the relation

$$\begin{aligned}\Sigma^2 J &\equiv \int_M \xi^* \mathcal{L}(\xi) r^2\, dm \\ &= \int_M \left\{16\pi^2 \Gamma_1 P \rho r^6 \left|\frac{d\xi}{dm}\right|^2 - 4\pi r^3 |\xi|^2 \frac{d}{dm}\left[(3\Gamma_1 - 4)P\right]\right\} dm, \end{aligned} \tag{9.44}$$

where in the second equality it has been assumed that P vanishes at the surface, and that $m = M;$ and where $\mathcal{L}$ is the linear operator defined in eq. (8.23). Finally, if $\delta(\epsilon - \partial L_r/\partial m)_{sp}$ vanishes at the surface, C is given by the relation

$$C = \int_M (\Gamma_3 - 1) \left(\frac{\delta\rho}{\rho}\right)^*_{sp} \delta\left(\epsilon - \frac{\partial L_r}{\partial m}\right)_{sp} dm$$

$$= C_r + iC_i \; (C_r, C_i \text{ real}). \tag{9.45}$$

Comparison of eqs. (9.45) with eqs. (9.3), (9.10), and (9.11) shows that the real part C_r of C is proportional to $\langle dW/dt \rangle$—the average, over one period, of the rate at which the pressure gradient and gravity forces do work on the whole star. As stated earlier, the main effect of C_i is, normally, to affect the pulsation period (see below).

Consider now the solution of eq. (9.43). In general there will be three roots of this equation. One, say ω_1, is normally very small and can be shown (see, e.g., Cox and Giuli 1968, Chap. 27) to correspond to the problem of secular instability of stars; this root is not of immediate interest to us here. Of much greater interest in connection with pulsating stars are the two large roots, say $i\omega_{2,3}$, of eq. (9.43). These two solutions are given, in terms of real and imaginary parts, by the expression (see, e.g., Cox and Giuli 1968, Chap. 27)

$$i\omega_{2,3} = i\left[\pm\Sigma + \frac{1}{2}\frac{C_i}{\Sigma^2 J} \pm \frac{3}{8}\frac{(C_r^2 - C_i^2)}{\Sigma^5 J^2}\right]$$

$$+ \left[\frac{1}{2}\frac{C_r}{\Sigma^2 J} \mp \frac{3}{4}\frac{C_r C_i}{\Sigma^5 J^2}\right] + O(\delta^3), \tag{9.46}$$

where $\delta \equiv |C/J\Sigma^3| \sim t_{ff}/t_k$ is normally small compared with unity (and has been assumed so in eq. [9.46]), and C_r and C_i denote, respectively, the real and imaginary parts of C (see eq. [9.45]). The two roots correspond to either all upper, or all lower, signs in eq. (9.46). These two roots are actually identical numerically, and describe only one kind of physical motion of the star (as was first pointed out by Iben 1971b; see also Lynden-Bell and Ostriker 1967). The two possible sets of signs merely correspond to $i\omega$ and its complex conjugate $(i\omega)^*$. It can be shown that if $i\omega$ is replaced by $(i\omega)^*$ in the differential equations, then all pulsation variables must be replaced by their complex conjugates. This replacement has the effect of reversing the sign of C_i. It is then obvious that $i\omega_3 = (i\omega_2)^*$, and both roots represent the same physical behavior of the star.

Equation (9.46) represents an oscillatory motion of period $\Pi = 2\pi/\Sigma$, where $\Sigma \approx \sigma$, the adiabatic angular pulsation frequency. The amplitude of this motion either grows or decays slowly in time (if $\delta \ll 1$), in accordance with eq. (9.14). According to eq. (9.46), κ, the stability coefficient (see §9.1), is given to first order in δ by eq. (9.13). Clearly, the largest effect of C_i is on the pulsation period of the model, as was stated previously.

9.4. EDDINGTON'S EVALUATION OF THE STABILITY COEFFICIENT

Eddington (1926, Chap. 8) was apparently the first to derive an expression for the stability coefficient defined in §9.1. His derivation was based on a clever entropy argument, and he considered only the thermodynamically reversible case in linear theory. We shall here sketch his derivation, but we shall at the same time generalize it to the nonlinear, irreversible case, as this turns out to be very easy.

Eddington immediately assumed that the stability coefficient κ was given by the expression

$$\kappa = -\frac{1}{2}\frac{\langle d(\delta\psi)/dt\rangle}{\langle\delta\psi\rangle}, \tag{9.47}$$

where $\delta\psi$ is the total pulsation energy, to be defined below (in general not the same as the $\delta\psi$ of §9.1), and angular brackets denote a time average over one complete period of duration Π. Eddington then evaluated the numerator of eq. (9.47) essentially by use of the energy theorem in the form of eq. (4.52).

We apply this theorem to both the pulsating and nonpulsating states of the star, the latter state being denoted by zero subscripts and assumed to be in complete equilibrium (see §5.2). Subtracting the two equations, we obtain an expression for $d\delta\psi/dt$, where

$$\delta\psi \equiv \int_M \left[\frac{1}{2}\dot{r}^2 - \left(\frac{Gm}{r} - \frac{GM}{r_0}\right) + (E - E_0)\right] dm \tag{9.48}$$

is a nonlinear expression for the total pulsation energy. The numerator of eq. (9.47) for κ is then obtained from the expression

$$\Pi\left\langle\frac{d(\delta\psi)}{dt}\right\rangle = \int_M dm \oint \frac{dq}{dt}\, dt, \tag{9.49}$$

where the time integration is to be carried out over a complete, *closed* cycle. By requiring that the cycle be closed, we are in effect interpreting the quantity on the left side of eq. (9.49) as the amount of energy that must be removed from or added to the star in one period if the motion is to be strictly periodic. This interpretation can be shown to be consistent with the results and discussion in earlier sections of this chapter if the damping time $|\kappa|^{-1}$ is large compared to the period Π.

Eddington wanted to evaluate the right side of eq. (9.49) by using only first-order quantities, known from the solution in some approximation of the linearized equations. However, this solution requires that dq/dt be sinusoidal in time; if this is true, the cyclic integral in eq. (9.49) obviously vanishes. Eddington thus sought a method which would yield the value of this cyclic integral to *second-order* accuracy, but which would involve quantities known only to first-order accuracy.

He adopted the following entropy argument, which we have generalized to the irreversible case. If the star is assumed to return precisely to its initial state after a complete period, then, clearly, the net change ds in the specific entropy s of each mass element will be zero around the period. Now, ds is given by the general thermodynamic relation

$$ds = \frac{dq}{T} + d\sigma, \tag{9.50}$$

where dq is the net heat gained per unit mass, T is the instantaneous temperature, and $d\sigma$ is the change in specific entropy resulting solely from irreversible processes such as viscosity. From the second law of thermodynamics, we have $d\sigma \geq 0$ always. However, we may write

$$\frac{1}{T} \equiv \frac{1}{T_0} - \frac{T - T_0}{TT_0}, \tag{9.51}$$

which is just an identity, where T_0 is any constant reference temperature which may be identified, if convenient, with the equilibrium temperature of the mass element. Using eq. (9.51) in eq. (9.50), and equating to zero the integral of ds around a closed cycle, we obtain for eq. (9.49)

$$\Pi\left\langle\frac{d(\delta\psi)}{dt}\right\rangle = \int_M dm \oint \frac{T - T_0}{T}\frac{dq}{dt}\,dt - \int_M dm\, T_0 \oint \frac{d\sigma}{dt}\,dt. \tag{9.52}$$

This is the desired generalization of Eddington's result to the nonlinear, irreversible case. Note that, according to this equation, irreversible processes always tend to damp pulsations, as expected. (See, for example, Cox and Giuli 1968, Chap. 27, for an alternative treatment of effects of viscosity.)

For the linear, reversible case ($[T - T_0]/T = \delta T/T$, $d\sigma = 0$), eq. (9.52) becomes

$$\Pi\left\langle\frac{d(\delta\psi)}{dt}\right\rangle = \int_M dm \oint \frac{\delta T}{T}\,\delta\left(\epsilon - \frac{\partial L_r}{\partial m}\right) dt. \tag{9.53}$$

which is Eddington's actual result.

An alternative and illuminating derivation which leads to Eddington's formula may be found in Ledoux (1958, §13).

It can easily be shown that eq. (9.53) is exactly the same as our results derived in previous sections of this chapter, obtained by computing directly the work done by the gravity and pressure gradient forces around a cycle, assumed to be closed.

10

Pulsational Stability of Actual Stars

In Chapter 9 we considered, from a fairly general standpoint, the effects of nonadiabaticity on (radial) stellar pulsations and the thermodynamic conditions which would lead to the existence of pulsations in stars. One rather obvious fact emerged from that discussion: that the total effect of all the driving regions had to be larger than that of all the damping regions in a star, if the star was to be pulsationally unstable, or overstable. These general conclusions have been well understood at least since the late 1930's and early 1940's, when Eddington was doing his important groundbreaking work on the causes of stellar pulsation. However, the successful identification of the actual mechanisms responsible for the pulsations of most kinds of variable stars has been effected only in the past 15–20 years. It is now fairly safe to say that these mechanisms are reasonably well understood, and that the causes of pulsation in common types of variable stars are accordingly known. While many details remain to be worked out, the overall picture seems to be correct in its essential features.

In general, these mechanisms are envelope ionization mechanisms, which basically involve the ionization of an abundant element such as hydrogen or helium at a critical depth below the stellar surface. More specifically, the partial ionization of He^{+} ($He^{+} \rightleftharpoons He^{++}$) (second helium ionization) seems to be the main agent responsible for the pulsations in most common types of variables stars. The relation between the abundance of helium and instability has been discussed by A.N. Cox, King, and Tabor (1973); see §10.3. Hydrogen ionization usually makes a fairly significant contribution to the instability, however, and may even be the main cause of the instability in the red variables. It is, in fact, hydrogen ionization that is mostly responsible for the phase lag of the luminosity variations in pulsating stars (see Chapter 11), according to current ideas.

The nuclear reactions that are the principal source of energy throughout most of the life of a star seem to play a totally negligible role (at least directly) in the phenomenon of pulsation in most known types of variable stars. The reason for this is that stellar pulsation is for the most part an envelope phenomenon. The pulsations, effectively, do not extend down to the deeper regions of a star where the nuclear reactions are occurring, and consequently, the direct effects of the nuclear reactions are negligible.

The effects of ordinary (molecular and radiative) viscosity have also been shown to be negligible in radial stellar pulsation (see, e.g., Cox and Giuli 1968, Chap. 27).

Since the present topics have been discussed in some detail in various review papers (e.g., J.P. Cox 1974a), we shall limit our remarks in this chapter to summarizing the basic ideas. Further elaboration can be found in the review papers.

A new kind of envelope ionization mechanism has been introduced by Stellingwerf (1978, 1979), in connection with δ Scuti stars and β Cephei variables. This mechanism is discussed more fully in §10.1 and in Chapter 13. See also J.P. Cox and Stellingwerf (1979), Saio and Cox (1979a,b), and J.P. Cox and Hansen (1979).

In §10.1 we shall summarize the important ideas involved in envelope ionization mechanisms. A brief historical sketch of the main steps leading to our present understanding of the causes of stellar pulsation will be given in §10.2. Finally, in §10.3 we shall consider a necessary condition for the pulsational instability in stars that arises from envelope ionization mechanisms.

10.1 ENVELOPE IONIZATION MECHANISMS

The ability of envelope ionization mechanisms to produce pulsational instability in stars arises essentially from the fact that the ionization of an abundant constituent of the stellar material can result in a modulation of the flux variations. As was pointed out in Chapter 9, this modulation must be such that the pulsating regions are gaining, or absorbing, heat when they are most compressed, and losing heat when most expanded. With this kind of phasing of the heat gains and losses, maximum pressure in the relevant regions will come *after* maximum density. Any incipient pulsations will then tend to be "pumped up," and so a destabilizing effect will be present here. Detailed considerations show that the phasing can indeed be precisely as described above, in regions where an abundant element is in the midstages of ionization. Moreover, these regions of partial ionization are located in the outer stellar layers where the pulsation amplitude is appreciable. In the inner regions the effect is normally exactly opposite to that described above, and so these inner regions tend to damp any incipient pulsations. However, the pulsation amplitude is relatively small in these inner regions, so that the damping effect here is diminished. Detailed calculations show that, under the proper conditions (see below), the effects of the driving can outweigh those of the damping, and the star can be unstable. This instability occurs, moreover, just in that part of the Hertzsprung-Russell diagram where common types of variable stars are located.

The effects of this modulation of the flux variations can be expressed quantitatively by writing the work integral (see Chapter 9) in the approximate form

$$C \approx -L \int_M (\Gamma_3 - 1)\left(\frac{\delta\rho}{\rho}\right) \frac{d}{dm}\left(\frac{\delta L_r}{L_r}\right) dm, \qquad (10.1)$$

where the indicated variations denote only the *space parts* of the corresponding quantities. In this expression we have neglected nuclear energy sources altogether, and have assumed that the equilibrium value of L_r is constant with interior mass m and equal to L, the total equilibrium stellar luminosity. Moreover, we have assumed (and shall assume, unless indicated otherwise, throughout the remainder of this section), that $\delta\rho/\rho > 0$ and is real in all regions of interest. This means, physically, that the regions in question are *compressed* at minimum stellar radius. This situation will obtain, approximately, throughout the star for the fundamental mode, and also exterior to the outermost node in the case of higher harmonics. These outermost regions are the only ones of much interest in the case of higher harmonics, because the pulsation amplitude is normally very small interior to this outermost node (see §8.12.c).

It is ordinarily true in most stars that $d(\delta L_r/L_r)/dm > 0$ everywhere in the important regions. In this case, then, according to eq. (10.1), $C < 0$, that is, the star is pulsationally stable. The magnitude of the luminosity variations in this case increases outward everywhere at the instant of minimum stellar radius. More heat therefore flows outward in unit time through the outside of every elementary mass shell than flows in through the inside, so that every mass shell is losing heat at this instant. According to the discussion in Chapter 9, then, every mass shell in this case is producing *damping*, and the pulsations of the star are accordingly damped. This kind of damping is often referred to as "radiative damping," although the term need not be restricted to radiative transfer alone, since L_r is the local luminosity due to all relevant transfer mechanisms.

This behavior of the luminosity variations is mainly a result of the opacity variations. Normally, the opacity *decreases* upon compression. The resulting increase in the transparency of the material during compression contributes strongly to the heat leakage at this instant. However, Eddington (1926, Chap. 8) pointed out that there would be some heat leakage even if the opacity were constant during the pulsations, provided that $\Gamma_1 \approx 5/3$. This last result is mainly a consequence of the fact that the local luminosity varies as some fairly high power (approximately the fourth) of the local temperature, which normally increases upon compression. The rapid outward increase in $\delta\rho/\rho$ is seen to contribute strongly to the damping in this case. These considerations show that, without some active destabilizing agent, most stars, particularly giants and supergiants, should be exceedingly stable against pulsations, as has been shown in the calculations of J.P. Cox (1955); Ledoux, Simon, and Bierlaire (1955); and Rabinowitz (1957). This result is satisfying in that it agrees with the fact that pulsation is indeed very rare among stars (see Chapter 1).

We consider now the effects on the luminosity variations of the ionization zones of hydrogen (H), neutral helium (He°), and once ionized helium (He^{+}) in the outer stellar layers. These are the only ionization zones that need be considered for stars of normal composition, because of the overwhelming abundance of hydrogen and helium; the present ideas, however, will apply in principle to *any* element, provided that it is sufficiently abundant. We also consider only radiative transfer in the present discussion, even though the material in the vicinity of such ionization zones is usually convectively unstable.

The middle of the He^{+} ionization zone is always characterized in the equilibrium state of the star by a temperature close to 4×10^{4}°K (corresponding to an ionization potential of 54.4 eV), almost independently of stellar parameters. The H and He° ionization zones are usually so close together in a star that their effect is nearly the same as that of a single ionization zone whose middle characteristically has a temperature in the approximate range $(1\text{–}1.5) \times 10^{4}$°K, again almost independently of stellar parameters. Since the effects of H ionization dominate the behavior of this combined ionization zone, it is usually referred to simply as the H ionization zone.

The most important effect of such an ionization zone on the material properties, for the present considerations, is that $\Gamma_3 - 1$ becomes rather small in such a zone. Most of the work of adiabatic compression in this case goes into ionization energy rather than into kinetic energy of thermal motion, so that the temperature does not increase very much upon compression. The small values of $\Gamma_3 - 1$ in such an ionization zone have a dramatic effect on the quasi-adiabatic luminosity variations, say $(\delta L_r/L_r)_a$, which are very sensitive to $\Gamma_3 - 1$. The overall effect is that $(\delta L_r/L_r)_a$ exhibits a marked dip in the ionization zone. The amount of this dip can be comparable to or larger than $(\delta L_r/L_r)_a$ itself.

The strong outward decrease of $(\delta L_r/L_r)_a$ in the inner parts of the ionization zone, where $\Gamma_3 - 1$ becomes smaller outward, clearly represents (at least when $[\delta L_r/L_r]_a \approx [\delta L_r/L_r]$) absorption of energy in this region at the instant of maximum compression of the star. This region is therefore a driving region (see Chapter 9).

Physically, the small values of $(\delta L_r/L_r)_a$ in a region of small $\Gamma_3 - 1$ result in part directly from the smaller temperature variations (that is, smaller $\delta T/T$) here. The reason is that, approximately, $L_r \propto T^4$ for given opacity κ. Hence, the flow of radiation is locally diminished upon compression; the radiation is dammed up and absorbed by the matter in the region of decreasing $\Gamma_3 - 1$ at the instant of greatest compression. This absorption of heat enhances the local *rate* of temperature increase and causes the temperature, and therefore also the pressure, to be slightly larger during the ensuing expansion than would be the case with adiabatic motion. As we saw in §9.1, it is this pressure excess during the expansion phase that effects the transformation of some of the heat absorbed upon compression into mechanical work and thence into pulsation energy. This direct effect of the temperature variations on the luminosity variations has been referred to as the "γ-mechanism"" (see J.P. Cox, Cox, Olsen, King, and Eilers 1966).

If the opacity law is of the form $\kappa \propto \rho^n T^{-s}$ $(n, s > 0)$, it is seen that the small values of $\delta T/T$ in regions of small $\Gamma_3 - 1$ may cause κ to *increase* upon compression in such a region; whereas, normally, κ *decreases* upon compression (see above). This local increase in κ upon compression enhances still further the damming up of the radiation in the region of outwardly decreasing $\Gamma_3 - 1$, and contributes still further to the driving in this region. This direct effect of the opacity variations on the luminosity variations has been called the "κ-mechanism" by Baker and Kippenhahn (1962).

Note that, if s is large and negative (as may be the case in the *H* ionization zone), there may be a damming up of radiation upon compression, and hence driving, even if Γ_3 has a value close to its normal value of 5/3. This fact has been found to be important in some cases by Stellingwerf (1978, 1979) in his calculations of the pulsational stability of models for δ Scuti-like stars and for β Cephei stars. In particular, at the temperature ($\sim 1.5 \times 10^5$°K) at which the photon energy in the maximum-energy peak in the radiation field is close to the ionization potential (54.4 eV) of He^+, there is an increase in the opacity. In the vicinity of this temperature, s becomes less positive than usual and produces driving. Stellingwerf has referred to this effect as the "bump mechanism."

Let us now return to the case of an ionization zone of an abundant element in the outer stellar layers. Ionization zones in stars are typically rather thin, at least in terms of the equilibrium temperature T, and $\Gamma_3 - 1$ *increases* rapidly outward in the outer regions of an ionization zone, where the degree of ionization is very small. (The minimum of $\Gamma_3 - 1$ occurs very nearly at the temperature corresponding to 50% ionization of the relevant element.)

Arguments similar to those above lead to the conclusion that $(\delta L_r/L_r)_a$ must *increase* rapidly outward in the outer regions of an ionization zone, such that the values of $(\delta L_r/L_r)_a$ immediately below and above an ionization zone are not greatly different. Hence, as long as $\delta L_r/L_r \approx (\delta L_r/L_r)_a$, we would expect that the strong damping of the outer portions of an ionization zone would largely cancel the strong driving of the inner portions, and the overall effect of the ionization zone on the stability of the star should be small if not negligible. However, because of nonadiabatic effects, $\delta L_r/L_r$ and $(\delta L_r/L_r)_a$ can be quite different in layers sufficiently far out in a star. It is this lack of equality between these two quantities (a nonadiabatic effect) that gives rise to the conditions referred to above for the occurrence of stellar pulsational instability brought about by envelope ionization mechanisms.

These important nonadiabatic effects can be discussed most simply in terms of a transition region which separates the quasi-adiabatic interior from the nonadiabatic exterior. In the quasi-adiabatic interior, the quasi-adiabatic approximation, exemplified by the relation $\delta T/T = (\Gamma_3 - 1)\delta\rho/\rho$, is valid, and $\delta L_r/L_r \approx (\delta L_r/L_r)_a$. On the other hand, in the nonadiabatic exterior the luminosity variation is effectively "frozen in," $(\delta L_r/L_r)_{sp} \approx$ constant (see Chapter 9), and bears practically no relation to the quasi-adiabatic luminosity variation. (However, it should be noted that $[\delta L_r/L_r]_{sp}$ is not strictly constant above the transition region. Actually, the hydrogen ionization zone farther out can introduce some spatial variation in $[\delta L_r/L_r]_{sp}$; see Chapter 11.) The transition region is actually fairly thick, so it is somewhat of an oversimplification to regard it as a single layer of zero thickness.

It can be shown from the energy equation (see, e.g., Cox and Giuli 1968, Chap. 27) that the location of the transition region in the stellar envelope is determined approximately by the order-of-magnitude relation

$$\frac{\langle c_V T_{TR} \rangle (\Delta m)_{TR}}{L\Pi} \sim 1, \tag{10.2}$$

where the numerator on the left side is approximately equal to the total internal energy in the stellar layers, of total mass $(\Delta m)_{TR} \equiv M - m_{TR}$, lying above the transition region; L is the equilibrium luminosity of the star; and Π is the pulsation period in the mode being considered. Here T and c_v are, respectively, the local values of temperature and specific heat per unit mass at constant volume, both in the equilibrium state of the model. Hence, the transition region defines that level above which the total internal energy of the material is of the order of magnitude of the total energy radiated by the star in a pulsation period. Note that the definition of the transition region does not assume any specific energy transport mechanism.

Given an equilibrium model of a stellar envelope, one can calculate the dependence of $(\Delta m)_{TR}$ or of T_{TR}, the equilibrium temperature in the transition region, on stellar parameters (such a calculation, for a simplified envelope model, is given in Cox and Giuli 1968, Chap. 27). Two things become evident from this calculation. (1) The location of the transition region in the envelope depends fairly strongly on the equilibrium radius R and only weakly on other stellar parameters such as mass M and L. It can be shown that, approximately, for radiative envelope models with given L, M, composition, and pulsation mode, $T_{TR} \propto R^{-1/2}$, and $(\Delta m)_{TR} \propto R^2$. Hence, T_{TR} is relatively large for stars of small radius R, and decreases as R increases. In other words, the transition region moves outward through mass as the radius increases (see below). (2) It turns out that, for stars in the oval region of the H-R diagram in Figure 3.1, $T_{TR} \sim 4 \times 10^4$°K, which is very nearly the temperature in the He^+ ionization zone. This near equality of T_{TR} and the above temperature suggests that the oval region in Figure 3.1 corresponds to the approximate coincidence in the stellar envelope of the transition region and the He^+ ionization zone. The following arguments show that this coincidence is a consequence of the interaction between nonadiabatic effects and ionization zones.

The main effects of nonadiabaticity on the outer stellar layers are seen to be to decouple the actual luminosity variations $\delta L_r/L_r$ from the quasi-adiabatic luminosity variations $(\delta L_r/L_r)_a$ in the regions exterior to the transition region. The density and temperature variations adjust themselves, in fact, so as to keep $\delta L_r/L_r$ approximately constant in space in these regions. Because of the strong dependence of the degree of ionization on temperature, the temperatures in the H and He^+ ionization zones are nearly constant and almost independent of stellar parameters. We may therefore consider a star of given composition, M, and L. At the same time, we may regard the equilibrium radius R as a free parameter; this procedure allows us to mimic approximately the computed evolution of certain real stars while they are crossing the instability strip.

If R is so small that both ionization zones lie *above* the transition region, the luminosity variations are effectively frozen in in space, and $\delta L_r/L_r$ (real part of $[\delta L_r/L_r]_{sp}$) $\approx$ constant in this region, in spite of the pronounced "dip" in $(\delta L_r/L_r)_a$ here. Hence $\delta L_r/L_r$ is positive and rises monotonically outward until the transition region is reached, and then remains essentially constant exterior to the transition region. Therefore, only damping exists in the envelope and the star is stable.

Consider next the case of a somewhat larger radius R. Since T_{TR} decreases as R increases, and since the temperatures in the ionization zones are nearly constant, there will be a critical value of R, which we shall call R_{crit}, at which the He^+ ionization zone (the deeper one) approximately coincides with the transition region. In this case the He^+ transition region effectively straddles the two regions—the quasi-adiabatic interior and the nonadiabatic exterior: the inner parts of the ionization zone lie essentially in the quasi-adiabatic interior where $\delta L_r/L_r \approx (\delta L_r/L_r)_a$ and where the outward decrease in $\Gamma_3 - 1$ produces a strong outward decrease in $(\delta L_r/L_r)_a$, and therefore also in $\delta L_r/L_r$. The outer parts of this ionization zone, however, lie essentially in the nonadiabatic exterior, where $\delta L_r/L_r$ has become frozen in at the small value it had near the middle of this ionization zone. The strong damping in the outer parts of the He^+ ionization zone has now been eliminated by nonadiabatic effects, and the H ionization zone still lies in the nonadiabatic exterior and so has only a small effect. The strong driving due to the rapid outward decrease in $\delta L_r/L_r$ in the inner parts of the He^+ ionization zone is now effective and the star may be unstable. The rapid outward increase in $\delta\rho/\rho$ weights the strong driving in the regions in and above the He^+ ionization zone more than the damping in the deeper regions (see, e.g., eq. [10.1]). This outward increase in $\delta\rho/\rho$, which is ordinarily a strong stabilizing influence, therefore now *enhances* instability. If *only* the He^+ ionization zone were present, the case of $R = R_{\text{crit}}$ would also correspond to *maximum* instability with respect to R, for given L, M, and composition. The low temperature boundary of the instability region, corresponding to $R > R_{\text{crit}}$, will be discussed below.

Detailed calculations (see, e.g., Baker and Kippenhahn 1962) show that in this case of $R \approx R_{\text{crit}}$ the driving in the outer regions of the envelope can indeed be more than strong enough to compensate for the damping in the interior, and the star can be unstable.

The present considerations offer an explanation of the fact, noted above, that the temperature T_{TR} in the transition region is close to the temperature (4×10^{4}°K) in the He^+ ionization zone, for stars lying in the oval region of Figure 3.1. In fact, the condition that the transition region and the He^+ ionization zone approximately coincide can even provide an approximate *necessary* condition for instability arising from the kinds of mechanisms under consideration here (see §10.3).

The Cepheid instability strip is observed to be fairly sharply bounded on the low temperature side. At fixed mean luminosity, its width is $\Delta \log T_e \approx 0.05$–$0.08$ (T_e = effective temperature), corresponding to $\Delta T_e \sim 600$–1100°K. Unfortunately, the envelope ionization mechanisms we have been discussing, which have been so successful in accounting for the high temperature edge of the instability region, do not on the basis of pure radiative transfer account for this return to stability at lower temperatures. In fact, detailed calculations, both linear and nonlinear (see, e.g., King, Cox, Eilers, and Davey 1973), show that the instability in the fundamental mode continues to increase as T_e is lowered at constant mean luminosity L.

This increasing instability in the fundamental mode, as R is increased, is probably for the most part a result of the H ionization zone, which is potentially a very powerful driving region (see above), moving into coincidence with the transition region as R increases.

However, it is well known that current theories of convection predict the onset of effective convective envelopes in the outer stellar layers at effective temperatures somewhat to the cooler side of the blue edge of the instability region. It therefore seems almost certain, as was first suggested by Baker and Kippenhahn (1965), that it is the onset of this efficient convective transfer that restores stability at low effective temperatures, that is, that determines the "red edge" of the Cepheid instability strip. However, because of the absence of reliable theories of time-dependent convection (see §§ 4.3 and 19.3), the exact mechanisms of this presumed stabilizing effect are very poorly understood.

Perhaps the most ambitious efforts to date to deal with this problem have been those of Deupree (1974; 1975a,b; 1976a,b,c; 1977a,b,c,d). His work is described in some detail in §19.3, where a physical discussion is presented as to how, according to his results, convection may terminate instability on the red side of the instability strip. A red edge of the Cepheid instability strip has also been found by Baker and Gough (1979).

10.2. HISTORICAL SKETCH

Eddington (1926) was apparently the first to consider seriously the problem of the maintenance of the pulsations and the related problem of the dissipation of pulsation energy. He obtained an expression for the total dissipation of pulsation energy for small radial oscillations; this expression, and Eddington's derivation of it, were summarized in §9.4. Eddington used this expression to calculate the decay time τ_d for the pulsations of a model of δ Cephei. Using the assumption that the equilibrium model could be represented by the polytrope $n = 3$ (Eddington's standard model), he obtained $\tau_d \approx 8000$ years. Since this is a relatively short time compared to evolutionary time scales, the probability of observing such a star pulsating would be vanishingly small. Later calculations, based on more realistic models, showed that τ_d would be more than a thousand times shorter than Eddington's value. These results emphasized the necessity of finding a strong and active destabilizing agent to maintain the pulsations.

Eddington evidently believed that the nuclear reactions near the stellar center provided the main driving agent for the pulsations of Cepheids. However, he also discussed another possible mechanism for exciting pulsations in a star, not based directly on nuclear reactions. As we have seen in the earlier parts of this chapter, this mechanism provides the basis of the currently accepted ideas on the causes of most stellar pulsations. He referred to this mechanism as a kind of valve mechanism whose operation is based on a modulation of the flux flowing through the stellar layers (see above). Eddington compared his valve mechanism to a thermodynamic heat engine, albeit of a perhaps rather bizarre kind by ordinary terrestrial standards. His valve mechanism achieved the desired effect by varying the leakage of heat, in contrast to the supply. It is the supply of heat which is varied in, for example, a conventional internal combustion engine. Thus, for example, *driving* of pulsations would result if the leakage of heat in a mass element were to be *diminished* during *compression*, and *increased* during *expansion* (see §9.1).

In Eddington's final work on the instability mechanism of pulsating stars, he suggested (Eddington 1941, 1942) that the seat of the above valve mechanism was in the hydrogen ionization region in the outer stellar layers. He apparently still believed that the nuclear reactions near the stellar center were the basic driving mechanism, and that the valve served only to decrease the dissipation in the outer regions to such an extent that the nuclear driving could predominate over the dissipation and render the star unstable.

The next important development was Epstein's (1950) work on the solutions of the LAWE for centrally concentrated models (see §8.13). These solutions were characterized by central pulsation amplitudes about 10^6 times smaller than surface amplitudes. A detailed quasi-adiabatic stability analysis of one of these models, based on Epstein's solutions, was next carried out by Cox (1955). The nuclear driving was found to be many orders of magnitude smaller than the radiative damping. The overall conclusion was that no effective or important sources of instability could be found interior to some 0.85 of the stellar radius, and that the driving mechanism, whatever its nature, would have to lie outside this level. Roughly similar conclusions were reached by Ledoux, Simon, and Bierlaire (1955) and Rabinowitz (1957).

These developments led investigators to look to the outer, nonadiabatic regions of the star in an attempt to locate the source of instability (see, e.g., J.P. Cox 1958). Nearly simultaneously with the work described above, Zhevakin (1953, 1954a,b) suggested that the region of second helium ionization ($He^+ \rightleftharpoons He^{++}$) might be a suitable site for a valve mechanism of the kind envisaged by Eddington. A similar suggestion was made by Cox and Whitney (1958). In view of the calculations described in the preceding paragraph, this valve mechanism would have to account not for *part*, but for *all*, of the instability. The effectiveness of He^+ ionization as a driving mechanism was first conclusively demonstrated by the linear but accurately nonadiabatic calculations of Baker and Kippenhahn (1962) and J.P.

Cox (1963). Since then, numerous detailed calculations, both linear and nonlinear (see, e.g., J.P. Cox 1974a for detailed references), have amply confirmed these conclusions, and leave little reasonable doubt that He^+ ionization is the main source of the instability of most kinds of pulsating stars lying in the oval region of the H-R diagram shown in Figure 3.1.

It was suggested by Christy (1962), and later confirmed by Baker and Kippenhahn (1965), that hydrogen ionization can in some cases be a significant contributor to the driving of pulsations. Subsequent calculations (see, e.g., J.P. Cox 1974a for detailed references), while confirming the qualitative correctness of this suggestion, have shown that, nevertheless, He^+ ionization is the major source of driving for the above kinds of pulsating stars. (Hydrogen ionization, may, however, be the most important destabilizing agent in many of the red variables; see, e.g., Langer 1971; Keeley 1970a,b; J.P. Cox 1974a; for a recent discussion of Mira variables, see Hill and Willson 1979).

Further details regarding the search for the destabilizing mechanism may be found in King and Cox (1968).

10.3. A NECESSARY CONDITION FOR INSTABILITY AND SOME OF ITS IMPLICATIONS

In this section we shall make use of the qualitative ideas considered earlier in this chapter to derive an order-of-magnitude necessary condition for pulsational instability by envelope ionization mechanisms of the kind we have been considering. This necessary condition, though very rough, is useful for orientation and also permits a number of qualitative features of at least the high-temperature boundary of the instability region (the oval region in Fig. 3.1), borne out by detailed calculations (see, e.g., J.P. Cox, 1974a), to be inferred from simple considerations. By restricting our attention to a necessary condition, we need not limit our considerations only to fundamental mode oscillations. Some of the ideas underlying this necessary condition have been applied to different kinds of stars by J.P. Cox and Stellingwerf (1979) and by J.P. Cox and Hansen (1979).

The arguments of §10.1 showed that the onset of instability on this side of the instability region is caused by the transition region moving into the He^+ ionization zone as the equilibrium radius R increases at constant M and L, i.e., as the mean effective temperature T_e decreases. Although, as mentioned above, the H ionization zone contributes to the driving, the major contributor is normally the He^+ ionization zone. This result is in part a reflection of the narrowness of the instability region on the H-R diagram (see Fig. 3.1): The instability is terminated on the low temperature side, presumably by effects of convection, before the H ionization zone can take over most of the driving. Hence, we may for simplicity consider only the He^+ ionization zone, and shall often refer to this as the "main driving region." We may then base our necessary condition for instability on the approximate coincidence in the stellar envelope of the transition region and the He^+ ionization zone (see §10.1). Letting an asterisk denote values of quantities at the midpoint of the He^+ ionization zone, we then have $T_{TR} \sim T^* \sim 4 \times 10^4$°K, and $(\Delta m)_{TR} \sim \Delta m^*$, where Δm^* is the mass lying above the He^+ ionization zone. The necessary condition is then given by eq. (10.2), with the replacements as just indicated:

$$\frac{\langle c_V T^* \rangle \Delta m^*}{L\Pi} \sim 1. \tag{10.3}$$

It can easily be shown that Δm^* increases sharply as R increases and is much less sensitive to other stellar parameters such as L and M. Thus, Δm^* increases sharply as T_e decreases. According to eq. (10.3), then, instability regions of the kind we are considering here are sharply bounded on the high-T_e side. Moreover, because of the strong dependence of the left side of eq. (10.3) on T_e, it follows that the high-T_e, or blue, boundaries are nearly *vertical* on the H-R diagram, as observed.

The necessary condition (10.3) for instability can also be expressed in an alternative form. Making simple assumptions about the envelope structure, assumed radiative, using the period-mean density relation, and so on, one can show (J.P. Cox 1959) that this equation can also be written as follows. Along the blue edge of an instability region, we have

$$L \propto Q^{-w} Y^{y} Z^{-z} M^{l} T_e^{-x}, \tag{10.4}$$

where Q is the pulsation constant for the mode under consideration (see §8.6); Y and Z are, respectively, the helium and heavy element (heavier than helium) mass fractions; and the exponents w, y, z, l, and x are all normally positive numbers whose values are determined by the envelope structure, the dependence of the opacity on density and temperature, and other factors. Ths simple model of the envelope structure can yield approximate values of the exponents, but it is better to determine them from detailed pulsation calculations. We only note here that x turns out to be quite large ($\sim$10–20), which confirms our general expectations that the blue edge of an instability strip should be nearly vertical on an H-R diagram.

The detailed pulsation calculations that have been carried out, both linear and nonlinear (see, e.g., J.P. Cox 1974a for references), confirm a relation of the form (10.4), at least over limited regions of the blue edges, and also provide numerical values of the various exponents. Typical values are the following:

$$\left.\begin{aligned} x &\approx 10\text{–}20 \\ l &\approx 0.5\text{–}1 \\ z &\approx (0.002\text{–}0.05)x \\ y &\approx (0.05\text{–}0.2)x \\ w &\approx (0.1\text{–}0.15)x \end{aligned}\right\} . \tag{10.5}$$

Equation (10.4) has a number of immediate applications. For example, it embodies the famous period-luminosity relation of classical Cepheids (see §3.1). If T_e is expressed as a function of L and R, and R in turn expressed as a function of Π, M, and Q from the period-mean density relation, then eq. (10.4) becomes, for given composition and pulsation mode, a relation between L, M, and Π. The assumption of a mass-luminosity relation $L = L(M)$, then yields $L = (\Pi)$, which is, schematically, the period-luminosity relation. The values of the exponents in eq. (10.5) lead to approximately the correct slope of this relation (see also J.P. Cox 1959). In the simplest terms, both the luminosity and the period increase together in ascending the instability strip, because the radius also then increases, and both of these quantities depend primarily on the stellar radius. It is therefore not too surprising that there should be a relation between period and luminosity for pulsating stars in the instability strip.

Because of the finite width of the instability strip, and for other reasons, there is some intinsic scatter in the period-luminosity relation, amounting to $\sim 1^m$ at given period.

Equation (10.4) is also useful for determining qualitatively how various factors can affect the blue edges of the instability strip. For example, for fixed L, an increase in Y moves the blue edges to *higher* T_e; an *increase* in Z moves the blue edges to *lower* T_e. However, the dependence on Z is normally rather weak. In regions where eq. (10.4) is applicable, a higher harmonic blue edge (smaller Q) should lie at *higher* T_e than a *lower* harmonic blue edge (larger Q). Thus, the first harmonic blue edge should lie, according to eq. (10.4), at *higher* T_e than the fundamental blue edge. This last prediction is not always confirmed (see, e.g., Tuggle and Iben 1973); the reason is that eq. (10.4) is only a *necessary,* not a *sufficient,* condition for instability (see, e.g., J.P. Cox, Castor, and King 1972).

The above dependences of the blue edges on Y and Z arise primarily through the *opacity*. An increase in Y, for example, implies a decrease in the mass fraction X of hydrogen, which is normally the main source of opacity in the envelope. The material in the envelope is then more transparent, and the temperature ($\sim 4 \times 10^4$°K) of the He^+ ionization is reached at a larger pressure. A larger pressure implies more mass Δm^* above this zone. Hence, T_e must be *increased* in order to restore the left side of eq. (10.3) to a value near unity. Similar considerations account for the dependence on Z. The dependence of the blue edges on Q arises essentially from the dependence of eq. (10.3) on period Π: A smaller Π (higher harmonic) implies a smaller Δm^*, which in turn implies a higher T_e.

The explanation of the dependence of the blue instability edges on mass M given by eq. (10.4) is more complicated than the explanation for its dependence on the other parameters, but nevertheless can be understood on the basis of the necessary condition (10.3). The argument, too lengthy to be given here, may be found in Cox and King (1971).

It should be cautioned that the above considerations regarding the implications of the necessary condition for instability are actually somewhat oversimplified. It is possible that complicating factors, such as curvature in the envelope (e.g., J.P. Cox 1963), effect of convection on the envelope structure (e.g., Cox and Giuli 1968, §27.7[b]), and so on, could in some cases alter certain conclusions drawn directly from eq. (10.3).

In Cox, King, and Tabor (1973) it was shown that a star would not be pulsationally unstable in the instability strip if the star possessed too small an abundance of helium, say less than some 20% by mass.

11

The Phase Lag

In this chapter we shall consider the well-known phase-lag discrepancy of pulsating stars of the classical Cepheid and RR Lyrae types (see §3.1). According to the usual interpretations of the observational results, maximum luminosity in these stars during a period occurs *not* at minimum radius, as might be expected on the basis of a naive application of adiabatic pulsation theory. Rather, maximum luminosity occurs roughly a quarter of a period *after* minimum radius, approximately when the stellar surface is moving most rapidly outward. This time lapse is called the "phase-lag discrepancy."

The origin of this phase lag has come to be rather well understood in recent years, as will be explained more fully in §11.1. A mostly qualitative discussion of the accepted theory will be presented in §11.2, and some quantitative details will be given in §11.3. Finally, some comparisons with detailed calculations, and some further remarks, will be presented in §11.4.

Throughout this chapter we shall adopt, unless stated otherwise, the basic assumptions made throughout this part (see, e.g., the beginning of Chapter 10).

11.1. OBSERVATIONS AND BRIEF HISTORY

The basic observations are derived from a simultaneous comparison of the light and velocity curves of pulsating stars of the types under consideration. Such curves were shown in Fig. 3.2 above for δ Cephei, the prototype of the classical Cepheids. These curves are seen to imply the existence of a phase lag, as stated above.

This phase lag has intrigued and, until recently, baffled astronomers since the earliest detailed observations of Cepheids were obtained, in about 1915. Many attempts to account for this phenomenon have been made down through the years, by such workers as Eddington, Rosseland, M. Schwarzschild, and Milne (see, e.g., Rosseland 1949 for a good discussion of this work). However, no attempts were successful until ten or fifteen years ago.

Linearized calculations, assuming radiative transfer, yield a phase lag of some 30 to 45° in the once-ionized helium (He^+) ionization zone, because of the absorption of heat in this region upon compression (see §9.1 for a qualitative discussion). However, in the earlier linearized calculations of

Cox (1963), in which the hydrogen (H) ionization zone was omitted entirely, the above phase lag was essentially eliminated in the regions between the He^{+} ionization zone and the stellar surface. Consequently, the computed phase lags in the emergent luminosity were very small, only a few degrees in magnitude. On the other hand, in the linearized calculations of Baker and Kippenhahn (1965), which included in detail the H ionization zone as well as the He^{+} ionization zone, the phase lags were found to be near 180°. As a result of the failure of the linearized calculations to yield the correct phase lag, workers in the field generally concluded that this lag must be a nonlinear effect.

The first successful calculation of the phase lag was carried out by Christy (1964) in the course of nonlinear machine calculations of stellar pulsation (see Chapter 12). Subsequently, the correct phase lag was calculated independently by Castor (1966) and by J. P. Cox, Eilers, and King (1967), using somewhat similar techniques. These nonlinear calculations, however, did not reveal the physical cause of the phenomenon. Moreover, since they were all nonlinear, their success strengthened the prevailing view that the phase lag must be a nonlinear effect. (However, in one set of nonlinear machine calculations by J. P. Cox, Cox, Eilers, and King 1967, the phase lag was found to persist down to the smallest amplitudes investigated, which corresponded to a relative radius semi-amplitude of $|\delta R/R| \sim 10^{-4}$. This result cast some doubt on the suspected nonlinearity of the effect, but was not followed up at that time.)

A simple physical picture of the cause of the phase lag, at least in RR Lyrae variables and possibly also in Cepheids, was proposed by Castor (1968b), and the approximations underlying this picture were later justified in detail by Castor (1971). According to his theory, the phase lag is basically a *linear* phenomenon (that is, it should appear in a strictly linear theory), although it is significantly affected by nonlinear effects. Detailed linearized calculations (see Cox 1974a for detailed references) have completely confirmed Castor's theory. We shall discuss in more detail below (see §11.4) why the linearized calculations of J. P. Cox (1963) and of Baker and Kippenhahn (1962, 1965) failed to reveal the correct phase lag.

The remainder of this chapter will be primarily devoted to consideration of Castor's theory. (A mostly qualitative discussion of this theory has been presented by King and Cox 1968.)

11.2. QUALITATIVE DISCUSSION OF PREVAILING THEORY

The essence of Castor's (1968b) simplified theory is that the phase lag is basically a consequence of the sweeping of the very thin H ionization zone back and forth through mass as the star pulsates. Now, it turns out that, for the classical Cepheids and RR Lyrae variables, the luminosity impinging on the bottom of the H ionization zone is largest near the instant when the stellar radius is smallest (the phase lag due to the He^+ ionization zone is nearly eliminated at the level of the H ionization zone; see remarks just after eq. [11.9] below). This large burst of luminosity on the bottom of this zone causes it to absorb energy most rapidly then. This absorbed energy ionizes the hydrogen in the inner parts of the H ionization zone, and so causes this zone to move farther out. The net result is that the H ionization zone is moving outward through mass most rapidly near the time of minimum stellar radius. Hence, in a roughly sinusoidal variation, this zone will find itself closest in mass to the surface roughly a quarter of a period *after* minimum stellar radius. Now, one can easily show (see §11.3) that with radiative transfer, the emergent luminosity varies (among other things) inversely as some power of the amount of mass lying above the H ionization zone. Since this mass will be smallest about a quarter period after minimum stellar radius, then the emergent luminosity will be largest near this instant.

An important, if not essential, element in Castor's theory is the extreme thinness of the H ionization zone under conditions of radiative equilibrium. This thinness is a result of the fact that the opacity κ is a strongly increasing function of temperature T in the regions within and exterior to the H ionization zone, for the kinds of stars to which the theory applies—say stars having effective temperature $T_e \lesssim (7 - 8) \times 10^3$°K. This strong increase of κ with increasing T requires (assuming radiative transfer) that the temperature gradient becomes extremely steep just exterior to and within the H ionization zone. Typically, T may increase from some 8000°K just outside to some 15,000°K just inside the H ionization zone in only about 1/20 of a pressure scale height. Hence, the H ionization zone behaves almost as a discontinuity and may be treated as such. Moreover, the amount of mass instantaneously lying above the H ionization zone is a well-defined quantity and may be reliably calculated from the theory of radiative transfer (see §11.3). This steep temperature rise is illustrated in J. P. Cox (1974a, Fig. 20).

11.3 FURTHER DETAILS OF THE SIMPLIFIED THEORY

We shall here, following Castor (1968b), treat the H ionization zone as a "front," or discontinuity. The luminosity impinging on the bottom of this zone will be denoted by L_1. This luminosity will be assumed known as a function of time t, from the interior solution, and will be assumed, for simplicity, to be largest when the stellar radius R has its minimum value during a cycle. The luminosity emerging from the top of the H ionization zone will be denoted by L_2. It will be assumed spatially (but not temporally) constant above the H ionization zone, as is consistent with the freezing-in effect discussed in §9.1. We shall also assume pure radiative transfer everywhere in the regions of interest.

First, the instantaneous derivative of the total pressure P with respect to interior mass m can be simply expressed by the "hydrostatic equilibrium" equation as a function of the effective gravity g_e, which is simply the sum of the ordinary gravitational acceleration of the surface layers, and the actual, instantaneous acceleration of these layers. Next, the instantaneous structure of the layers above the H ionization zone can easily be obtained from the theory of radiative transfer. This theory gives, in the diffusion approximation, an expression for the instantaneous derivative of the local temperature T with respect to m in terms of T itself, the emergent luminosity L_2 (assumed spatially constant in the relevant regions), the radial distance r of these regions (assumed equal to the instantaneous stellar radius R), and the opacity κ of these layers. By eliminating m between these expressions, we can obtain a first-order differential equation for $\partial P/\partial T$. This equation can easily be integrated if a simple expression for κ as a function of P and T is assumed. We adopt the relation

$$\kappa = \text{const. } P^n T^\sigma \tag{11.1}$$

where typical values in and above the H ionization zone lie in the approximate ranges $1/2 \lesssim n \lesssim 1$, $4 \lesssim \sigma \lesssim 12$. This integrated expression shows that, if $\sigma > 4$ and if T is at least a few times the boundary temperature of the star, P approaches the nearly spatially constant value P_{H}, the pressure in the H ionization zone. The fact that P approaches constancy with increasing T is just a manifestation of the very steep temperature gradient referred to earlier. We can express P_{H} in terms only of g_e, R, and L_2 if we express the boundary temperature in terms of L_2 and R. This expression is

$$P_{\mathrm{H}} \approx \text{const. } \frac{g_e^{1/(n+1)} R^{\sigma/[2(n+1)]}}{L_2^{\sigma[4(n+1)]}}. \tag{11.2}$$

However, P_{H} is also given by instantaneous integration of the hydrostatic equilibrium equation referred to above. This integrated expression gives P_{H} in terms of g_e, R, and Δm, the total stellar mass instantaneously lying exterior to the H ionization zone. Equating P_H as given by these two expressions and solving for L_2, we obtain

$$L_2 = \text{const. } \frac{R^{2(1+1/q)}}{g_e^{\beta}(\Delta m)^{1/q}}, \tag{11.3}$$

where

$$q \equiv \frac{\sigma}{4(n+1)}, \tag{11.4}$$

$$\beta \equiv \frac{n}{q(n+1)}. \tag{11.5}$$

Typical values are $q \approx 1$–2, $\beta \approx 1/2$–$1/6$. Also, the value of the constant in eq. (11.3) depends on total stellar mass M, as well as on the values of n and σ. Equation (11.3) gives the emergent luminosity L_2 as a function of R, g_e, and Δm, according to the theory of radiative transfer.

The next and only remaining essential requirement of the present simplified picture is to obtain an expression for the mass velocity, $dm/dt = -d(\Delta m)/dt$, of the H ionization zone. This velocity will tell us the rate at which this zone sweeps back and forth through mass as the star pulsates.

The method used by Castor (1968b) makes use of the extreme thinness of the H ionization zone, and treats this zone as a front, or discontinuity, applying Rankine-Hugoniot-type relations across the front. (For further discussion of this approach, see Adams and Castor 1979.) The front conditions derive ultimately from the mass conservation equation (4.4) and the equation (4.26) of conservation of thermal and mechanical energy, both written in conservation form. We assume a steady state in a coordinate system moving with the front. Denoting by "Δ" jumps in the respective quantities across the front (i.e., the values outside minus inside the front, where outside here means closer to the stellar surface), neglecting the gravitational force, and assuming that there is no thermonuclear energy generation, we may combine these front conditions to obtain

$$\left(\frac{dm}{dt}\right)\Delta[h + (1/2)v^2] = -\Delta L. \tag{11.6}$$

Here L is the total local luminosity, $h = E + P/\rho$ is the specific enthalpy of the stellar material, and v (assumed purely radial) is the velocity of the front with respect to the local material.

Now, the jump in $[h + (1/2)v^2]$ across the front will be approximately equal to the ionization energy of hydrogen per unit mass, say χ, assuming hydrogen to be the main constituent in the regions of interest. This statement is approximately justified by noting that, if $v = 10$ km s^{-1} (a typical value), then $(1/2)v^2 \approx 0.05 \times 10^{13}$ ergs gm^{-1}; whereas $\chi \approx 1.4 \times 10^{13}$ ergs gm^{-1} for pure hydrogen. Thus normally, we expect that $\chi \gg (1/2)v^2$. Also $\Delta h \approx \Delta E$ (E = internal energy per unit mass) $\approx \chi$. We therefore obtain, from eq. (11.6),

$$\frac{dm}{dt} \approx \frac{L_1 - L_2}{\chi}, \tag{11.7}$$

where subscripts 1 and 2 denote, respectively, values immediately inside and outside the front.

Equation (11.7) just says that the H ionization zone is moving outward through mass when $L_1 > L_2$. If $L_1 > L_2$, energy is being absorbed in this zone and used to ionize hydrogen; this absorption of energy forces the H ionization zone to move outward then.

The key equations of Castor's (1968b) simplified theory are eqs. (11.3) and (11.7). If Δm as given by eq. (11.3) is substituted into eq. (11.7), a first order differential equation in L_2, the exterior luminosity, results, whose solution would give L_2 as a function of time t, interior luminosity L_1, stellar radius R, and effective gravity g_e. Rather than consider the general nonlinear problem, we shall immediately linearize the equations, in part for simplicitly and in part to show explicitly that the phase lag does indeed follow from a linear theory.

We accordingly linearize eqs. (11.3) and (11.7), and eliminate dm/dt between the resulting linearized equations. We also replace Δm by its value, say Δm_0, in the equilibrium model, and drop many zero subscripts from equilibrium quantities. We finally obtain the (linearlized) differential equation for $(\delta L/L)_2$:

$$\frac{d}{dt}\left(\frac{\delta L}{L}\right)_2 + \frac{L_0}{q\chi\Delta m_0}\left(\frac{\delta L}{L}\right)_2 = \frac{L_0}{q\chi\Delta m_0}\left(\frac{\delta L}{L}\right)_1 + 2\left(1+\frac{1}{q}\right)\frac{d}{dt}\left(\frac{\delta R}{R}\right) - \beta\frac{d}{dt}\left(\frac{\delta g_e}{g_e}\right), \quad (11.8)$$

where β and q were defined in eqs. (11.4) and (11.5).

We now assume that all pulsation variables have a time dependence of the form $e^{i\sigma t}$, where $\sigma = 2\pi/\Pi$ is assumed purely real, and Π denotes the pulsation period. With this assumption the relation between $\delta g_e/g_e$ and $\delta R/R$ is easily obtained, with Ω^2 as a parameter, where Ω is the usual dimensionless frequency (see, e.g., §8.6). We also assume, with Castor, that

$$\left(\frac{\delta L}{L}\right)_1 = \alpha\left(\frac{\delta g_e}{g_e}\right), \quad (11.9)$$

where α is assumed purely real. This assumption is equivalent to neglecting the small phase lag of $(\delta L/L)_1$ relative to minimum radius, resulting from the He^+ ionization zone further in (see §11.1). A typical value of α for pulsating stars of the kind of interest here is $\alpha \approx 1/2$ (Castor 1968b).

The solution of eq. (11.8) may be written in the form

$$\left(\frac{\delta L}{L}\right)_2 = -i\left(\frac{\delta L}{L}\right)_1 \cdot \frac{\Theta}{q} \cdot \frac{1 + A + i[(\Theta/q) - A(\Theta/q)^{-1}]}{1 + (\Theta/q)^2}, \quad (11.10)$$

where

$$A \equiv \frac{1}{\alpha}\left[\frac{1 + 1/q}{1 + \frac{1}{2}\Omega^2} + \frac{n}{q(n+1)}\right], \tag{11.11}$$

and

$$\Theta \equiv \frac{L_0}{\sigma\chi\Delta m_0}. \tag{11.12}$$

It is seen that Θ is equal to the ratio of the energy radiated by the star at its equilibrium luminosity L_0 in $1/(2\pi)$ pulsation periods, to the energy required to ionize all the hydrogen above the H ionization zone.

We see from eq. (11.10) that for $\Theta/q \to \infty$,

$$\left(\frac{\delta L}{L}\right)_2 \to \left(\frac{\delta L}{L}\right)_1. \tag{11.13}$$

Thus the phase lag of $(\delta L/L)_2$ behind $(\delta L/L)_1$, or behind minimum radius, approaches *zero:* the instantaneous luminosity of the star is largest when its radius is smallest. Since $\Theta \propto (\Delta m_0)^{-1}$, the present case corresponds to Δm_0's being too small. So little mass lies above the H ionization zone that the freezing-in effect discussed, for example, in §9.1 is operative even in the H ionization zone. The heat content in the layers lying within and above this zone is so small that it cannot introduce a significant difference between $(\delta L/L)_2$ and $(\delta L/L)_1$. It is easy to see from eq. (11.3) that, for constant L_0 and M and for typical values of the opacity exponents, $\Delta m_0 \propto T_e^{-12}$ and $\Theta \propto T_e^{+12}$, where T_e denotes the equilibrium effective temperature of the star. Hence, the present case corresponds to high T_e; this case is, in fact, realized only for T_e so large that the star very likely lies well to the blue side of the blue edge of the instability region.

Consider next the case $\Theta/q \to 0$. Equation (11.10) then shows that

$$\left(\frac{\delta L}{L}\right)_2 \to -A\left(\frac{\delta L}{L}\right)_1, \tag{11.14}$$

which corresponds to a phase of $(\delta L/L)_2$ relative to $(\delta L/L)_1$ of $-180°$; that is, maximum L now comes at maximum R. This case corresponds to Δm_0's being too large. The H ionization zone now lies so deep in the star that the variations of Δm during a period are relatively small and can be neglected. In this case $(\delta L/L)_2$ is essentially uncoupled from $(\delta L/L)_1$ and is smallest when $\delta g_e/g_e$ is largest. This dependence of $(\delta L/L)_2$ on $\delta g_e/g_e$ is seen to arise mostly from the opacity variations (see eq. [11.3]). When Δm_0 is this large, however, T_e is so small that the star very likely lies well to the cool side of both edges of the instability strip, that is, is stable.

Consider, finally, the intermediate case where $\Theta/q \sim A^{1/2}$. In this case eq. (11.10) shows that

$$\left(\frac{\delta L}{L}\right)_2 \approx -iA^{1/2}\left(\frac{\delta L}{L}\right)_1, \tag{11.15}$$

which corresponds to a phase lag of $(\delta L/L)_2$ behind $(\delta L/L)_1$ of just 90°: maximum luminosity now comes approximately a quarter period after minimum radius. According to eq. (11.10), the condition that the phase lag be 90° is that

$$\Theta/q = A^{1/2}. \tag{11.16}$$

For the reasonable values $\alpha = 1/2$ and $\Omega^2 = 10$ and for representative values of n and σ, $A^{1/2}$ varies between ~0.91 and ~1.10. Hence eq. (11.16), the condition for a 90° phase lag, may be written to adequate accuracy as

$$\Theta/q \approx 1. \tag{11.17}$$

In this case Δm_0 is intermediate, as is T_e; the variations of Δm during a period are now appreciable and yet the layers within and above the H ionization zone have sufficient heat capacity that this zone can introduce an appreciable difference between $(\delta L/L)_1$ and $(\delta L/L)_2$. In this case it can be shown that $d(-\Delta m)/dt$ is approximately in phase with $(\delta L/L)_1$. Then the H ionization zone is moving outward through mass most rapidly at about the time that $(\delta L/L)_1$ is largest and R is smallest. The minimum separation in mass between the H ionization zone and the surface, and

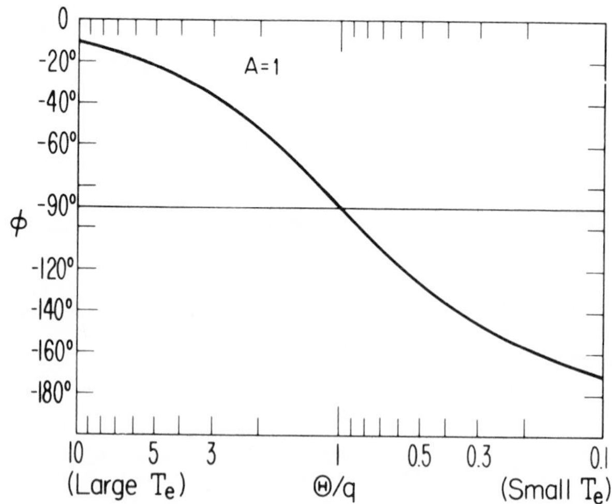

Figure 11.1. Phase lag ϕ versus Θ/q from eq. (11.10), for $A = 1$. See text for further explanation.

hence maximum luminosity, occurs roughly when the star is expanding most rapidly, as interpretations of observations show (see §3.1).

A plot of ϕ—the phase of $(\delta L/L)_2$ relative to $(\delta L/L)_1$, and therefore very nearly relative to minimum radius—computed from eq. (11.10) versus Θ/q is shown in Figure 11.1, for the case $A = 1$. It is seen that ϕ varies between $\sim 35°$ for $\Theta/q = 3$ and $\sim -140°$ for $\Theta/q = 0.3$. This range of Θ/q corresponds to a total range (at fixed L_0 and M) of $|\Delta \log T_e| \approx 0.08$. i.e., $\Delta T_e \approx 1200°$K for stars in the oval region in Figure 3.1. This range of T_e is only slightly larger than the empirical width of the Cepheid instability strip.

11.4. COMPARISON WITH DETAILED CALCULATIONS AND FURTHER REMARKS

Detailed calculations show that Θ lies in the approximate range $0.5 \lesssim \Theta \lesssim 5$ for most Cepheids and RR Lyrae variables; this is just the range of Θ values required for the existence of the approximate 90° phase lag. It is a remarkable fact that the condition for the appearance of the phase lag and the necessary condition for instability (see §10.3) are both satisfied for parameter values corresponding to stars lying in the oval region in Figure 3.1 where many common types of pulsating stars are found. Because these two phenomena (instability and the phase lag) are caused by the action of two different ionization zones, it appears that the occurrence of the phase lag in pulsating stars is more or less an accident of nature; attributing both phenomena to a single physical mechanism, which was Eddington's method (Eddington 1941, 1942), is evidently not entirely correct. However, partial substantiation for Eddington's view lies in the fact that the existence of the phase lag implies that a non-negligible fraction of the total driving must be due to the H ionization zone. This statement is based on the fact that, if there is enough heat capacity in this ionization zone to introduce a significant difference between $(\delta L/L)_1$ and $(\delta L/L)_2$, then this zone should contribute substantially to the driving. Detailed calculations, both linear and nonlinear (see, e.g., J. P. Cox 1974a for detailed references), show that this is indeed the case.

A plot of ϕ versus equilibrium effective temperature, at fixed L_0 and M, for a sequence of classical Cepheid models with radiative envelopes, according to unpublished linearized calculations of J. P. Cox and King (1970), is shown in Figure 21 of J. P. Cox (1974a). This figure shows that the phase lag is close to 90° for both the fundamental and first harmonic modes near the blue instability edges; also, that the variation of ϕ with T_e is in the same sense as, and in qualitative agreement with, Castor's (1968b) simplified picture (cf. Fig. 11.1). Qualitatively similar results have been obtained for radiative models of RR Lyrae variables in the linearized calculations of Castor (1971) and of Iben (1971b).

Castor (1968b) presents semiquantitative arguments concerning the effects of nonlinearity on the phase lag. He concludes that these nonlinear effects should conspire to cause the luminosity maximum to tend to occur near the instant of maximum outward surface velocity, over a wide range of stellar parameter values. Hence, nonlinear effects should cause the actual phase lag to be relatively insensitive to the location of a pulsating star within the instability strip. This expectation is in agreement both with observations and with detailed nonlinear calculations.

Many qualitative features of Castor's theory have been confirmed by the nonlinear calculations of Keller and Mutschlecner (1971).

Effects of nonlinearity on the H ionization zone are discussed further by Castor (1971). He shows that, although in a pulsating star with small but finite pulsation amplitude the linearized equations may break down seriously *in* the H ionization zone, the overall behavior of this zone is nevertheless described by the linearized equations to about the same accuracy as are the other pulsation variables. An analogy is the use of Rankine-Hugoniot relations across a shock front; such a procedure yields accurate relations between conditions far upstream and downstream from the shock, although this procedure does not accurately describe conditions in the shock front itself. This result provides further justification for the use of a linear theory to describe the phase lag.

It has been pointed out by Karp (1975b) that in his nonlinear Cepheid pulsation calculations the phase lag seems to be caused more by the He^{+} ionization zone than by the H ionization zone. The relation between this result and Castor's theory is not clear at present.

It is also emphasized by Castor (1971) that the phase lag would be too large—near 180°—if convective transfer were important in the regions within and exterior to the H ionization zone. This conclusion is in agreement with the earlier linearized calculations of Baker and Kippenhahn (1965), which revealed a phase lag of nearly 180° for a model in which effective convection in the equilibrium model was present in the relevant regions. Qualitatively similar conclusions were also reached by Iben (1971b). Physically, this result is a consequence of the fact that convection in the equilibrium model has the effect of increasing the amount of mass Δm_0 lying above the H ionization zone, and also of increasing the thickness (in mass) of this ionization zone, for a star of given L_0, M, and T_e. Hence, convection in the relevant regions of the equilibrium model tends to force the star toward the limit $\Theta/q \rightarrow 0$ discussed above, which corresponds to $\phi \rightarrow 180°$. Castor implies that the existence of the phase lag in actual pulsating stars suggests that convection in the relevant regions of the envelope is smaller than is predicted by conventional theories of convection.

The two-dimensional calculations of convection by Deupree (1977a) in pulsating stars show that, indeed, the amount of convection that is needed to stabilize a pulsating star is so small that the phase lag should not be appreciably affected by convection.

Finally, the reason the linearized calculations of Cox (1963) did not reveal the correct phase lag is evidently that the H ionization zone was omitted entirely, for expediency, from the calculations.

12

Nonlinear Theory

Actual pulsations of real stars must certainly involve *nonlinear* effects, which cannot possibly be given directly by a linear theory. Some specific evidence that such effects are present in nature is presented by J. P. Cox (1974a). Thus, for example, with the surface amplitudes that are inferred from the observations, it is shown there that the relative pressure, density, and temperature variations at the surface would be much too large to be given correctly by a linear theory. Also, the light and velocity curves, even in a single mode, are ordinarily decidedly not simple sine waves (see, e.g., Fig. 3.2). Finally, the fact that the observed pulsation amplitudes of pulsating stars of a given type do not normally show enormous variations from star to star suggests the existence of a (nonlinear) limit-cycle type of behavior.

However, the full set of nonlinear equations is so complicated that there are apparently no cases of realistic stellar models for which analytic solutions exist. Only two specific (and not very realistic) cases are known that may be exceptions to this statement, and these are considered briefly in §12.1. Also, there does not appear to be any very helpful underlying mathematical framework for nonlinear oscillations. Accordingly, most recent theoretical studies of stellar pulsation have proceeded through purely numerical methods with fast digital computers, as described briefly in §12.2. However, a few new attempts to apply analytic techniques to the phenomenon have recently been made, and these items are discussed briefly in §12.3.

Unless specifically stated otherwise, we shall adopt in this chapter the same basic assumptions as in the remainder of this part (see, e.g., the opening remarks in Chapter 10).

12.1. EXACTLY INTEGRABLE CASES

The only exactly integrable cases are two in number. They are both for *adiabatic* motion with constant $\Gamma_1 = (d \ln P/d \ln \rho)_{\text{ad}}$. These are a general stellar model with $\Gamma_1 = 4/3$, and the *homogeneous* model (density ρ not a function of position in the star) with *any* (constant) value of $\Gamma_1 > 4/3$. So we consider here radial, nonlinear, adiabatic pulsations of a stellar model with constant Γ_1.

The equation of motion is eq. (6.11), with $f(m,t)$ given by Gm/r^2, m

being the mass interior to radial distance r. By using the mass equation (e.g., eq. [6.8]) and assuming adiabatic oscillations, one can express the acceleration $\ddot{r}$ of a mass element in terms only of r, r_0 (value of r of the same mass element in the nonpulsating model), m, and Γ_1. We now assume that the variables are separable:

$$r(r_0,t) = x(r_0)w(t), \tag{12.1}$$

where $x(r_0)$ is a function only of r_0, and $w(t)$ is a function only of time t. Substituting this equation into the equation for $\ddot{r}$ and rearranging somewhat, we obtain

$$w^2\ddot{w} = -\frac{x(r_0)}{r_0^{\,2}}\frac{1}{w^{3\Gamma_1-4}}\frac{1}{\rho_0}\frac{d}{dr_0}\left\{P_0\left[\frac{r_0^{\,2}}{x^2(r_0)}\frac{1}{dx(r_0)/dr_0}\right]^{\Gamma_1}\right\} - \frac{Gm}{x^3(r_0)}, \tag{12.2}$$

where all quantities having zero subscripts are functions only of r_0.

In the case of a general model with $\Gamma_1 = 4/3$, we see immediately that the right side of eq. (12.2) is a function only of r_0, while the left side is a function only of t. Hence, separation of variables is possible in this case, and both sides of eq. (12.2) must be equal to the same constant, say A. Perhaps the simplest way of satisfying this requirement is to have $x(r_0) = r_0$. This condition, in turn, requires, since the unperturbed model is in hydrostatic equilibrium, that $A = 0$. We therefore obtain

$$w = Bt + C, \tag{12.3}$$

where B and C are constants of integration.

Thus, for this case a solution is (with proper choice of constants)

$$r = r_0\left(\frac{t}{t_0}\right), \tag{12.4}$$

or $r \propto t$, with $\ddot{r} = 0$ (hydrostatic equilibrium). From the mass equation, we have

$$\frac{\rho}{\rho_0} = \left(\frac{t}{t_0}\right)^{-3} = \left(\frac{r}{r_0}\right)^{-3}. \tag{12.5}$$

From the momentum equation,

$$\frac{P}{P_0} = \left(\frac{t}{t_0}\right)^{-4} = \left(\frac{r}{r_0}\right)^{-4}. \tag{12.6}$$

But eqs. (12.4)–(12.6) are just the equations of *homologous motion* in hydrostatic equilibrium. We may therefore conclude that *an arbitrary stellar model with* $\Gamma_1 = 4/3$ *is neutrally stable under adiabatic motion,* and can be in hydrostatic equilibrium at any radius.

As the other exactly integrable case, we assume that Γ_1 is constant and not equal to 4/3. We then see from eq. (12.2) that, for our assumed separation of variables to be possible, the unperturbed model must be such that $x \propto m^{1/3}$. It then follows that

$$\frac{P_0}{\rho_0^{\Gamma_1}} = B\, m^{2/3} + C, \tag{12.7}$$

where B and C are constants. However, realistic stellar models do not obey eq. (12.7), from which it follows that separation of variables is not possible for such models. But this equation is obeyed precisely for the *homogeneous* model in hydrostatic equilibrium. Therefore, separation of variables is possible for this model, and the solution for radial, nonlinear, adiabatic pulsations with constant Γ_1 can be obtained exactly. This case is not at all realistic, but is instructive anyway. It is discussed in some detail by Cox (1974a) and by Ledoux and Walraven (1958, §85), and accordingly will not be considered further here. We may only note that this analysis shows that oscillatory behavior is possible for this model only if $\Gamma_1 > 4/3$, just as for the linear case (see §8.1).

12.2. NUMERICAL METHODS

Since analytic solutions cannot generally be obtained for the pulsations of realistic stellar models, one must normally resort to purely numerical methods for solutions. Two approaches have been tried and used in the past: (a) initial value techniques and (b) seeking periodic solutions of the nonlinear equations. Method (b) is highly promising and has already produced some preliminary results (see below), but is still, to some extent, in the developmental stages. Quite a number of calculations have by now been carried out with approach (a). We shall therefore confine most of our remarks to this approach.

The initial value techniques involve conceptually dividing the star or stellar envelope into a number (roughly 50) of discrete, spherical mass zones or shells, and replacing the equations of hydrodynamics and heat flow by finite difference equations, somewhat after the manner described by Richtmyer and Morton (1967). The model is then followed step by step in time on a fast digital computer from certain initial conditions. Published calculations of this kind were pioneered by Christy (1962); J. P. Cox, Cox, and Olsen (1963); and Aleshin (1964). More detailed descriptions of such techniques may be found in Christy (1964); A. N. Cox, Brownlee, and Eilers (1966); Cox, Cox, Olsen, King, and Eilers (1966); and Stobie (1969a,b,c). More recent discussions and further developments have been provided, for example, by Fraley (1968); Papaloizou (1973b); and Kutter and Sparks (1972). See also Castor, Davis, and Davison (1977); Davis and Davison (1978); and Adams, Davis, and Keller (1978). Some results of applications of these methods, and numerous references, are found in J. P.

Cox (1974a), Fischel and Sparks (1975), and A. N. Cox and Deupree (1976).

These techniques make it possible, provided that sufficient computer time is available, to examine the growth or decay of pulsations, the transition from small to large amplitudes, the approach to limiting amplitude, limiting amplitude characteristics, the mode behavior, and so on, of realistic stellar models. These calculations have confirmed and extended the linearized results, and have answered certain questions, but have raised others, about real pulsating stars. Such calculations, at least in their present state of development, involve a certain amount of artistry, and different workers develop different detailed techniques for handling different parts of the calculations. This difficulty was alleviated, to some extent, by a comparison of results of a number of different investigators for one model by Fischel and Sparks (1975). There was surprisingly good overall agreement among the various results. We may therefore conclude that such calculations can mimic, at least in rough fashion, the temporal behavior of real stars.

In recent years, some efforts have been devoted to attempts to obtain more-or-less directly periodic solutions of the nonlinear equations of hydrodynamics and heat flow. To a large extent, these efforts have been motivated by a desire to circumvent many of the difficulties inherent in the initial-value approaches (such as extremely long computer runs and all the disadvantages that accompany them).

The first such attempts along these lines were due to Baker and von Sengbusch (1969, 1970) and von Sengbusch (1973) (see also Baker 1973). In these methods the zoning of the model, the finite difference equations, the inclusion of pseudoviscosity pressure, and so forth, are very similar to those used in the initial value approach. In this latter approach the finite difference forms of the four basic equations (mass, momentum, energy, and flux) are solved iteratively for four physical variables (say $r_i, \dot{r}_i, T_{i-1/2}$, and $V_{i-1/2}$) in each of, say, J zones, at each time step. Thus, taking into account the boundary conditions, $4J$ coupled, nonlinear equations are solved for $4J$ unknowns, at each time step. If there are K time steps in a period (for a discussion of the factors that determine K, see, for example, Cox, Brownlee, and Eilers 1966), the total number of equations and unknowns to be solved in a period is $4JK$ (which, in typical cases, may amount to some 40,000–50,000). In the method of seeking periodic solutions of the nonlinear equations, the same $4JK$ unknowns are involved, but there are now $4J$ additional unknowns because the *initial* values (at the beginning of the period) are not known *a priori*. Since the value of the period Π is also not known in advance, it must be included as one of the unknowns. Thus, there are $4J(K + 1) + 1$ unknowns in all. There are $4JK$ equations and $4J$ periodic boundary conditions. In this method all of the unknowns in a period are determined essentially simultaneously, by a kind of sophisticated Newton-Raphson scheme. The above periodicity condition makes Π an eigenvalue of the nonlinear equations, so that its value is also determined as part of the solution. These periodic solutions may represent not only the fundamental mode but also higher modes as well. Besides the periodic, nonlinear behavior of the model, the method also yields information regarding the stability of the nonlinear limit cycles, and, in general, shows a great deal of promise.

More recently, a method has been developed by Stellingwerf (1974a,b; 1975) which achieves the same goals as the Baker-von Sengbusch method, but by a combination of the two techniques described above. In this method, values of all dependent physical variables are guessed at some initial time, as well as the period itself, and the number of time steps in a period is chosen. The model is then integrated forward in time through the specified number of time steps by use of the initial value techniques (or a variation thereof) described above. The final values of the physical variables at the end of a period are then compared with the initial values, and the latter are then varied iteratively, as well as the period itself, until strict periodicity is achieved. As in the Baker-von Sengbusch method, this method also yields complete stability information for the nonlinear limit cycles for the various modes.

These methods may be the only feasible methods for studying the limiting amplitude behavior of stellar models with very long *e*-folding times for growth or decay of pulsations, very slow switching rates among various modes, etc. Some preliminary results obtained by use of these methods have been summarized by J. P. Cox (1975). Further exploration and development of these methods and their potentialities is being carried out by A. N. Cox and collaborators (see, e.g., A. N. Cox, Hodson, and Davey 1976). Another new approach, due to Buchler (1978), is based on a two-time formalism (see, e.g., Cole 1968) and appears very promising.

12.3. ANALYTICAL TECHNIQUES

Numerous attempts have been made to obtain nonlinear solutions in certain special cases via expansions in terms of the linear, adiabatic eigenfunctions, or by other means, but we do not consider these in detail in this book. A summary of some of these attempts, as well as references to the earlier literature on this subject, may be found in Ledoux and Walraven (1958, §§86–90); and in Rosseland (1949, Chap. 7). More recent work along these lines has been carried out by van der Borght and Murphy (1966); van der Borght (1969, 1970); Murphy (1968); Murphy and Smith (1970); Simon (1970, 1971, 1972a,b, 1976); Simon and Sastri (1972); Mohan (1972); Papaloizou (1973a); and Melvin (1977).

13

Simple Models of Stellar Pulsation

It is clear that nonadiabatic (i.e., real) stellar pulsation is a complicated phenomenon. This is true even for very small amplitudes, where the linear theory is applicable, and for purely radial oscillations. For larger amplitudes, where nonlinear effects are important, the situation is almost hopelessly complicated. For these reasons a number of attempts have been made to devise models of stellar pulsation that are simple enough to understand in terms of familiar physical concepts, but which nevertheless retain, it is hoped, the essential physics of the phenomenon. Most of these attempts have been based on a "one-zone" type of model which may be visualized as a single, perhaps relatively thin, spherical mass shell concentric with the stellar center, lying somewhere in the star. Perhaps the best known, and historically the first, of these simple models is the "Baker one-zone model" (Baker 1966), discussed in some detail in §13.1 below. This model was designed primarily for very small, radial oscillations, that is, for elucidating some of the complicated physics underlying pulsations describable by the linear theory. This model has received considerable attention in the literature, and has been extended to cases not discussed in Baker's original paper (Unno and Kamijo 1966; Gough 1967; Unno 1967; Okamoto and Unno 1967; Ishizuka 1967; Zahn 1968). We shall not consider these extensions in detail in this book. Also, unless we state otherwise, we shall consider only purely radial oscillations.

The linear and nonlinear oscillations of a thermally excited oscillator, which was supposed to represent a buoyant cell in a stratified stellar envelope, were discussed by Moore and Spiegel (1966). While this model was not intended to be directly applicable to pulsating stars, some of the solutions nevertheless are strikingly reminiscent of the temporal behavior of certain types of real pulsating stars.

The oscillations of a one-zone model similar in many respects to the Baker one-zone model were discussed by Usher and Whitney (1968). They considered both linear and nonlinear oscillations. The latter were treated primarily by an asymptotic method in which only the first-order nonlinear terms were retained. This model and the one studied by Moore and Spiegel (1966) will not be considered explicitly here.

A model designed specifically for nonlinear, nonadiabatic pulsations was proposed by Rudd and Rosenberg (1970). This model is also essentially a one-zone model, and will be discussed in §13.2.

A one-zone model designed primarily to study the emergent luminosity variations, in both linear and nonlinear theory, was suggested by Stellingwerf (1972). This model was a generalization of the Baker one-zone model, and will be discussed in §13.3.

Finally, a model whose purpose was to elucidate some of the features of the attainment of a limiting amplitude, the stability of limit cycles, and so forth, was proposed by Castor (1970, unpublished notes). This model will be described in §13.4 (with Castor's kind permission).

These models have helped considerably to clarify some of the complicated physics involved in stellar pulsations, and the roles played by different physical mechanisms in determining or influencing certain aspects of the phenomenon. However, it is almost certainly true that the actual situation in real pulsating stars is considerably more complicated than that in these one-zone models. For example, the distributed (that is, variable in space) nature of real stellar envelopes, incorporated only to a limited extent (if at all) in these models, may play a crucial role in the phenomenon. Simplified models for *distributed* stellar envelopes would no doubt be useful for our eventual understanding of the physics of stellar pulsation.

13.1 BAKER ONE-ZONE MODEL

In this section we shall discuss the Baker one-zone model (Baker 1966) in linear theory only, for purely radial oscillations. Nonlinear oscillations of a generalization of the Baker one-zone model will be discussed briefly in §13.3.

It may be noted that Jean's (1928) linearized treatment, based on the assumption of *homologous* motion of the mass layers (i.e., $\delta r/r$ = constant in space), yields results identical to those of Baker's one-zone model in linear theory.

The essential assumptions of the Baker model are that it represents a single spherical shell of mass Δm, somewhere in the star concentric with the center, and throughout which all physical variables (with the exception of the luminosity variation, see below) are assumed to be constant in space. The basic equations are then linearized (see Chapters 5 and 7), and all spatial derivatives, with the exception of $\partial(\delta L_r/L_r)/\partial m$ ($\delta L_r/L_r$ is the relative luminosity variation), are dropped. Because Baker intended his one-zone model primarily for applications to common types of pulsating stars, for which nuclear energy sources are not important (see Chap. 10), he omitted these sources from his model. Hence, with ϵ (rate per unit mass of thermonuclear energy production) = 0, essential physics would be lost in dropping the $\partial(\delta L_r/L_r)/\partial m$ term.

Baker handles this term as follows. He assumes that the luminosity incident on the *inner* boundary of the zone is constant and equal to the local equilibrium luminosity; hence $\delta L_r/L_r = 0$ on this boundary. However, he allows $\delta L_r/L_r$ on the outer boundary, say $(\delta L_r/L_r)_U$, to be variable in time, thus providing a mechanism for the zone to gain or lose heat through a modulation of the flux flowing through it. He also defines $\delta L_r/L_r$ *in* the zone to be equal to $(1/2)(\delta L_r/L_r)_U$, that is, to the instantaneous average of $\delta L_r/L_r$ on the two boundaries. Replacing $\partial(\delta L_r/L_r)/\partial m$ by $(\delta L_r/L_r)_U/\Delta m$, Baker obtains the approximation

$$\frac{\partial}{\partial m}\left(\frac{\delta L_r}{L_r}\right) = \frac{2}{\Delta m}\left(\frac{\delta L_r}{L_r}\right). \tag{13.1}$$

Baker assumes radiative transfer only, and that the opacity variations $\delta\kappa/\kappa$ are given by the expression (in our notation)

$$\frac{\delta\kappa}{\kappa} = n\frac{\delta\rho}{\rho} - s\frac{\delta T}{T}, \tag{13.2}$$

where $\delta\rho/\rho$ and $\delta T/T$ are, respectively, the relative density and temperature variations, and n and s are constants. The linearized equation of state is also written (again in our notation) in the form

$$\frac{\delta P}{P} = \chi_\rho\frac{\delta\rho}{\rho} + \chi_T\frac{\delta T}{T}, \tag{13.3}$$

where $\delta P/P$ is the relative pressure variation and χ_ρ and χ_T are constants (see §4.2c).

With all spatial derivatives thus eliminated, the pulsation equations become a set of *ordinary* differential equations of the third order in time. Combining these equations into a single differential equation and assuming a time dependence for all pulsation variables of the form $e^{\mathscr{s}t}$, we obtain the following cubic equation for the eigenfrequency $\mathscr{s}$:

$$\mathscr{s}^3 + K\sigma_0 A\mathscr{s}^2 + B\sigma_0{}^2\mathscr{s} + K\sigma_0{}^3 D = 0, \tag{13.4}$$

where

$$A \equiv (\Gamma_3 - 1)(s + 4)/\chi_\rho, \tag{13.5}$$

$$B \equiv 3\Gamma_1 - 4, \tag{13.6}$$

$$D \equiv (\Gamma_3 - 1)[3n\chi_T - s(4 - 3\chi_\rho) + 4(\chi_T + 3\chi_\rho - 4)]/\chi_\rho, \tag{13.7}$$

where Γ_1 and Γ_3 are defined in §4.2. The quantities σ_0 and K are defined as follows:

$$\sigma_0{}^2 \equiv g/r = Gm/r^3, \tag{13.8}$$

$$K \equiv \frac{\chi_\rho}{\Gamma_3 - 1} \cdot \frac{2L}{\Delta m c_V T} \cdot \frac{1}{\sigma_0}, \tag{13.9}$$

where g, m, r, L, c_V, and T denote, respectively, local gravitational acceleration, interior mass, radial distance, luminosity, specific heat per unit mass at constant volume, and temperature, all regarded as equilibrium values and therefore constant in time.

It will be seen presently that $\sigma_0 \approx 2\pi/\Pi$, where Π is the adiabatic pulsation period of the one-zone model. Hence, eq. (13.9) shows that K is approximately equal to the ratio of the energy flowing through the model in $1/(2\pi)$ periods, to its internal energy. Roughly, K may be thought of also as the ratio of the free-fall time to the Kelvin time for the zone. Hence, K is seen to be a kind of coupling constant between the adiabatic and nonadiabatic behaviors of the model: small K corresponds to nearly adiabatic motion, large K to large departures from adiabaticity.

For $K = 0$, corresponding to perfectly adiabatic motion, the only nonvanishing roots of eq. (13.4) are

$$\mathscr{s} = \pm i(3\Gamma_1 - 4)^{1/2}\sigma_0 \tag{13.10a}$$

$$= \pm i[(3\Gamma_1 - 4) \cdot \tfrac{4}{3}\pi G\bar{\rho}(r)]^{1/2} \tag{13.10b}$$

where we have defined $\bar{\rho}(r)$ by the relation $\bar{\rho}(r) \equiv m/[(4/3)\pi r^3]$, where m is the mass interior to radial distance r. If $\Gamma_1 > 4/3$, these roots are purely imaginary, corresponding to pulsations of constant amplitude; the pulsation period is $\Pi = 2\pi/[(3\Gamma_1 - 4)^{1/2}\sigma_0]$. Note that, if $\bar{\rho}(r)$ is replaced by $\bar{\rho}$, the mean density of the whole star, this result is precisely the same as that for the adiabatic pulsations of the homogeneous model (see §8.1) or of the shell model of §2.1. For the same reasons as given in §9.3, the two alternative signs of $\mathscr{s}$ have no physical significance.

It is clear that, if $\Gamma_1 < 4/3$, both of the above roots are purely real, corresponding to *dynamical* instability. Hence, the condition for dynamical stability of the one-zone model is that

$$B > 0 \tag{13.11}$$

(see eq. [13.6]).

Consider now the case $K \neq 0$. Baker discusses the roots of eq. (13.4) and shows that there are two necessary conditions, besides (13.11), that all roots have negative real parts; that is, that the one-zone model shall be stable with respect to all possible types of (radial) motion. These conditions are that

$$D > 0, \tag{13.12}$$

$$AB - D > 0, \tag{13.13}$$

where A, B, and D are defined in eqs. (13.5)–(13.7).

The condition (13.12) may easily be seen to be the condition for *secular stability* (see §19.6). For, clearly, there exists a "small" root, say $\mathscr{s}_1$, of eq. (13.4), given approximately by

$$\mathscr{s}_1 \approx -\frac{K\sigma_0 D}{B} = -\frac{K\sigma_0 D}{3\Gamma_1 - 4}. \tag{13.14}$$

This root is purely real, corresponding to aperiodic motion which is stable if $D > 0$ (assuming that $\Gamma_1 > 4/3$). The time scale associated with this motion is of the order of $(L/c_V T\Delta m)^{-1}$, which is clearly the thermal relaxation or Kelvin time for the zone. Note that, for $\chi_\rho = \chi_T = 1$ (simple perfect gas), eq. (13.12) yields the condition

$$3n - s > 0 \tag{13.15}$$

for secular stability. This condition is just the Jeans (1928) condition without the nuclear energy generation terms (see, e.g., Ledoux 1963, pp. 420–421; and §19.6 of this book). Physically, it states that $\delta L_r/L_r$ must decrease upon compression. If this is the case, then radiation is being trapped in the zone when it is compressed; this trapped radiation will cause the zone, subsequently, to expand and cool, and thus to return, eventually, to its original configuration.

The condition (13.13) is the condition for *pulsational* stability. This may be seen most easily, at least for $K \ll 1$, by obtaining solutions of eq. (13.4) for the two "large" roots, say $\mathscr{s}_{2,3}$, as expansions in terms of K. We obtain

$$\mathscr{s}_{2,3} \approx \pm iB^{1/2}\sigma_0 - \frac{\sigma_0(AB - D)}{2B}K + \ldots. \tag{13.16}$$

Equation (13.16) makes it obvious that eq. (13.13) is indeed the necessary condition for pulsational stability.

Using eqs. (13.5)–(13.7), we may also write, after some manipulation and the use of some thermodynamic identities, the condition (13.13) for pulsational stability of the one-zone model in the form

$$4(\Gamma_3 - 1) + [s(\Gamma_3 - 1) - n] - 4/3 \equiv \Lambda > 0, \tag{13.17}$$

which defines the quantity Λ. Note that this relation is valid even if $\chi_\rho \neq 1$ and $\chi_T \neq 1$. Each term in eq. (13.17) has a distinct physical significance, as Baker (1966) points out. The first term, $4(\Gamma_3 - 1)$, is always positive and therefore always tends to stabilize. This term represents cooling upon compression as a result of enhanced radiation flow at this instant ($L \propto T^4$

and, approximately, $T \propto \rho^{\Gamma_3 - 1}$ in the adiabatic approximation). In an ionization zone of a dominant element, we have seen that $\Gamma_3 \rightarrow 1$, so that the damping effects represented by this term are diminished in such a zone. Physically, the ionization prevents the temperature from rising very much upon compression, and hence inhibits the loss of radiation at this instant. These effects of ionization on this first term represent (in part) the operation of the gamma mechanism (see §10.1).

The second term, $s(\Gamma_3 - 1) - n$, in eq. (13.17) gives the effect of the opacity variations, and can be of either sign. For the normal values $n \approx 1$, $s \approx 3$, $\Gamma_3 - 1 \approx 2/3$, this term is positive and therefore contributes to the damping. Physically, the opacity in this case decreases upon compression and thus allows radiation to leak out of the zone at this instant, thus producing a tendency for damping. However, in an ionization zone of a dominant element, $\Gamma_3 \rightarrow 1$, and this term may then be negative, representing driving of pulsations. Physically, the ionization keeps the temperature variations relatively small, so that the density variations now dominate the opacity variations; and the opacity always increases with increasing density (at constant temperature). Hence, the opacity variations now produce a damming up of the radiation during compression, thus resulting in driving. These destabilizing effects of the opacity variations on stability represent the operation of the kappa mechanism (see §10.1).

If s were large and negative (or at least less positive than usual), this term involving s would represent damming up of the radiation during compression, and hence driving, even if Γ_3 had its normal value of about 5/3. This circumstance (that is, s less positive than usual and $\Gamma_3 \simeq 5/3$) is realized exactly in Stellingwerf's (1979) bump mechanism, referred to in §10.1.

The third term, $-4/3$, in eq. (13.17), being negative, always tends to destabilize. This term arises from the assumed spherical symmetry, and would be absent for plane-parallel stratifications. This term merely reflects the dependence of the local luminosity on the area of the spherical surface bounding the mass level of interest ($L_r \propto r^4$). The area of this surface diminishes upon compression, thus producing a tendency for a diminution, or trapping, of the radiation at this instant, and hence a destabilizing tendency. This effect of curvature has been called by Baker (1966) the "radius effect."

Another interpretation of eq. (13.17) is obtained by the following observation. If in the linearized radiative transfer equation for the one zone model, that is,

$$\frac{\delta L_r}{L_r} = 4\xi - n\frac{\delta\rho}{\rho} + (s + 4)\frac{\delta T}{T}, \tag{13.18}$$

where $\xi \equiv \delta r/r$, $\delta L_r/L_r$, $\delta\rho/\rho$, and $\delta T/T$ denote *the spatial* parts of, respectively, the relative radius, luminosity, density,and tempeature variations, we make the quasi-adiabatic approximation, $\delta T/T = (\Gamma_3 - 1)\,\delta\rho/\rho$, we obtain

$$\frac{\delta L_r}{L_r} \equiv \left(\frac{\delta L_r}{L_r}\right)_{\text{ad}} = \Lambda\left(\frac{\delta\rho}{\rho}\right). \tag{13.19}$$

Here Λ is defined in eq. (13.17) and $(\delta L_r/L_r)_{\text{ad}}$ means $(\delta L_r/L_r)$ as computed in the quasi-adiabatic approximation. Hence, the condition (13.17) for pulsational stability of the one-zone model may alternatively be expressed in the form

$$\left(\frac{\delta L_r}{L_r}\right)_{\text{ad}} > 0, \tag{13.20}$$

which perhaps makes the physical content of the condition somewhat clearer: upon compression, the zone must *lose* heat as a result of enhanced radiation from its surface. This statement is completely consistent with the physical discussion in §9.1.

The above statement appears at first sight not to be strictly correct, since only $(\delta L_r/L_r)_{\text{ad}}$, and not $(\delta L_r/L_r)$, the *actual* luminosity variation, appears in eq. (13.20). However, it can be shown that $(\delta L_r/L_r)$ essentially always has the same sign as $(\delta L_r/L_r)_{\text{ad}}$. Hence, eq. (13.20), without the subscript "ad," is also a completely equivalent statement of the condition for pulsational stability of the one-zone model under the assumed conditions.

Another viewpoint, which provides some interesting insights into the one-zone model and immediately generalizes it quite considerably, is to apply the integral expressions (9.45) and (9.46) to the model. The first expression gives the value of the quantity C, the real part C_r of which is proportional to the average rate, over a period, at which the pressure gradient and gravity forces do work on the star. Hence, in terms of C, the condition for pulsational stability is $C_r < 0$.

In applying this condition to the one-zone model, we may assume that only the one zone of interest, of mass Δm, is pulsating, so that the integration is carried out only over Δm. Also, by the very nature of the one-zone model, the spatial part of the relative density variation $\delta\rho/\rho$ may always be taken as purely real. Since all physical quantities are assumed spatially constant in the one-zone model, we obtain as the condition for pulsational stability

$$C_r = (\Gamma_3 - 1)\left(\frac{\delta\rho}{\rho}\right)\delta\left(\epsilon - \frac{\partial L_r}{\partial m}\right)_r \Delta m < 0, \tag{13.21}$$

where $\delta(\epsilon - \partial L_r/\partial m)_r$ denotes the real part of $\delta(\epsilon - \partial L_r/\partial m)$, which henceforth in this subsection shall denote only the spatial part of the complete variation. We have retained ϵ here for generality and to illustrate how easily the Baker one-zone model may be generalized with the present approach. If we take $(\delta\rho/\rho)$ as positive, then the condition (13.21) for pulsational stability of the one-zone model becomes

$$\delta\left(\epsilon - \frac{\partial L_r}{\partial m}\right)_r < 0. \qquad (13.22)$$

This condition states simply that, in order to be pulsationally stable, the zone must suffer a net *loss* of heat upon compression, in full agreement with our physical discussion in §9.1.

Because of the generality of our present approach, eq. (13.22) must comprise eq. (13.20) as a special case, as can in fact be shown (but we shall not do so here). Note, for example, that no assumptions as to the transfer mechanism are contained in eq. (13.22), whereas radiative transfer was explicitly assumed in obtaining eq. (13.20). Also, it was pointed out above that eq. (13.20) involved only the *quasi-adiabatic* luminosity variations, whereas eq. (13.22) involves the real parts of the *nonadiabatic* variations of energy production rate ϵ and interior luminosity L_r.

When $\delta(\delta L_r)/\partial m$ is evaluated as in Baker's approximation (see above), ϵ is set equal to zero, and one makes the quasiadiabatic approximation, one obtains eq. (13.20), as expected. It may be noted that this equation may be applied straightaway to the many extensions or elaborations of the original Baker one-zone model. For example, L_r in eq. (13.22) may be the sum of the radiative and convective luminosities. Also, ϵ need not be set equal to zero. Indeed, if we set $\delta\epsilon/\epsilon = \lambda\delta\rho/\rho + \nu\delta T/T$ and use eq. (13.18) for $\delta L_r/L_r$, in both cases making the adiabatic approximation, we obtain as the condition for stability

$$\lambda + \nu(\Gamma_3 - 1) - (s + 4)(\Gamma_3 - 1) + (n + 4/3) < 0. \qquad (13.23)$$

This condition differs from Baker's (1966) condition (13.17) only in the presence of the terms containing λ and ν. The condition (13.23) is, in fact, identical to Jeans' (1928) condition for the pulsational stability of a continuously distributed star executing homologous pulsations and obeying the perfect gas law. Equation (13.23) may also be derived, with considerable algebra, directly from Baker's approach as outlined near the beginning of this subsection.

13.2. RUDD-ROSENBERG MODEL

The model of Rudd and Rosenberg (1970) is a one-zone model similar in its basic features to that studied by Usher and Whitney (1968). That model is also similar to the shell model introduced in §2.1 of this book, except that the radius of the rigid core of the Rudd-Rosenberg model may be any fraction of the equilibrium radius of the shell, which may be identified with the stellar surface. The main purpose Rudd and Rosenberg hoped to accomplish with their model was a clarification of some of the mathematics underlying fully nonlinear stellar pulsations and the occurrence of limit cycle-like phenomena of the kind observed in real pulsating stars. By restricting their considerations to a one-zone model, they could study true limit cycle behavior. More specifically, they wanted to devise a model which would exhibit overstability at small amplitudes and a stable limit cycle at large amplitudes.

If M is the mass of the rigid core, $\mathcal{M}(\ll M)$ is the mass of the gas in the zone, and r is the instantaneous radius of the zone, the equation of motion of the zone is taken to be

$$\mathcal{M}\frac{d^2r}{dt^2} = 4\pi r^2 P - \frac{GM\mathcal{M}}{r^2}, \tag{13.24}$$

as in Usher and Whitney (1968), where P is the pressure of the gas in the zone. It is assumed in eq. (13.24) that the gas in the zone behaves as if its entire mass $\mathcal{M}$ were concentrated into an infinitely thin shell of radius r; this shell may be thought of as a thin, freely distensible membrane that contains the gas. The equilibrium radius r_0 of the zone is given by the relation which results from setting d^2r/dt^2 equal to zero in eq. (13.24).

The energy equation (e.g., eq. [4.30a]) is introduced not by differentiating eq. (13.24) and then eliminating $\dot{P}$ (a dot denotes d/dt) with the help of the energy equation, as was done in obtaining eq. (6.13). Rather, the energy equation is introduced in integrated form. We consider here only thermodynamically reversible processes, for which $dq/dt = T\dot{s}$ (s = specific entropy). Denoting equilibrium quantities by zero subscripts, the formal integral of the energy equation (4.30a) is

$$\left(\frac{P}{P_0}\right) = K\left(\frac{\rho}{\rho_0}\right)^{\Gamma_1}, \tag{13.25}$$

where

$$K = \rho_0^{\Gamma_1-\Gamma_{1,0}} \exp\left\{-\int_t \dot{\Gamma}_1 \ln\rho dt + \int_t \frac{\chi_T \dot{s}}{c_V} dt\right\}. \tag{13.26}$$

For adiabatic motion ($\dot{s} = 0$) with constant Γ_1, K would be unity. The authors allow K to be variable, but do not relate the specific functional form they assume for it (see below) to detailed physical processes operating in the zone.

The exact relation between ρ/ρ_0 and r/r_0 for the present model may be written in the form

$$\frac{\rho}{\rho_0} = \left(\frac{r}{r_0}\right)^{-m}, \tag{13.27}$$

where

$$m \equiv \frac{\ln \{[(r/r_0)^3 n_0^3 - 1]/(n_0^3 - 1)\}}{\ln (r/r_0)}, \tag{13.28}$$

in which $n_0 = r_0/r_c$ (r_c = radius of rigid core) ≥ 1. If $r = r_0$ in eq. (13.28) ($r \approx r_0$ in eq. [13.27]), then m becomes a constant whose value depends only on n_0. The value of m in this case is

$$m = \frac{3n_0^3}{n_0^3 - 1}. \tag{13.28$'$}$$

If the core were of zero radius ($n_0 = \infty$), we would have $m = 3$, as in the homogeneous model (§8.1) or in the shell model of §2.1. It is easily seen from eq. (13.28′) that $m > 3$ if $n_0 > 1$ and $r/r_0 \approx 1$ (say $0.95 \leq r/r_0 \leq 1.1$, as in Cepheids and RR Lyrae variables). A value of $m \approx 12$ corresponds to $n_0 \approx 1.2$; the authors argue that such values might be appropriate to real pulsating stars for which only the outermost layers are effectively pulsating (see §8.12c). Physically, in a thin layer of gas overlying a rigid sphere, a small change in the outer radius of the layer will have a relatively large effect on the mean density of the gas in the layer.

Combining eqns. (13.25) and (13.27) with eq. (13.24) yields the final equation of motion adopted for the model:

$$\ddot{w} = \frac{GM}{r_0^3}\left[\frac{K}{w^n} - \frac{1}{w^2}\right], \tag{13.29}$$

where $w \equiv r/r_0$, $n \equiv m\Gamma_1 - 2$, and K is assumed to be expressible as a function only of the small quantity $1 - w$. Note that $n = 3\Gamma_1 - 2$ for $m = 3$ (the homogeneous model), and $n = 18$ (the value eventually adopted in the numerical applications) for $m = 12$ and $\Gamma_1 = 5/3$. From the discussion in §8.1, it is seen that the condition for dynamical stability, i.e., for the existence of oscillatory solutions, is that $n > 2$.

The novel feature introduced by Rudd and Rosenberg is to regard the expansion and contraction phases of the pulsations as quantitatively different. Some support for this viewpoint is provided by observations of the phase-plane diagram of δ Cephei, as is pointed out by the authors: the expansion and contraction phases appear to occur about two different equilibrium radii, that for expansion being larger than that for contraction.

Different parameters for the two phases are thus used in the functional forms for K. The specific forms finally adopted are the following:

$$K_e = E_0 + E_2(1 - w)^2 \qquad (13.30a)$$

$$K_c = C_0 + C_2(1 - w)^2, \qquad (13.30b)$$

where subscripts e and c refer, respectively, to the expansion and contraction phases. Some arguments in support of these functional forms and discussion of the values of the constants will be given presently.

The transitions between eqs. (13.30) with $K = K_e$ and $K = K_c$ are assumed to occur discontinuously at the ends of successive phases of expansion and contraction. However, w and $\dot{w}$ are assumed to be continuous throughout the motion. Hence, $\ddot{w}$, and thus the total force acting on the zone, change discontinuously at the end of each phase of expansion and of contraction.

The above statements regarding the different equilibrium radii for the two halves of a cycle imply that

$$E_0 > 1 > C_0. \qquad (13.31)$$

The justification for eq. (13.31) can easily be seen by setting $\ddot{w} = 0$ in eq. (13.29), and then solving for the corresponding value, say w_0, of w. If we neglect the quadratic terms in eqs. (13.30), then we are effectively assuming small oscillations about $w = 1$. Noting that w_0 is larger for the expansion phase than for the contraction phase, eq. (13.31) follows. Hence, the positive outward forces act over a greater distance during the expansion phase immediately after minimum radius (when the velocities are positive) than during the contraction phase immediately preceding minimum radius (when the velocities are negative). The quadratic forms in $(1 - w)$ for K_e and K_c were chosen for consistency with the requirement (see eq. [13.26]) that K never be negative. The same considerations require that E_0, E_2, C_0, and C_2 in eqs. (13.30) all be positive numbers.

The authors show that, in the limit of small amplitudes, where the quadratic terms in K_e and K_c can be neglected, the condition $E_0 > C_0$ leads to a build-up of the amplitude, that is, to self-excitation of the pulsations. This conclusion may be most readily understood in terms of a phase-plane diagram, as in Figure 13.1. During the expansion phase, $E_0 > 1$ is the "center" for that half-cycle; during the contraction phase, $C_0 < 1$ is the "center" for that half-cycle. Clearly, the above successive switching from one center to the other leads to an increase in the oscillation amplitude.

We may note that the above condition can also be interpreted in terms of an oscillator in which the restoring force lags maximum displacement by a small amount. Treating the K's as constants in this limit is equivalent to

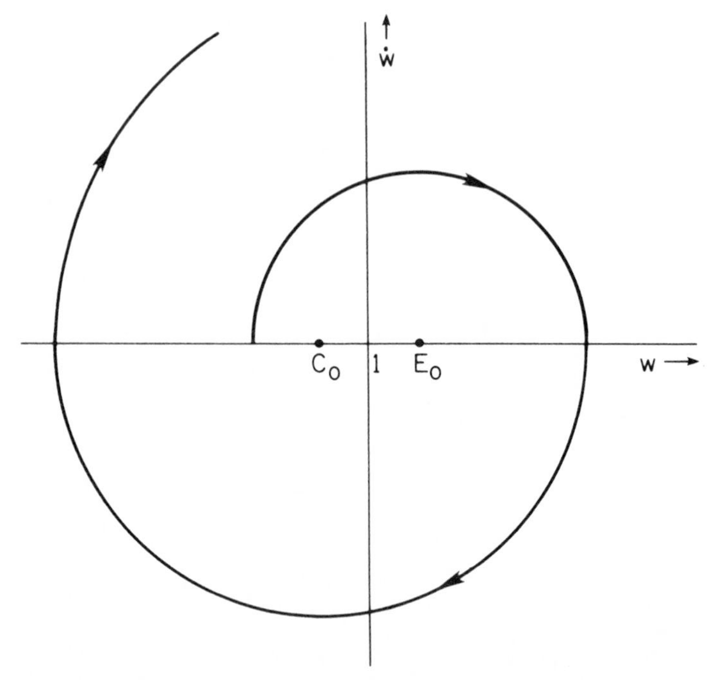

Figure 13.1. Schematic phase-plane diagram for an oscillator that switches discontinuously from one "center" ($E_0 > 1$) during the expansion phase to another "center" ($C_0 < 1$) during the contraction phase.

assuming that the motion is *adiabatic* during each half-cycle; but nonadiabatic for an infinitesimal time at each turn-around point, with more energy inserted into the zone at minimum radius than is extracted at maximum radius. Hence, this model is, in the words of the authors, "piecewise conservative," but "globally nonconservative."

The limitation of the amplitude and the occurrence of a limit cycle are produced, in this model, by the nonadiabatic, quadratic terms in the K's (see eqs. [13.30]). It is shown by a rather intricate argument that the necessary and sufficient condition for the existence of a stable limit cycle for the present model is that $C_2 > E_2$. Physically, this condition asserts, approximately, that nonadiabatic effects during large inward excursions, for example, must produce a relative strengthening of the outward restoring forces when the gas is moving *inward* toward minimum radius, and a relative *weakening* of the inward restoring force when the gas is expanding away from minimum radius (the converse is true for a large outward excursion), if a limit cycle is to exist.

The occurrence of a limit cycle for the present model can also be understood in terms of the potential curves defined by Rudd and Rosenberg, which are essentially first integrals of the equation of motion (13.29). These first integrals may be written in the form

$$\tfrac{1}{2}\,\dot{w}^2 = -V + \text{const.}, \tag{13.32}$$

where

$$V \equiv -\int \left(\frac{K}{w^n} - \frac{1}{w^2}\right) dw. \tag{13.33}$$

Explicit relations, corresponding to the assumed functional forms of K (see eqs. [13.30]), are given by the authors. The intersection of one of the curves $V(w)$ with a line parallel to the w-axis (corresponding to a given total energy) gives the turn-around points where $\dot{w} = 0$. The corresponding potential curves, say $\overline{V}_e$ and $\overline{V}_c$, for small amplitudes can be obtained from these relations by setting $E_2 = C_2 = 0$ therein.

The potential curves $\overline{V}(w)$ for small amplitudes are illustated by the authors. Note that $\overline{V}_e$ is always greater than $\overline{V}_c$, and that $\overline{V}_e - \overline{V}_c$ becomes larger as w decreases below unity, and smaller as w increases above unity. These figures show clearly the growth of the oscillation amplitude. Clearly, there is no limit cycle for these kinds of potential curves, and the oscillation amplitude would increase without limit.

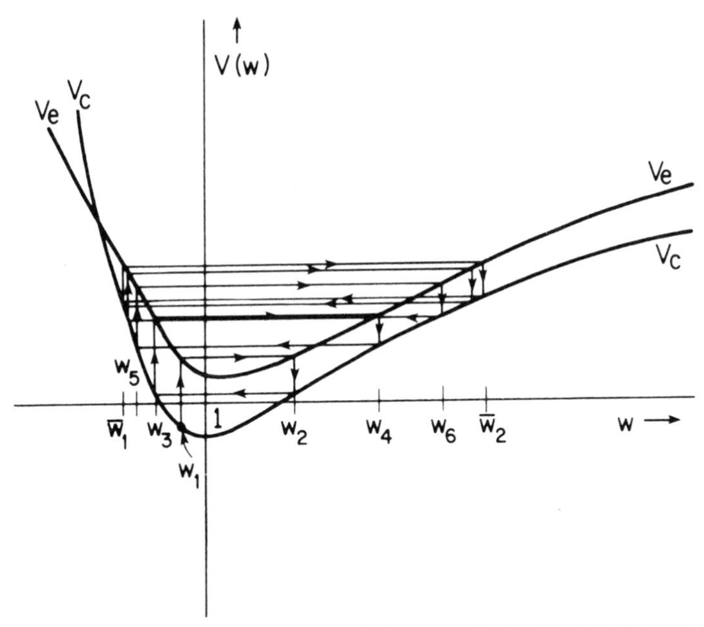

Figure 13.2. Potential diagram (schematic) for large amplitudes ($E_2 \neq 0$, $C_2 \neq 0$), for the case where a limit cycle exists ($C_2 > E_2$).

The potential curves V_e and V_c, illustrated schematically in Figure 13.2 (adapted from Rudd and Rosenberg 1970), show the effects of the quadratic, nonadiabatic terms at large amplitudes, for the case $C_2 > E_2$. Upon compression, the V_c curve is distorted more by these terms than is the V_e curve. The distortion, if $C_2 > E_2$, is such that the V_c curve becomes steeper, implying a larger restoring force during compression, at large inward excursions, than during the ensuing expansion phase, which confirms the qualitative statements made above. The straight lines show the motion of the oscillator and the successive turn-around points w_0, w_1, $w_2, \ldots$. It is clear now that a limit cycle exists, having successive turn-around points $\overline{w}_1$ (<1) and $\overline{w}_2$ (>1) which do not change with time. In this limit cycle the same amount of energy is removed at maximum expansion as is inserted at maximum compression.

By a suitable choice of values for E_0, C_0, E_2, and C_2, the authors can get good agreement with the phase-plane diagram of δ Cephei.

Overall, this model is very enlightening, particularly from a mathematical standpoint.

13.3. STELLINGWERF MODEL

The Stellingwerf model (Stellingwerf 1972) is, essentially, an extension and generalization of the Baker one-zone model, and contains some features of the Rudd-Rosenberg model. Stellingwerf (1972) has carried out both linear and nonlinear calculations of the oscillations of his model.

The Stellingwerf model differs in two major respects from the Baker one-zone model. First, the relation between the density ρ of the gas in the shell and the radius r of the thin, freely distensible membrane which holds the gas in, is the same as in the Rudd-Rosenberg model (see eqs. [13.27] and [13.28]). Thus, in the linear theory the relation between the relative density variations $\delta\rho/\rho$ of the gas (we shall hereafter drop zero subscripts from equilibrium quantities) and the relative radial displacements $\delta r/r$ of the membrane here is $\delta\rho/\rho = m\,\delta r/r$, where $m(\geq 3)$ is determined only by n_0, the ratio of the equilibrium radius of the membrane to the radius of the rigid core (see eq. [13.28′]). The value of m in the Baker one-zone model is 3.

Second, Stellingwerf allows explicitly for the possibility of a luminosity variation δL_i incident on the lower boundary of the shell. He also neglects thermonuclear energy production ($\epsilon = 0$), and makes the approximation

$$\delta\left(\epsilon - \frac{\partial L_r}{\partial m}\right) = -\frac{L_0}{\Delta m}\left(\frac{\delta L_r}{L_r} - \frac{\delta L_i}{L_0}\right), \tag{13.34}$$

where Δm is the mass of the shell, L_0 is the equilibrium luminosity of the model, and $\delta L_r/L_r$ is the relative luminosity variation at the top of the shell. He adopts the same opacity law as in the Baker one-zone model, so that the relative opacity variations are given in linear theory by eq. (13.2). Only radiative transfer is considered, so that $\delta L_r/L_r$ is given in the linear theory by eq. (13.18). He also assumes a perfect gas equation of state, that is, takes $\chi_\rho = \chi_T = 1$ in eq. (13.3).

He then neglects the spatial derivatives of the remaining pulsation variables, combines the linearized equations, and obtains a cubic equation for the eigenfrequency $\mathscr{s}$ (time dependence of all pulsation variables: $e^{\mathscr{s}t}$). This cubic equation is entirely analogous to that obtained by Baker (1966), except generalized as noted above. Stellingwerf introduces a nonadiabaticity parameter, which he calls ζ, that is related to Baker's nonadiabaticity parameter K_{Baker} (see eq. [13.9]) by

$$K_{\text{Baker}} = \frac{(3\Gamma_1 - 4)^{1/2}}{\pi(\Gamma_3 - 1)}\zeta \approx \zeta. \tag{13.35}$$

The factor of proportionality is normally near unity, so that $\zeta \ll 1$ corresponds to the adiabatic limit; $\zeta \sim 1$ to the partially adiabatic limit; and $\zeta \gg 1$ to the extreme nonadiabatic limit.

On the basis of this cubic equation, Stellingwerf then obtains the necessary conditions for stability by requiring that $\text{Re}(\mathscr{s}) < 0$. In this discussion he assumes that $\delta L_i = 0$. The condition for *dynamical* stability is found to be

$$m\Gamma_1 - 4 > 0, \tag{13.36}$$

which is a not unexpected generalization of Baker's condition, eq. (13.11). The condition for *secular* stability is

$$4 + mn + (m - 4)(s + 4) > 0, \tag{13.37}$$

which reduces to eq. (13.15) for $m = 3$ (recall that Stellingwerf has taken $\chi_\rho = \chi_T = 1$). The condition for *pulsational* stability is

$$4(\Gamma_3 - 1) + [s(\Gamma_3 - 1) - n] - 4/m \equiv \Lambda' > 0, \tag{13.38}$$

which defines the quantity Λ'. Note that this condition reduces to Baker's condition, eq. (13.17), for $m = 3$. Note also that eqs. (13.36) and (13.38) are not affected by the assumption that $\chi_\rho = \chi_T = 1$. Finally, we note that the luminosity variations, in the quasi-adiabatic approximation, are given from eq. (13.18) by

$$\left(\frac{\delta L_r}{L_r}\right)_{\text{ad}} = \Lambda'\left(\frac{\delta\rho}{\rho}\right), \tag{13.39}$$

where the subscript "ad" denotes this approximation.

The physical interpretation of the three main terms on the left side of eq. (13.38) is exactly as in the Baker one-zone model. However, note that, since $m \geq 3$, the "radius effect" may be smaller in magnitude in the Stellingwerf model than in the Baker model.

Stellingwerf (1978, 1979) points out that the relatively small values of m appropriate for δ Scuti-like stars and β Cephei stars render the bump mechanism more effective for these stars than for highly centrally concentrated stars such as the Cepheids, for which m would be relatively large.

It is clear, moreover, that the discussion in §13.1 of the Baker one-zone model in terms of the C integral also applies exactly to the Stellingwerf model. Hence, the appropriate generalization of the condition (13.23) for pulsational stability (if $\delta L_i = 0$) is exactly the same as in this equation, except that the 3 therein is replaced by m.

Perhaps one of the most interesting features of Stellingwerf's paper (Stellingwerf 1972) is his discussion of the phase lag in his model. This discussion can be paraphrased in our notation as follows.

Using the equation (13.18) for the variations in the radiative luminosity, we can express $\delta L_r/L_r$ in terms only of m, n, $\delta\rho/\rho_0$, and $\delta T/T$. On the other hand, $\delta T/T$ is given by the energy equation (5.34b), once one assumes that all pulsation variables possess a time dependence of the form $e^{i\omega t}$. Using the approximation (13.34) in the energy equation, and setting $\omega = 2\pi/\Pi$ (Π = period), we can express $\delta T/T$ in terms only of Γ_3, ζ (Stellingwerf's nonadiabaticity parameter), $\delta L_r/L_r$, and $\delta L_i/L_0$. Eliminating $\delta T/T$ between these last two equations and solving for $\delta L_r/L_r$, we obtain

$$\frac{\delta L_r}{L_r} = \frac{2\pi i}{2\pi i + (s+4)\zeta}\left(\frac{\delta L_r}{L_r}\right)_{\text{ad}} + \frac{(s+4)\zeta}{2\pi i + (s+4)\zeta}\left(\frac{\delta L_i}{L_0}\right). \tag{13.40}$$

This equation relates the relative luminosity variations $\delta L_r/L_r$ of the radiation emerging from the top of the shell, to the quasiadiabatically computed value $(\delta L_r/L_r)_{\text{ad}}$ of that quantity (see eq. [13.39]) and the relative luminosity variations $(\delta L_i/L_0)$ of the radiation incident on the lower boundary of the shell; these latter are also the relative luminosity variations of the radiation emerging from the rigid core.

Consider first the case $\delta L_i = 0$ (constant luminosity incident on the lower boundary of the shell). It is then clear from eq. (13.40) that in the adiabatic limit, $\zeta \to 0$,

$$\frac{\delta L_r}{L_r} \to \left(\frac{\delta L_r}{L_r}\right)_{\text{ad}}. \tag{13.41}$$

If, moreover, the shell is *driving*, then $\Lambda' < 0$ (see above), and $(\delta L_r/L_r)_{\text{ad}}$ will attain maximum when $\delta\rho/\rho$ attains minimum (the radiation is trapped in the shell when it is most compressed; see §9.1 and eq. [13.39]). If by "phase lag" we mean the phase of the maximum of $\delta L_r/L_r$ behind the minimum of $\delta r/r$, then the present situation represents a 180° phase lag.

For an intermediate value of ζ, say $\zeta = 2\pi/(s + 4)$ (~ 1), we have

$$\frac{\delta L_r}{L_r} = \frac{i}{i + 1}\left(\frac{\delta L_r}{L_r}\right)_{\text{ad}}. \tag{13.42}$$

If $\Lambda' < 0$ (the shell is pulsationally unstable), then $(\delta L_r/L_r)_{\text{ad}} < 0$, and the present case corresponds to a 135° phase lag.

Finally, in the extreme nonadiabatic limit, $\zeta \gg 1$, we have

$$\frac{\delta L_r}{L_r} \longrightarrow \frac{2\pi}{(s + 4)\zeta}\, i\left(\frac{\delta L_r}{L_r}\right)_{\text{ad}} \longrightarrow 0. \tag{13.43}$$

If $\Lambda' < 0$, then the present case corresponds to a 90° phase lag but with vanishingly small luminosity variations (this last item is a manifestation of the freezing-in effect discussed in §9.1).

Thus, in the present case of $\delta L_i = 0$ and driving in the shell ($\Lambda' < 0$), the emergent luminosity variations, when represented as a vector in the complex plane, with $\delta\rho/\rho$ (or $-\delta r/r$) pointing along the positive real axis, lie in the third quadrant. The vector $\delta L_r/L_r$ thus lies, in the present case, when $\zeta \ll 1$, almost along the negative real axis, and is nearly of magnitude $|(\delta L_r/L_r)_{\text{ad}}|$. It rotates counterclockwise and shortens as ζ increases, and asymptotically approaches the negative imaginary axis, with vanishing length, as ζ becomes large.

We may note that $\zeta \ll 1$ corresponds, for given mass and equilibrium luminosity, to high effective temperatures; and $\zeta \gg 1$ to low effective temperatures.

Consider now the case $\delta L_i \neq 0$. If we suppose that all the damping occurs interior to the shell, then $\delta L_i/L_0$ should be largest approximately when $\delta\rho/\rho$ is largest. We shall assume, with Stellingwerf, that this is exactly true, so that $\delta L_i/L_0$ lies exactly along the positive real axis in the complex plane discussed above. Then the vector represented by the last term on the right side of eq. (13.40) lies, when $\zeta \ll 1$, nearly along the negative imaginary axis and is of vanishing length ($\{[s + 4]\zeta/[2\pi i]\}[\delta L_i/L_0]$). It rotates counterclockwise and grows in length as ζ increases, and asymptotically approaches the positive real axis, with length $(\delta L_i/L_0)$, as ζ becomes large. This vector therefore always lies in the fourth quadrant (see Fig. 13.3 below).

Considering the vector sum of the two terms on the right side of eq. (13.40), then, we see that, in the adiabatic limit, $\zeta \ll 1$, $\delta L_r/L_r \approx (\delta L_r/L_r)_{\text{ad}}$ and is negative (the luminosity variations on the inner boundary do not affect the values of the emergent luminosity variations). In the extreme nonadiabatic limit, $\zeta \gg 1$, $\delta L_r/_r \approx (\delta L_i/L_0)$ and is positive (the emergent luminosity variations simply follow the interior luminosity variations; the shell possesses so little heat capacity that the gas within it cannot modulate the luminosity flowing through it), and $\delta L_r/L_r$ is given, for intermediate values of $\zeta(\sim 1)$, by the vector sum of the two vectors, one lying in the third quadrant and one in the fourth quadrant. In this last case the resultant vector will clearly lie closer to the negative imaginary axis, i.e., will have a more nearly 90° phase lag, than will either of the above two individual vectors alone, as shown in Figure 13.3. If the relative lengths of these two vectors were suitable, the resultant vector could easily lie quite close to the negative imaginary axis in this case of intermediate values of ζ. In fact, Stellingwerf argues that these lengths of the above vectors will indeed be such that the resultant will correspond very nearly to a 90° phase lag for $\zeta \sim 1$, if the length of the vector $(\delta L_i/L_0)$ is so chosen that the interior damping roughly balances the driving in the shell.

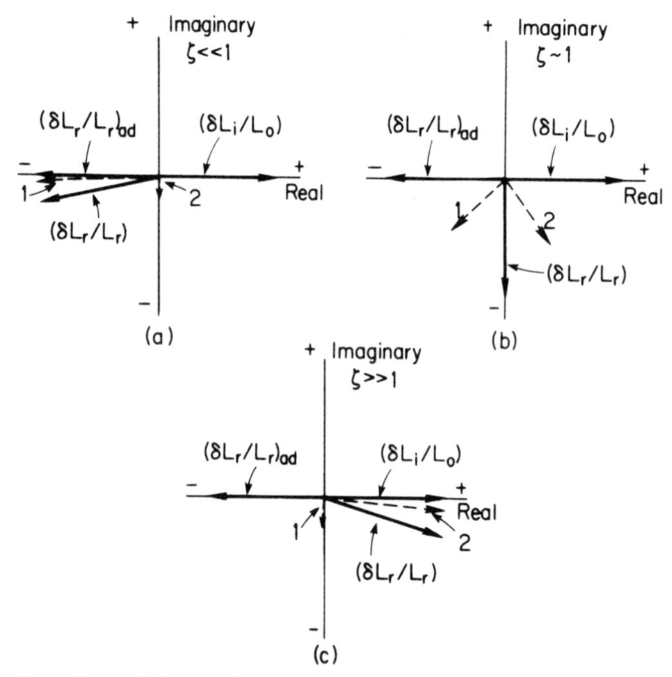

Figure 13.3. The relative luminosity variation $\delta L_r/L_r$ emerging from the Stellingwerf one-zone model. The relative density variation $\delta\rho/\rho$ (or $-\delta r/r$) is assumed to be represented by a vector (not shown) pointing along the positive real axis. The quasi-adiabatic luminosity variation $(\delta L_r/L_r)_{ad}$ and the relative luminosity variation $(\delta L_i/L_0)$ incident on the lower surface of the model, are shown as solid vectors pointing along, respectively, the negative and positive real axes. The situation corresponds to driving in the shell and damping in the interior (see text for further explanation). The symbols 1 and 2 represent, respectively, the first and second terms on the right side of eq. (13.40), denoted by dashed vectors. The solid vector denotes $\delta L_r/L_r$, or the vector sum of the dashed vectors. All of the above vectors are shown schematically for (a) small nonadiabaticity, $\zeta \ll 1$; (b) intermediate nonadiabaticity, $\zeta \sim 1$; and (c) large nonadiabaticity, $\zeta \gg 1$; where ζ is Stellingwerf's nonadiabaticity parameter (see eq. [13.35]).

Thus, in the Stellingwerf model the approximate 90° phase lag in the emergent luminosity variations is caused by the combination of damping in the interior and driving in the shell. These are exactly the conditions demanded in Castor's simplified picture of the phase lag (see Chap. 11).

Stellingwerf (1972) has also carried out nonlinear calculations of his model. These nonlinear calculations confirmed to a surprising degree the expectations obtained on the basis of a linearized treatment (see above) and also revealed other properties which we shall not consider in detail here. However, we may note that there is no saturation of the driving mechanism in the Stellingwerf model—all the damping in this model is supposed to take place in the regions interior to the shell. Therefore, in contrast to the Rudd-Rosenberg model, the Stellingwerf model does not possess a built-in amplitude-limiting mechanism.

Nevertheless, the light curves obtained by Stellingwerf are in some cases qualitatively very similar to actual light curves of many Cepheids and RR Lyrae variables (see, for example, Payne-Gaposchkin and Gaposchkin 1965).

It is not clear that details of actual light curves can in all cases be related (as in the Stellingwerf model) in a straightforward manner to physical properties in a simple, one-zone model. Nevertheless, the model is instructive and enlightening, and provides some possibly useful insights.

13.4. CASTOR MODEL

The main purpose of the Castor model (J. Castor, unpublished notes, 1970; used with his kind permission) was to investigate the problem of amplitude limitation in soft self-excited systems (for the meaning of the term "soft," see §5.1). Detailed numerical calculations (see, e.g., Cox 1974a for references) have shown such amplitude limitation to be a common (if not usual) feature of such calculations, as it is, presumably, in real stars.

Castor first notes from the general, nonlinear energy equation (see, e.g., eq. [4.52]), that for adiabatic motion (right side of eq. [4.52] equal to zero), pulsations of *any* amplitude are possible. He thus reasons that the main physical causes of amplitude limitation are likely to be found in nonlinearities in the *nonadiabatic* part of the motion. This conclusion is consistent with the amplitude-limiting mechanism in the Rudd-Rosenberg model (§13.2).

He then notices that the general, nonlinear, nonadiabatic, third-order partial differential equation for the motion of a spherical surface containing interior mass m—eq. (6.13)—has all explicit nonadiabatic effects contained in the right side. He then proposes, in essence, linearizing only the *left* side of this equation. The right side is to be expanded into a linear term plus one higher-order term, which Castor chooses to be a cubic in small quantities.

In order to get a simple equation, Castor drops the spatial derivatives on the left side of eq. (6.13). He is therefore effectively adopting a one-zone model. The third order (in time) differential equation that he obtains is of the same form as eq. (13.49) below. He studies his equation by use of a "two-time" variable perturbation method such as those described in Cole (1968).

Following Castor, we shall also adopt a one-zone model here, and shall choose a model of the kind studied by Usher and Whitney (1968) and by Rudd and Rosenberg (1970). The equation of motion of the model is identical to eq. (13.24) above.

We then take the time derivative of eq. (13.24) and eliminate dP/dt by use of the energy equation, for example, eq. (4.30a) with dq/dt as given by eq. (4.40). We express $d\rho/dt$ ($\equiv \dot{\rho}$) in terms of dr/dt ($\equiv \dot{r}$) in the above equation by use of the relation

$$\frac{\dot{\rho}}{\rho} = -m(r)\frac{\dot{r}}{r}, \tag{13.44}$$

where

$$m(r) \equiv \frac{3n_0^{\,3}}{n_0^{\,3} - 1}\left(\frac{r}{r_0}\right)^3\left(\frac{\rho}{\rho_0}\right), \tag{13.45}$$

and $n_0 = r_0/r_c$ is the ratio of the equilibrium, or static, radius of the freely distensible outer membrane of the model, to the radius of the rigid core underlying the model (we are here using exactly the same notation as in §13.2, on the Rudd-Rosenberg model). We finally obtain the nonlinear equation

$$\dddot{r} + \frac{\dot{r}\ddot{r}}{r}[m(r)\Gamma_1 - 2] + [m(r)\Gamma_1 - r]\frac{GM}{r^2}\frac{\dot{r}}{r} = \frac{4\pi r^2}{\mathcal{M}}\rho(\Gamma_3 - 1)\left(\epsilon - \frac{\partial L_r}{\partial m}\right), \tag{13.46}$$

where $\mathcal{M}$ is the mass of gas in the zone (cf. 13.2), and m on the right side (and here only) denotes interior mass. Equation (13.46) is the analogue of eq. (6.13) for our one-zone model. In this form we see that all explicit nonadiabatic effects are contained on the right side of the equation.

In accordance with Castor's reasoning (see above), we now linearize only the left side of eq. (13.46). In so doing, we assume that the unperturbed model is in hydrostatic equilibrium ($\ddot{r}_0 = 0$) and that Γ_1 is constant. At the same time, we define a characteristic angular frequency ω_0:

$$\omega_0 \equiv \left[(m_0\Gamma_1 - 4)\frac{GM}{r_0^3}\right]^{1/2}, \tag{13.47}$$

where m_0 is the value of m in the unperturbed model. We assume that $\Gamma_1 \geq 4/m_0$, so that ω_0 is purely real. We define, finally, a new time scale by the relation

$$t' \equiv \omega_0 t. \tag{13.48}$$

In terms of the new time variable we are therefore measuring time in units of $\Pi/2\pi$ ($\Pi = 2\pi/\omega_0$ = period); in other words, we are considering unit angular frequency.

Again following Castor's reasoning, we now expand the right side of eq. (13.46) into a linear term in $\xi \equiv \delta r/r$ and a cubic term in ξ. Henceforth dropping primes from the new time variable, we may finally write eq. (13.46) in the form

$$\frac{d^3\xi}{dt^3} + \frac{d\xi}{dt} + \epsilon(\xi - Q\xi^3) = 0, \tag{13.49}$$

where ϵ, a pure number (not to be confused with the rate per unit mass of thermonuclear energy production), is assumed to be small, and Q is a parameter (set equal to 1 in Castor's treatment) whose value governs the importance of the cubic term. Equation (13.49) is of the same form as the one studied by Castor (see above). We shall now examine the solutions, limit cycles, and so on, of eq. (13.49).

We consider first the *linear* regime, characterized by $Q = 0$ and $|\xi| \ll 1$. Assuming that $\xi \propto e^{i\omega t}$, we obtain from eq. (13.49) a cubic equation in ω whose solution is

$$\omega = \pm 1 - \tfrac{1}{2} i\epsilon + O(\epsilon^2). \tag{13.50}$$

The solution of eq. (13.49) in this case shows that the zone is self-excited (overstable) if $\epsilon > 0$ (as we already know is the case; see Chap. 9).

The important thing to note here is that we have oscillations on one time scale, i.e., $t \sim 1$; and *growth* of oscillations on another time scale, i.e., $\epsilon t \sim 1$. Thus, following Castor (see above), we explicitly introduce *two* time variables, as in Cole (1968), and treat them as two independent variables. One of these will be the ordinary pulsation time, t. The other will measure the growth of amplitude, and will be denoted by $\tilde{t}$ and defined by the relation

$$\tilde{t} \equiv \epsilon t. \tag{13.51}$$

We then assume that ξ can be represented by the following series:

$$\xi = F_0(t,\tilde{t}) + \epsilon F_1(t,\tilde{t}) + O(\epsilon^2), \tag{13.52}$$

where all the $F_i(t,\tilde{t})$ are assumed to be *periodic* in t. Using eq. (13.52), we may express ξ, $d\xi/dt$, etc., in terms of F_0, $\partial F_0/\partial t$, $\partial F_0/\partial\tilde{t}$, ... and F_1, $\partial F_1/\partial t$, $\partial F_1/\partial\tilde{t}$, ..., etc. The differential equation (13.49) may then be written in the form

$$\frac{\partial^3 F_0}{\partial t^3} + \frac{\partial F_0}{\partial t} + \epsilon\left(3\frac{\partial^3 F_0}{\partial t^2 \partial\tilde{t}} + \frac{\partial F_0}{\partial\tilde{t}} + F_0 - QF_0^{\,3} + \frac{\partial^3 F_1}{\partial t^3} + \frac{\partial F_1}{\partial t}\right) + O(\epsilon^2) = 0. \tag{13.53}$$

Equating, now, successively, the coefficients of ϵ^0, ϵ^1, and so on, to zero yields the following set of equations:

$$\frac{\partial^3 F_0}{\partial t^3} + \frac{\partial F_0}{\partial t} = 0, \tag{13.54}$$

$$\frac{\partial^3 F_1}{\partial t^3} + \frac{\partial F_1}{\partial t} = -3\frac{\partial^3 F_0}{\partial t^2 \partial\tilde{t}} - \frac{\partial F_0}{\partial\tilde{t}} - F_0 + QF_0^{\,3}, \tag{13.55}$$

and so on.

We note that eq. (13.54) just describes *linear adiabatic* oscillations. Its general solution is

$$F_0 = A(\tilde{t})\cos t + B(\tilde{t})\sin t + C(\tilde{t}), \tag{13.56}$$

where A, B, and C may, in general, be functions of $\tilde{t}$ alone. We may also write eq. (13.56) as

$$F_0 = R^{1/2}(\tilde{t})\cos\,[t + \phi(\tilde{t})] + C(\tilde{t}), \tag{13.57}$$

where

$$R \equiv A^2 + B^2 \tag{13.58}$$

and ϕ is a phase angle, without significance, whose value depends on the values of A and B. The idea of the method is then to use eq. (13.56) to evaluate the right side of eq. (13.55), etc.

Thus, F_0, $F_0^{\,3}$, and the various partial derivatives of F_0 can be calculated, and they will involve products of $\cos^0 t$, $\cos^1 t$, $\cos^2 t$, or $\cos^3 t$ and $\sin^0 t$, $\sin^1 t$, $\sin^2 t$, or $\sin^3 t$. We express these powers and products in terms of trigonometric functions of multiples of t by means of trigonometric identities (given by Cole 1968, p. 92). Using these expressions in the differential equation (13.55), we obtain

$$\frac{\partial^3 F_1}{\partial t^3} + \frac{\partial F_1}{\partial t} = 2\frac{dA}{d\tilde{t}}\cos t + 2\frac{dB}{d\tilde{t}}\sin t - \frac{dC}{d\tilde{t}}$$

$$- A\cos t - B\sin t - C + Q(\tfrac{3}{4}A^3 + \tfrac{3}{4}AB^2$$

$$+ 3AC^2)\cos t + Q(\tfrac{3}{4}B^3 + \tfrac{3}{4}A^2B + 3BC^2)\sin t$$

$$+ Q(\tfrac{3}{2}A^2C + \tfrac{3}{2}B^2C + C^3) + (\tfrac{1}{4}A^3$$

$$- \tfrac{3}{4}AB^2)\cos 3t + Q(\tfrac{3}{4}A^2B - \tfrac{3}{4}B^3)\sin 3t$$

$$+ Q(\tfrac{3}{2}A^2C - \tfrac{3}{2}B^2C)\cos 2t + Q3ABC\sin 2t. \qquad (13.59)$$

Now, F_1 would not be periodic in t, as assumed, if the coefficients of the $\cos t$, $\sin t$, and 1 terms in eq. (13.59) did not all vanish. On the other hand, the coefficients of the $\cos 3t$, $\sin 3t$, $\cos 2t$, and $\sin 2t$ terms on the right side of eq. (13.59) do not need to vanish.

Accordingly, setting the sum of the coefficients of the $\cos t$, $\sin t$, and 1 terms on the right side of eq. (13.59) equal to zero, yields three conditions for the periodicity of F_1 with respect to t. These three conditions are three first-order differential equations in $\tilde{t}$ for the three quantities $A(\tilde{t})$, $B(\tilde{t})$ and $C(\tilde{t})$ (see eq. [13.56]). Two of these differential equations can easily be combined into one, in terms of the quantity $R(\tilde{t})$, defined by eq. (13.58). The resulting two differential equations are

$$\frac{dR}{d\tilde{t}} - R + Q(\tfrac{3}{4}R^2 + 3RC^2) = 0, \qquad (13.60)$$

$$\frac{dC}{d\tilde{t}} + C - Q(\tfrac{3}{2}RC + C^3) = 0, \qquad (13.61)$$

These equations describe, to first-order accuracy in ϵ, the growth of the oscillations.

The oscillator will have a limit cycle if $dR/d\tilde{t} = 0$, $dC/d\tilde{t} = 0$, for then the motion will be strictly periodic, and the growth will have become limited. There are four solutions of eqs. (13.60) and (13.61) with $dR/d\tilde{t} = 0$, $dC/d\tilde{t} = 0$, corresponding to (in principle) four limit cycles:

I: $R = 0, C = 0;$

II: $C = 0, R = 4/(3Q);$

III: $R = 0, C = \pm 1/Q^{1/2};$

IV: $R = 8/(15Q), C = \pm 1/(5Q)^{1/2}.$

We note that limit cycle I merely represents the static case ($\xi = 0$). This case is clearly a limit cycle, though a degenerate one. Limit cycle III just represents a constant displacement, as may also be seen directly from the original differential equation (13.49).

But are all four of these limit cycles stable? If any one of them is unstable, then that particular limit cycle would be realized only transiently, if at all, in nature. To determine the stability of the limit cycles, we must go back to the differential equations (13.60) and (13.61) for the time rates of change of the amplitude parameters R and C, and linearize these equations about the respective limit-cycle values of R and C, which we henceforth denote with zero subscripts. We write the linearized forms of these equations as a single matrix equation:

$$\frac{dX}{d\tilde{t}} = \mathcal{N}X, \tag{13.62}$$

where

$$X \equiv \begin{pmatrix} \delta R \\ \delta C \end{pmatrix} \tag{13.63}$$

and

$$\mathcal{N} \equiv \begin{pmatrix} 1 - \tfrac{3}{2}QR_0 - 3QC_0^2 & -6QR_0C_0 \\ \tfrac{3}{2}QC_0 & -1 + \tfrac{3}{2}QR_0 + 3QC_0^2 \end{pmatrix}. \tag{13.64}$$

Here δR, for example, is the difference between the actual value of R and the limit-cycle value R_0.

The question of the stability of the limit cycles is then answered by determining the eigenvalues of the matrix $\mathcal{N}$. Thus, if V_i denotes an eigenvector of $\mathcal{N}$ corresponding to the eigenvalue λ_i, we have

$$\mathcal{N}V_i = \lambda_i V_i. \tag{13.65}$$

The solution of eq. (13.62) at any time $\tilde{t}$ is

$$X(\tilde{t}) = \sum_i \alpha_i V_i e^{\lambda_i \tilde{t}}, \tag{13.66}$$

where the α_i are constant coefficients (we are here ignoring the possibility that several of the λ_i might be equal to one another, in which case the relevant $\alpha_i V_i$ would have to be replaced by polynomials in $\tilde{t}$). The eigenvalues of $\mathcal{N}$ are obtained by solving the characteristic equation:

$$|\mathcal{N} - \lambda\mathcal{I}| = 0, \tag{13.67}$$

where vertical bars denote the determinant of the indicated matrix, λ is a number, and $\mathcal{I}$ denotes the unit matrix. If we apply eq. (13.67) to the matrix $\mathcal{N}$ as defined in eq. (13.64), the former equation becomes a simple quadratic in the eigenvalue λ. The solution of this quadratic is

$$\lambda = \pm (1 - 3QR_0 + \tfrac{9}{4}Q^2R_0^2 - 6QC_0^2 + 9Q^2C_0^4)^{1/2}. \quad (13.68)$$

The roots of eq. (13.67) are therefore either *purely real* or *purely imaginary*. It should be noted that the existence of a positive real value of λ immediately signifies instability,

It can be shown that only the limit cycle IV is not unstable, and is presumably the one which could be realized in nature (if $\epsilon > 0$). To first order in ϵ, this limit cycle is

$$\xi = \left(\frac{8}{15Q}\right)^{1/2} \cos(t + \phi) \pm \left(\frac{1}{5Q}\right)^{1/2} + \epsilon F_1(t,\tilde{t}), \quad (13.69)$$

where the value of F_1 is determined, as above, from the next higher order terms in ϵ.

We note that ϵ itself does not appear at all in the *zero*-order limit cycle; ϵ only determines the time ($\sim\epsilon^{-1}$ periods) required to reach this limit cycle. We also note that, as is to be expected, the limit cycle *does* depend on Q (the factor whose value determines the importance of the third-order terms in ξ), the amplitude of the limit cycle increasing as Q decreases. Finally, we note that, to zero order in ϵ, there is no *period* change. This result just obtains because, as we remarked above, the zero-order solution is just the solution of the linear adiabatic wave equation, only at a certain amplitude.

Limit cycles of other oscillators have been discussed by Melvin in a number of unpublished manuscripts (see, e.g., Melvin 1975).

In the real stellar case, of course, the situation is vastly more complicated. In fact, it is not clear that the above one-zone model (or, for that matter, *any* one-zone model) is even very relevant for the actual stellar case. The only point of contact with reality is probably that actual models of real pulsating stars often seem to possess limit cycle types of behavior. The stability of these types of behavior has been investigated by Stellingwerf (1974a,b).

III

NONRADIAL OSCILLATIONS OF STARS

In this part we shall be concerned with more general types of stellar oscillations than the purely radial ones considered in Part II. The displacement of a typical mass element from its unperturbed position may now be in any direction at all, and need not be parallel to a radius vector from the stellar center. Consequently, this displacement must now be described by a vector in three dimensions. Three numbers are accordingly required for the complete specification of this displacement, instead of just one as in the simpler case of purely radial oscillations. As a result, the theory of nonradial oscillations is considerably more intricate than the corresponding theory of purely radial oscillations. Largely for this reason, the former theory is much less highly developed than the latter. However, because of the generality of the theory of nonradial oscillations, it forms potentially a far closer approximation to actual stellar oscillations, and is thus much more powerful than the simpler and more restricted theory of purely radial oscillations. For example, effects can be treated in the present theory which may give rise to a preferential direction in the star, such as stellar rotation and stellar magnetic fields.

In the present part, however, we shall be concerned for the most part with stars in which such effects are absent. (Some of the effects of rotation, magnetic fields, and so on, will be considered in Chapter 19.) Thus, we consider in the present part mainly stars which in their unperturbed (nonoscillating) states are spherically symmetric, static and in hydrostatic equilibrium, and do not possess large-scale magnetic fields. We shall also neglect in this part molecular and radiative viscosity and turbulence, and shall assume the applicability of nonrelativistic mechanics and Newtonian gravitation theory.

In Chapters 14 through 16 we shall be concerned with rather general features of nonradial stellar oscillations. In Chapter 14 we shall derive a third order (in time) partial, nonlinear vector differential equation that describes nonlinear, nonradial, nonadiabatic motions. In the remaining chapters of this part, except in portions of Chapter 19, we shall consider only oscillations which are so small that a linear theory (see Chapter 5) is applicable.

When the oscillations are assumed to be adiabatic as well, the equations can be combined into a linear adiabatic wave equation (LAWE), which is a generalization to nonradial oscillations of the LAWE for radial oscillations (Chapter 8). It will be shown in Chapter 15 that this equation can be written in operator form, and it will also be shown there that the operator which appears in this equation is *Hermitian,* just as in the case of radial oscillations. The LAWE, which is and has been in the past the basis for most discussions of nonradial oscillations in stars, will be discussed in considerable detail in Chapter 17. As this chapter was written mostly in 1976–1977, it does not reflect the sophistication contained, for example, in the papers of Christensen-Dalsgaard (1979), Shibahashi (1979), and Wolff (1979), or in the monograph by Unno, Osaki, Shibahashi, and Ando (1979).

In Chapter 16 we shall show that the linearized energy equation (see Chapter 9) applies to nonradial, as well as to radial, stellar oscillations. We shall also show in this chapter that integral expressions, such as those derived in Chapter 9 for radial oscillations, also apply to nonradial oscillations. In this chapter it is necessary, of course, to restore the nonadiabatic terms which are dropped in Chapters 15 and 17. The considerations in Chapter 16 provide the justification for commonly used methods of estimating the vibrational stability of nonradial stellar oscillations.

In Chapter 18 we shall consider the problem of nonradial, *nonadiabatic* stellar pulsations in linear theory. Finally, in Chapter 19 we shall examine, rather briefly, certain miscellaneous topics in the theory of stellar pulsation, both radial and nonradial.

Recently, an excellent treatise on nonradial stellar oscillations has been published by Unno, Osaki, Ando, and Shibahashi (1979).

14

General Nonlinear Partial Differential Equation

In this chapter we shall combine the mass, momentum, and energy equations into a single third-order (in time) partial, nonlinear vector differential equation that describes general, nonlinear, nonadiabatic motions in an inviscid fluid medium. Although this equation is not very useful for studying stellar oscillations, its very existence is interesting. This equation is the nonradial analogue of eq. (6.13) for purely radial motions.

We start with the momentum equation in Eulerian form (see eq. [4.17]), in which we have assumed that the pressure tensor reduces to a pure hydrostatic, thermodynamic pressure (see §4.2c). We now take the Stokes derivative of the equation of motion (4.17) (after dividing through by the density ρ), to obtain

$$\frac{d^2\mathbf{v}}{dt^2} = \frac{1}{\rho^2}\left(\frac{d\rho}{dt}\right)\nabla P - \frac{1}{\rho}\frac{d}{dt}(\nabla P) + \frac{d\mathbf{f}}{dt}. \tag{14.1}$$

However, we may use the operator relation (4.1), to write

$$\frac{d}{dt}(\nabla P) = \frac{\partial}{\partial t}\nabla P + \mathbf{v}\cdot\nabla(\nabla P) = \nabla\frac{\partial P}{\partial t} + \mathbf{v}\cdot\nabla(\nabla P). \tag{14.2}$$

The last term of this equation may be eliminated by making use of the identity

$$\nabla(\mathbf{v}\cdot\nabla P) = \mathbf{v}\cdot\nabla(\nabla P) + (\nabla P)\cdot\nabla\mathbf{v} + (\nabla P)\times(\nabla\times\mathbf{v}). \tag{14.3}$$

One of the terms in this identity can be combined with the first term in the last equality in eq. (14.2), so that this equation, again with use of the operator relation (4.1), now involves dP/dt rather than $\partial P/\partial t$. Next, the quantity dP/dt can be eliminated by making use of the energy equation in the form (4.30a) (see also eq. [4.40]). We therefore obtain

$$\frac{d^2\mathbf{v}}{dt^2} = \frac{1}{\rho^2}\left(\frac{d\rho}{dt}\right)\nabla P - \frac{1}{\rho}\nabla\left(\frac{\Gamma_1 P}{\rho}\frac{d\rho}{dt}\right) - \frac{1}{\rho}\nabla\left[\rho(\Gamma_3 - 1)\left(\epsilon - \frac{1}{\rho}\nabla\cdot\mathbf{F}\right)\right]$$
$$+ \frac{1}{\rho}(\nabla P)\cdot\nabla\mathbf{v} + \frac{1}{\rho}(\nabla P)\times(\nabla\times\mathbf{v}) + \frac{d\mathbf{f}}{dt}. \tag{14.4}$$

The quantities $\rho^{-1}\nabla P$ and $\rho^{-1}d\rho/dt$ can be eliminated by use of, respectively, the momentum equation (4.17) and the mass equation (4.5). We may therefore finally write eq. (14.4) in the form

$$\frac{d^2\mathbf{v}}{dt^2} - \left(\frac{d\mathbf{v}}{dt} - \mathbf{f}\right) \cdot [(\nabla \cdot \mathbf{v})\mathbf{I} - \nabla\mathbf{v}]$$
$$+ \left(\frac{d\mathbf{v}}{dt} - \mathbf{f}\right) \times (\nabla \times \mathbf{v}) - \frac{1}{\rho}\nabla[\Gamma_1 P \nabla \cdot \mathbf{v}] - \frac{d\mathbf{f}}{dt}$$
$$= -\frac{1}{\rho}\nabla\left[\rho(\Gamma_3 - 1)\left(\epsilon - \frac{1}{\rho}\nabla \cdot \mathbf{F}\right)\right], \tag{14.5}$$

where $\mathbf{I}$ denotes the unit tensor. This is the desired generalized (nonradial) version of the earlier eq. (6.13) for purely radial motion. Note that only $\mathbf{v}$ and $\mathbf{f}$ (and their time derivatives) appear on the left side, and only nonadiabatic terms appear on the right side.

It can be shown that, if $\mathbf{f}$ is the force per unit mass due to self-gravitation and for purely radial motion, eq. (14.5) reduces, as it should, to eq. (6.13).

15

Hermiticity of the Linear Adiabatic Wave Equation

In this chapter we shall be concerned with nonradial oscillations which are so small that a *linear* theory (Chapter 5) is applicable. We also restrict ourselves in this chapter to oscillations that are *adiabatic* as well. We shall see that for these assumptions the linearized equations can be combined into a single *linear adiabatic wave equation* (LAWE). This equation is the nonradial analogue of the LAWE for purely radial oscillations discussed at considerable length in Chapter 8. This nonradial LAWE possesses a number of interesting and useful properies which will be discussed in this chapter.

The nonradial LAWE will be obtained in §15.1, where certain other useful results will also be derived. Some interesting and useful properties of this equation will be obtained and discussed in §15.2 for an important special case. This case embraces almost all that is of interest in the theory of nonradial stellar oscillations. A still more general case will be considered briefly in §15.3. Some of the conclusions reached in this section will find applications in the remainder of this part.

15.1. INTRODUCTION

We first recall the linearized forms of the mass, momentum, and energy equations, where the last equation will be specifically for the case of adiabatic oscillations. The mass equation will be taken as eq. (5.29b), and the momentum equation is eq. (5.32). When the only body force is due to self-gravitation (as we shall assume throughout this chapter is the case unless we state otherwise), we have for the net body force per unit mass $\mathbf{f} = -\nabla\psi$, where ψ is the gravitational potential (see §4.5). Finally, the linearized energy equation, for adiabatic oscillations, is eq. (5.36a).

What is the expression for the Eulerian variation ψ'? First, we recall that the gravitational potential ψ itself is a solution of Poisson's equation, eq. (4.42). The integral solution of this equation is given in eq. (4.43). If, now, each mass element is displaced to a new position $\mathbf{x} + \delta\mathbf{r}$, then the contribution to the potential at $(\mathbf{r},t)$ due to dm' will be $-Gdm'/|\mathbf{x} + \delta\mathbf{r} - \mathbf{r}|$, and the new potential at $(\mathbf{r},t)$ will be as given by eq. (4.43), except that $\mathbf{x}$ in the denominator will now be replaced by $\mathbf{x} + \delta\mathbf{r}$.

Remembering the meaning of the Eulerian variation and that $\delta\mathbf{r}$ is infinitesimally small, we have

$$\psi'(\mathbf{r},t) = -G\int_V \left\{\delta\mathbf{r}(\mathbf{x},t)\cdot\nabla_{\mathbf{x}}\left(\frac{1}{|\mathbf{x}-\mathbf{r}|}\right)\right\}\rho_0(\mathbf{x},t)\,d\tau', \qquad (15.1)$$

where $\nabla_{\mathbf{x}}$ means that the spatial derivatives in the gradient operator are to be taken with respect to the coordinates of the point $\mathbf{x}$. Integrating eq. (15.1) by parts and replacing one of the resulting volume integrals by a surface integral by use of the divergence theorem, we obtain

$$\psi'(\mathbf{r},t) = -G\oint_S \frac{\rho_0(\mathbf{x},t)[\delta\mathbf{r}(\mathbf{x},t)]\cdot d\mathbf{S}'}{|\mathbf{x}-\mathbf{r}|} + G\int_V \frac{\nabla_{\mathbf{x}}\cdot[\rho_0(\mathbf{x},t)\delta\mathbf{r}(\mathbf{x},t)]}{|\mathbf{x}-\mathbf{r}|}\,d\tau', \qquad (15.2)$$

where the first integral is to be carried out over the entire surface S of the configuration. We must be certain, to be consistent with our original integral solution, eq. (4.43), of Poisson's equation, that $\rho_0 = 0$ everywhere *exterior* to the surface S.

The important relation (15.2) can also be obtained as follows. Taking the Eulerian variation of Poisson's equation, we have

$$\Delta^2\psi' = 4\pi G\rho'. \qquad (15.3)$$

The formal solution of this equation is

$$\psi'(\mathbf{r},t) = -G\int \frac{\rho'(\mathbf{x},t)d\tau'}{|\mathbf{x}-\mathbf{r}|}, \qquad (15.4)$$

where $d\tau'$ is an element of volume at $\mathbf{x}$ and the integration is to be extended over all regions of space at which ρ' exists. However, in the more general case where the density does not vanish at the surface of the configuration, and yet where a definite surface exists exterior to which there is no matter (as in the case of a liquid globe, or of the homogeneous stellar model, for example), careful attention must be given to eq. (15.4).

In this case the density must have a discontinuity, jumping from some finite value immediately interior to the (imaginary) bounding surface, say S, to zero just outside S. Hence the Eulerian density variation ρ', while it may be small (say of first order in smallness) everywhere sufficiently *interior* to S, must be relatively large, say of *zero* order, in the immediate vicinity of S. This "immediate vicinity" of S must, in fact, comprise a thin shell-like volume (similar to the eggshell around an egg) containing S and wrapped around the volume, V, within which the density is greater than zero. The thickness of the shell-like region is determined at each instant of time by the extent to which S departs from its unperturbed shape.

Clearly, ρ' at any point $\mathbf{x}$ within this shell-like region (but not interior to it!) will have a maximum magnitude nearly equal to the value, ρ_s, of the unperturbed fluid at its surface.

The integral in eq. (15.4) must therefore consist of two terms: one, an integral over the entire volume V where $\rho_0(\mathbf{x},t) \neq 0$, and an integral over the shell-like volume discussed above containing the surface S. In a linear theory, $\rho'(\mathbf{x},t)$ is of first order in smallness everywhere within V, and so the linearized mass equation (5.29a) can be used therein; moreover, V may be taken as the volume of the *unperturbed* system. This first integral then yields the volume integral in eq. (15.2). In the second integral we may use, to first-order accuracy, the expression

$$\rho'(\mathbf{x},t)d\tau' = \rho_0(\mathbf{x}_0,t)\delta\mathbf{r} \cdot d\mathbf{S}'. \tag{15.5}$$

Here $\mathbf{x}_0$ denotes the unperturbed position of a fluid particle that lies on the surface of the configuration, and $\delta\mathbf{r} = \mathbf{x} - \mathbf{x}_0$. In this expression it is to be understood that $\mathbf{x}_0$ actually denotes a point immediately *interior* to the surface S, so that ρ_0 will have a definite, finite value at $\mathbf{x}_0$. Also, $\delta\mathbf{r} \cdot d\mathbf{S}'$ is the volume of a "pillbox" about $\mathbf{x}$, with $d\mathbf{S}'$ a surface element directed toward the *outward normal* of the shell-like region discussed above. Moreover, S may to first-order accuracy be taken as the surface of the unperturbed configuration, and the second integral above becomes a surface integral over the (closed) surface S. This surface integral is just the surface integral in eq. (15.2).

In many cases—in fact, in practically all cases considered in the present Part III, and specifically in §15.2—the density ρ_0 will vanish at the surface of the configuration. In this event the surface integral in eq. (15.2) vanishes, and we have the simpler result

$$\psi'(\mathbf{r},t) = G\int_V \frac{\nabla_{\mathbf{x}} \cdot [\rho_0(\mathbf{x},t)\delta\mathbf{r}(\mathbf{x},t)]d\tau'}{|\mathbf{x} - \mathbf{r}|}, \tag{15.6}$$

which may also be obtained simply by taking the Eulerian variation of the solution (4.43) of Poisson's equation and using in it the linearized mass equation (5.29a).

Another consequence of taking ρ_0 to be zero at the surface of the configuration is the following. We shall essentially always assume that the pressure vanishes at the surface of the configuration. Hence $P_0 = 0$ and

$$\delta P = 0. \tag{15.7}$$

We now use the relation (5.16) between Lagrangian and Eulerian variations and express ∇P_0 in terms of $d\mathbf{v}_0/dt$ and $\nabla\psi_0$ by use of the unperturbed momentum equation. We obtain

$$P' = \rho_0 \delta\mathbf{r} \cdot \left(\frac{d\mathbf{v}_0}{dt} + \nabla\psi_0\right) = 0 \tag{15.8}$$

at the surface of the configuration if ρ_0 vanishes here and if the quantity in parentheses is everywhere finite. Hence, the vanishing of both P and ρ at the surface implies that both δP and P' are also zero here. This implication will be made use of in §15.2 below.

We see, then, that for small adiabatic oscillations the right side of the momentum equation (say eq. [5.32]) can be expressed in the present case (all body forces due to self-gravitation) entirely in terms of the field of displacements $\delta\mathbf{r}$ of mass elements. This equation, with the right side so expressed, is then the linear adiabatic wave equation (LAWE) for nonradial oscillations, and will form the basis for most of the discussion in this part. The linearized momentum equation is

$$\frac{d^2\delta\mathbf{r}}{dt^2} = \left(-\frac{1}{\rho}\nabla P - \nabla\psi\right)' + \delta\mathbf{r} \cdot \nabla\left(-\frac{1}{\rho_0}\nabla P_0 - \nabla\psi_0\right). \tag{15.8'}$$

15.2. PROPERTIES OF THE LAWE FOR AN IMPORTANT SPECIAL CASE

The special case to be considered in this section (and, in fact, throughout most of this part) is characterized by the following assumptions, in addition to those mentioned at the beginning of this part:

(a) The pressure P and density ρ both vanish at the surface of the configuration.

(b) The unperturbed (nonoscillating) state of the system is *static* ($\mathbf{v}_0 \equiv 0$) and in *hydrostatic equilibrium* ($d\mathbf{v}_0/dt \equiv 0$). Hence, the system is, for example, not rotating.

(c) The unperturbed system is *spherically symmetric*. (This assumption is actually a corollary of assumption *b*, if the only body force is due to self-gravitation.)

In this case, because of assumption *b*, the Lagrangian variation of the right side of the linearized momentum equation is identical to the Eulerian variation. Thus, the right side can be written as a linear operator $\mathcal{L}$, operating on $\delta\mathbf{r}$. We have (see eq. [15.8'])

$$\frac{d^2\delta\mathbf{r}}{dt^2} = -\mathcal{L}(\delta\mathbf{r}), \tag{15.9}$$

where the specific form for $\mathcal{L}(\delta\mathbf{r})$ is, if ρ_0 vanishes at the surface of the configuration (assumption *a* above),

$$\mathcal{L}(\boldsymbol{\zeta}) = \frac{1}{\rho^2}(\nabla P)\nabla \cdot (\rho\boldsymbol{\zeta}) - \frac{1}{\rho}\nabla(\boldsymbol{\zeta} \cdot \nabla P) - \frac{1}{\rho}\nabla(\Gamma_1 P \nabla \cdot \boldsymbol{\zeta}) + \nabla\left\{G\int_V \frac{\nabla_{\mathbf{x}} \cdot [\rho(\mathbf{x},t)\boldsymbol{\zeta}(\mathbf{x},t)]\,d\tau'}{|\mathbf{x} - \mathbf{r}|}\right\}. \quad (15.10)$$

Here we have dropped zero subscripts from equilibrium (or unperturbed) quantities, and have written $\boldsymbol{\zeta} \equiv \delta\mathbf{r}$ for short. The linearity of $\mathcal{L}$, for example, is now evident.

The operator $\mathcal{L}$ has the important property that it is *Hermitian* with respect to the element of mass $\rho d\tau$, as was first shown by Chandrasekhar (1963, 1964d). This property may be expressed by the relations

$$\int_V \boldsymbol{\eta}^* \cdot \mathcal{L}(\boldsymbol{\zeta})\rho d\tau = \left[\int_V \boldsymbol{\zeta}^* \cdot \mathcal{L}(\boldsymbol{\eta})\rho d\tau\right]^* = \int_V \boldsymbol{\zeta} \cdot [\mathcal{L}(\boldsymbol{\eta})]^*\rho d\tau, \quad (15.11)$$

where $\boldsymbol{\zeta}$ and $\boldsymbol{\eta}$ are any two (mathematically well-behaved) vectors and the asterisk denotes the complex conjugate.

The Hermiticity of $\mathcal{L}$ can be proved by forming the quantity in the integrand of the first integral in eq. (15.11) and using the form (15.10) for $\mathcal{L}(\boldsymbol{\zeta})$, in which we recognize the quantities ρ', P', and ψ'. We then perform parts integrations on the integrals involving P' and ψ' and make use of the divergence theorem. We note that the two resulting surface integrals are zero since, in accordance with our assumptions, both P' and ρ vanish on the surface of the configuration (see above). Replacing ρ', P', and ψ' by expressions derived earlier, the integral on the left side of eq. (15.11) can be written as the sum of five volume integrals, one of which is a double integral. Recognizing the ∇P and $\nabla\rho$ are both directed radially (assumption *c* above), it is evident $\boldsymbol{\eta}^*$ and $\boldsymbol{\zeta}$ may be freely interchanged in this last equation. Since $\mathcal{L}$ is a real operator, $[\mathcal{L}(\boldsymbol{\eta})]^* = \mathcal{L}(\boldsymbol{\eta}^*)$, and so the Hermiticity of $\mathcal{L}$ as expressed in eq. (15.11) immediately follows. We shall see in §15.3 that the right side of the linearized momentum equation can be written as a Hermitian operator for adiabatic oscillations, even when the above three assumptions are dropped.

The Hermiticity of $\mathcal{L}$ has a number of important consequences, which we shall now discuss. First, we note that, if $\mathbf{v}_0 = 0$ as assumed in this section, $d^2/dt^2 = \partial^2/\partial t^2$, and so the LAWE may be written as

$$\frac{\partial^2 \delta\mathbf{r}}{\partial t^2} = -\mathcal{L}(\delta\mathbf{r}). \quad (15.12)$$

Hence, if

$$\delta\mathbf{r}(\mathbf{r},t) = \mathbf{u}(\mathbf{r})e^{i\sigma t}, \tag{15.13}$$

then eq. (15.12) becomes the eigenvalue equation,

$$\mathcal{L}(\mathbf{u}) = \sigma^2\mathbf{u}. \tag{15.14}$$

Here σ^2 and $\mathbf{u}(\mathbf{r})$ are, respectively, the eigenvalue (square of the angular oscillation frequency) and eigenfunction, or eigenvector, of $\mathcal{L}$. The consequences of the Hermiticity of $\mathcal{L}$ are the following.

(i) *The eigenvalues are all real.* The reality of σ^2 follows by using eq. (15.14) and its complex conjugate, as applied to the kth eigenvalue, say σ_k^2, and eigenfunction, say $\mathbf{u}_k$, in the Hermiticity condition, eq. (15.11). In this way we see that σ_k^2 must be equal to its complex conjugate, and this condition requires that σ_k^2 be real. Hence, just as in the purely radial case, the motion in the nonradial case must be either oscillatory and of constant (in time) amplitude, or exponentially growing in time (dynamically unstable): complex frequencies do not exist for adiabatic oscillations for vanishing surface pressure.

It should be pointed out that this conclusion regarding the reality of the eigenvalues depends explicitly on our assumptions that the system in its unperturbed state is static ($\mathbf{v}_0 = 0$) and in hydrostatic equilibrium ($d\mathbf{v}_0/dt = 0$), and also that the pressure vanishes at the surface. The conclusion might not follow if motions or accelerations were present in the unperturbed system, or if the pressure did not vanish at the surface. For example, as is well known (see below and §19.1), the eigenvalues are not necessarily all real if the unperturbed system is rotating.

(ii) *The eigenfunctions are orthogonal to one another with respect to the element of mass $\rho d\tau$.* Consider two modes, say k and l, of nonradial oscillations, with eigenfunctions $\mathbf{u}_k$ and $\mathbf{u}_l$ and eigenvalues σ_k^2 and σ_l^2. Then we have $\mathcal{L}(\mathbf{u}_l) = \sigma_l^2\mathbf{u}_l$ and $(\mathcal{L}[\mathbf{u}_k])^* = \sigma_k^2\mathbf{u}_k^*$, making use of the reality of σ_k^2. Using these relations in the Hermiticity condition (15.11), we see that, if $\sigma_k^2 \neq \sigma_l^2$,

$$\int_V \mathbf{u}_k^* \cdot \mathbf{u}_l\,\rho d\tau = 0\ (k \neq l), \tag{15.15}$$

or the eigenfunctions $\mathbf{u}_k$ and $\mathbf{u}_l$ are orthogonal to one another with respect to $\rho d\tau$. In the case of *degeneracy* ($\sigma_k^2 = \sigma_l^2$), it is well known that an orthogonal set of functions can always be constructed (see, e.g., Schiff 1955, §10). Hence, even in this case the above assertion is true.

(iii) *The eigenvalues obey a variational principle.* This fact was first shown by Chandrasekhar (1964d). The present proof (different from Chandrasekhar's) is along the lines given by Lynden-Bell and Ostriker

(1967) and proceeds very much as in the purely radial case discussed in detail in §8.10. We define the real number Σ^2 by the relation

$$\Sigma^2 \int_V \mathbf{u}^* \cdot \mathbf{u} \rho d\tau \equiv \int_V \mathbf{u}^* \cdot \mathcal{L}(\mathbf{u}) \rho d\tau, \tag{15.16}$$

where $\mathbf{u}(\mathbf{r})$ is any sufficiently regular vector function of $\mathbf{r}$. We now subject $\mathbf{u}(\mathbf{r})$ to a small variation $\Delta\mathbf{u}$ at each point $\mathbf{r}$, keeping $\mathbf{r}$ and all unperturbed quantities fixed during the variation. We obtain from eq. (15.16) an expression involving $\Delta\mathbf{u}$, $\Delta\mathbf{u}^*$, $\Delta\mathcal{L}(\mathbf{u})$, and $\Delta\Sigma$, where $\Delta\Sigma$ is the corresponding variation in Σ. However, it follows from the linearity of $\mathcal{L}$ that $\Delta\mathcal{L}(\mathbf{u}) = \mathcal{L}(\Delta\mathbf{u})$. Also, because of the Hermiticity of $\mathcal{L}$, $\mathbf{u}^*$ and $\Delta\mathbf{u}$ can be interchanged in this expression. Finally, the facts that $\mathcal{L}$ is a real operator and Σ^2 is a real number permit considerable simplification, and so we obtain

$$2\Sigma(\Delta\Sigma) \int_V \mathbf{u}^* \cdot \mathbf{u} \rho d\tau + 2Re\left\{\int_V (\Delta\mathbf{u}^*) \cdot [\Sigma^2 \mathbf{u} - \mathcal{L}(\mathbf{u})] \rho d\tau\right\} = 0, \tag{15.17}$$

where $Re\{\cdot\ \cdot\ \cdot\}$ means the real part of the indicated quantity.

If, now, $\mathcal{L}(\mathbf{u}) = \Sigma^2\mathbf{u}$, then $\Delta\Sigma = 0$, even if $\Delta\mathbf{u}^* \neq 0$. Conversely, if $\Delta\Sigma = 0$ and $\Delta\mathbf{u}^*$ is arbitrary (but small), then $\mathcal{L}(\mathbf{u}) = \Sigma^2\mathbf{u}$. But this is just the LAWE for nonradial oscillations, from which we must have $\mathbf{u}(r) = \mathbf{u}_k(\mathbf{r})$ and $\Sigma^2 = \sigma_k^2$, that is, $\mathbf{u}(\mathbf{r})$ and Σ^2 must be, respectively, an eigenfunction and eigenvalue of this equation. We have therefore proved the following: those solutions of the equation $\mathcal{L}(\mathbf{u}) = \Sigma^2 \boldsymbol{u}$ for which Σ is an extremum ($\Delta\Sigma = 0$) with respect to arbitrary small variations of $\boldsymbol{u}(\mathbf{r})$, are the *eigenfunctions* $\mathbf{u}_k(\mathbf{r})$, and the associated eigenvalues are the corresponding *adiabatic eigenvalues* σ_k^2, both for the kth mode.

However, the present variational principle is not as useful as in the case of purely radial oscillations. The reason is mainly that the equations of linear, adiabatic, nonradial oscillations do not form a Sturm-Liouville system, as will be emphasized in Chapter 17. Consequently, a unique, simple ordering of the eigenvalues is not possible, and hence it is not possible to say in advance, for any given mode, whether the relevant extremum is a minimum or a maximum. In contrast, in the case of purely radial oscillations, where the Sturm-Liouville character of the equations permits a unique ordering of modes, it is possible to prove that the extremum for the fundamental mode is a *minimum* (see §8.10).

It should be pointed out that, in case there are motions (such as rotation) in the unperturbed state of the system, the above variational principle must be modified somewhat (see §19.1).

It may also be noted that a period-mean density relation, analogous to that which applies to radial oscillations, also applies to nonradial oscillations of stars. This conclusion follows most simply from a dimensional analysis of, say, the integral expression (15.16) for the periods of nonradial oscillations. However, as we shall see (Chap. 17), the *value* of the product $(\text{period})^2 \cdot (\text{mean density})$ may be much more sensitive to the nondimensional structure (for example, relative mass concentration) of the star than for the case of radial oscillations.

15.3. A MORE GENERAL CASE

The more general case to be considered in this section is one characterized by linear, adiabatic oscillations, but in which the three assumptions made in §15.2 are dropped. Hence, the systems considered here need not have vanishing surface density (but the surface pressure is still assumed to vanish), may have arbitrary velocities (and accelerations), and need not be spherically symmetric. A special case of this more general case would be a rotating star, where the rotation may be slow or fast, uniform or differential.

The first discussion of the small, adiabatic oscillations of a rotating star was by Clement (1964). It was noticed by Chandrasekhar and Lebovitz (1964) that the Hermiticity property persisted even if the assumption of vanishing surface density was dropped. Smeyers (1973) showed that the variational principle discussed by Chandrasekhar and Lebovitz (1964) was equivalent to Hamilton's principle of classical mechanics, even for nonradial oscillations. A more general and more rigorous discussion than Clement's (1964) was given by Lynden-Bell and Ostriker (1967), who considered specifically the more general case described above. In our treatment here we shall not present the details of Lynden-Bell and Ostriker's (1967) discussion, as their paper is quite clear and may be consulted by the interested reader.

In this case the Lagrangian and Eulerian variations of the right side of the linearized momentum equation (say eq. [5.32]) are not equal. We write this equation in the form

$$\frac{d^2\delta\mathbf{r}}{dt^2} = -\mathcal{P}(\delta\mathbf{r}) - \mathcal{V}(\delta\mathbf{r}), \tag{15.18}$$

where $\mathcal{P}$ and $\mathcal{V}$ are linear vector operators. We have, if the only body force is self-gravitation,

$$\mathcal{P}(\boldsymbol{\zeta}) \equiv \left(\frac{1}{\rho}\nabla P\right)' + \boldsymbol{\zeta}\cdot\nabla\left(\frac{1}{\rho}\nabla P\right), \tag{15.19a}$$

$$\mathcal{V}(\zeta) \equiv \nabla\psi' + \zeta \cdot \nabla(\nabla\psi), \tag{15.19b}$$

where the variations of P are assumed to be adiabatic. Note that the right side of eq. (15.18) is equal to the right side of eq. (15.8′). Now, we have the important result, first proved by Lynden-Bell and Ostriker (1967), that both $\mathcal{V}$ and $\mathcal{P}$ are Hermitian (in the sense of eq. [15.11]), where the Hermiticity of $\mathcal{P}$ requires that the pressure vanish at the surface of the configuration.[1]

It should, however, be emphasized that, although the right side of the linearized momentum equation (15.18) can thus be written as a Hermitian operator for small adiabatic oscillations even in this more general case, the eigenvalues of the LAWE need not all be real, as was shown by Lynden-Bell and Ostriker (1967) (see also §19.1).

[1]The operators actually introduced by Lynden-Bell and Ostriker (1967) differ from the ones above by the factor ρ, the density in the unperturbed (nonoscillating) configuration. More specifically, we have, for example, $\mathcal{P}_{\text{Lynden-Bell and Ostriker}} = \rho\mathcal{P}_{\text{Cox}}$, where $\mathcal{P}_{\text{Cox}}$ is as defined in eq. (15.19a) above. However, this difference in definition makes no difference in the final results, essentially because the definition of Hermiticity adopted by Lynden-Bell and Ostriker (1967) differs from ours (see eq. [15.11]) by just this factor of ρ.

16

Growth or Decay of Small, Nonradial Oscillations

In this chapter on the growth or decay of nonradial stellar oscillations in linear theory, we shall employ two approaches: one largely physical (§16.1), the other more formal (§16.2). Both of these approaches lead to the same result, which is an expression for the stability coefficient (see §9.1) for nonradial oscillations. This expression is formally identical to the one (see eq. [9.13]) that applies to purely radial oscillations.

We adopt throughout this chapter the three basic assumptions stated in §15.2.

16.1. "LINEARIZED" ENERGY EQUATION

In this section we shall show that the "linearized" (second-order) energy equation that was derived in §9.1 (see eq. [9.3]) for purely radial oscillations of stars, is precisely valid also for *general* (nonradial) oscillations. This demonstration provides a physical basis behind the expressions used for calculating the growth or decay of oscillations, and is also generally illuminating.

In view of the three assumptions referred to above, we have $\mathbf{v} = \delta\mathbf{v}$, and $d\mathbf{v}/dt = \partial\mathbf{v}/\partial t \equiv \dot{\mathbf{v}}$. The linearized equation of motion is then eq. (5.31), with $\mathbf{f}' = -\nabla\psi'$, where ψ' is the Eulerian variation of the gravitational potential. We shall usually omit zero subscripts from unperturbed quantities, and our notation will be standard (see the earlier chapters).

In this equation we express P' in terms of δP by use of the relation (5.16) between Eulerian and Lagrangian variations. We have

$$\dot{\mathbf{v}} = -\frac{1}{\rho}\nabla\delta P + \frac{1}{\rho}\nabla(\delta\mathbf{r}\cdot\nabla P) + \frac{\rho'}{\rho^2}\nabla P - \nabla\psi'. \qquad (16.1)$$

We now take the scalar (dot) product of $\rho\, d\tau\, \mathbf{v}$ with eq. (16.1) ($d\tau$ = a volume element) and integrate over the entire volume V of the configuration. The first two terms on the right side of the resulting equation can then be integrated by parts, and the divergence theorem used, so that the right side of this equation consists of six integrals, two of which are surface integrals. But these surface integrals vanish since, in accordance with our assumptions, $\delta P = 0$ at the surface, and also $\nabla P = 0$ because $\rho = 0$ here.

Hence this equation is

$$\int_V \rho\mathbf{v}\cdot\dot{\mathbf{v}}d\tau = -\int_V \frac{\delta P}{\rho}\frac{d\delta\rho}{dt}d\tau + \int_V \frac{1}{\rho}(\delta\mathbf{r}\cdot\nabla P)\frac{d\delta\rho}{dt}d\tau$$
$$+\int_V \frac{\rho'}{\rho}\mathbf{v}\cdot\nabla P d\tau - \int_V \rho\mathbf{v}\cdot\nabla\psi' d\tau. \quad (16.2)$$

Consider now the second integral on the right side of this equation. We integrate it by parts (with respect to time) and take the d/dt operator outside of the integral sign, as is allowed because of mass conservation ($d[\rho d\tau]/dt = 0$, see Chapter 4). Also, we use the relations $d(\delta\mathbf{r})/dt = \delta\mathbf{v} = \mathbf{v}$ and the fact that the unperturbed quantities are not functions of time in the other integral resulting from the parts integration. We now write $\delta\rho = \rho' + \delta\mathbf{r}\cdot\nabla\rho$ in one of these integrals and $\delta\rho = -\rho\nabla\cdot\delta\mathbf{r}$ in the other. We do another parts integration (with respect to time) and make use of the fact that $\nabla\rho$ and ∇P are purely radial in a spherical star. We finally obtain

$$\int_V \frac{1}{\rho}(\delta\mathbf{r}\cdot\nabla P)\frac{d\delta\rho}{dt}d\tau = -\frac{d}{dt}\int_V (\nabla\cdot\delta\mathbf{r})(\delta\mathbf{r}\cdot\nabla P)\,d\tau$$
$$-\int_V \frac{\rho'}{\rho}(\mathbf{v}\cdot\nabla P)\,d\tau - \frac{d}{dt}\left[\frac{1}{2}\int_V \frac{1}{\rho}(\delta\mathbf{r}\cdot\nabla\rho)(\delta\mathbf{r}\cdot\nabla P)\,d\tau\right]. \quad (16.3)$$

Consider now the last integral in eq. (16.2). We do a parts integration, using the divergence theorem, the fact that $\rho = 0$ at the surface, and the linearized mass equation. We also make use of the results of §15.1. We obtain

$$-\int_V \rho\mathbf{v}\cdot\nabla\psi' d\tau = \frac{d}{dt}\left[\frac{1}{2}G\int_V\int_V \frac{\rho'(\mathbf{r},t)\rho'(\mathbf{x},t)d\tau' d\tau}{|\mathbf{x}-\mathbf{r}|}\right], \quad (16.4)$$

where $d\tau'$ is a volume element located at $\mathbf{x}$.

Consider, finally, the first integral on the right side of eq. (16.2). We use the thermodynamic identity (5.35a) to eliminate δP from this integral. Integrating (with respect to time) by parts, and using the linearized mass equation in the form $\delta\rho = -\rho\nabla\cdot\delta\mathbf{r}$, we obtain

$$-\int_V \frac{\delta P}{\rho}\frac{d\delta\rho}{dt}d\tau = -\frac{d}{dt}\left[\frac{1}{2}\int_V \Gamma_1 P(\nabla\cdot\delta\mathbf{r})^2\,d\tau\right]$$
$$-\int_V (\Gamma_3 - 1)T\,\delta s\left[\frac{d}{dt}\left(\frac{\delta\rho}{\rho}\right)\right]\rho d\tau. \quad (16.5)$$

Assembling the above results, and comparing with eq. (15.10), we see that we may write

$$\frac{d}{dt}\left\{\int_V \frac{1}{2}\rho v^2 d\tau + \frac{1}{2}\int_V \delta\mathbf{r} \cdot \mathcal{L}(\delta\mathbf{r})\rho d\tau\right\}$$

$$= -\int_V (\Gamma_3 - 1)T\delta s\left[\frac{d}{dt}\left(\frac{\delta\rho}{\rho}\right)\right]\rho d\tau, \qquad (16.6)$$

where $\mathcal{L}$ is the (Hermitian) linear operator introduced in Chapter 15, and s denotes specific entropy.

In order to complete the derivation, we must obtain the interpretation of the quantity $\frac{1}{2}\int_V \delta\mathbf{r} \cdot \mathcal{L}(\delta\mathbf{r})\rho d\tau$. This we do now.

First, we note (see eq. [15.9]) that $-\mathcal{L}(\delta\mathbf{r})$ is the Lagrangian variation of the total force per unit mass at $(\mathbf{r},t)$ resulting from a field of adiabatic Lagrangian displacements $\delta\mathbf{r}(\mathbf{x},t) \equiv \boldsymbol{\eta}(\mathbf{x},t)$ of mass elements throughout the system (if the unperturbed accelerations are zero, this is also the Eulerian variation of this force). Now consider further increments $d\boldsymbol{\eta}(\mathbf{x},t)$ in the displacements of all mass elements. The total work done by the above forces over these increments for the entire system is

$$dW = -\int_V (d\boldsymbol{\eta}) \cdot \mathcal{L}(\boldsymbol{\eta})\rho d\tau. \qquad (16.7)$$

But we know that $\mathcal{L}$ is Hermitian. Making use of this fact, taking the d operator outside the integral ($d[\rho d\tau] = 0$), and integrating over the total displacements $\delta\mathbf{r}$, we have for the total work done by these forces over these displacements,

$$W = -\frac{1}{2}\int_0^{\delta\mathbf{r}} d\int_V \boldsymbol{\eta} \cdot \mathcal{L}(\boldsymbol{\eta})\rho d\tau = -\frac{1}{2}\int_V \delta\mathbf{r} \cdot \mathcal{L}(\delta\mathbf{r})\rho d\tau. \qquad (16.8)$$

But this work is conservative in the sense that its value is independent of the paths by which the final displacements were reached. This work therefore has the properties of a potential energy. In fact, noting that a potential energy is equal to the work done *against* the forces exerted by the system, we may write $W = -\delta\Phi$, where

$$\delta\Phi \equiv \frac{1}{2}\int_V \delta\mathbf{r} \cdot \mathcal{L}(\delta\mathbf{r})\rho d\tau \qquad (16.9)$$

is the total potential energy of the system (for adiabatic displacements) in its perturbed state, taken as zero when the displacements are zero everywhere. Thus, the quantity on the right side of eq. (16.9) has exactly the same physical interpretation in the general case of nonradial oscillations of stars as in the case of purely radial displacements. The fact that $\delta\Phi$

is of *second* order in small quantities is a result of the circumstance that its first-order variation vanishes identically if the unperturbed system is in hydrostatic equilibrium.

We note that we can now supply the physical interpretation of the integral expression (15.16) for the eigenvalue, say σ^2, for nonradial, adiabatic oscillations (Σ^2 in eq. [15.16] becomes σ^2 if $\mathbf{u}$ therein denotes the eigenfunction). If we write $\delta\mathbf{r} = \mathbf{u}\cos(\sigma t + \phi)$, and so on, the same procedures as were used in Chapter 8 for purely radial oscillations lead to the results that

$$\overline{\delta\mathcal{T}} = \frac{1}{4}\sigma^2 \int_V \mathbf{u}^* \cdot \mathbf{u}\rho d\tau, \tag{16.10}$$

$$\overline{\delta\Phi} = \frac{1}{4}\int_V \mathbf{u}^* \cdot \mathcal{L}(\mathbf{u})\rho d\tau, \tag{16.11}$$

where $\delta\mathcal{T}$ is the variation of the total kinetic energy of the system, and bars denote averages over one period in the mode considered. Hence eq. (15.16) for σ^2 means that

$$4\overline{\delta\mathcal{T}} = 4\overline{\delta\Phi}, \tag{16.12}$$

just as in the case of radial oscillations.

Returning now to our linearized energy equation, we have

$$\frac{d\delta\Psi}{dt} = -\int_V (\Gamma_3 - 1)T\delta s\left[\frac{d}{dt}\left(\frac{\delta\rho}{\rho}\right)\right]\rho d\tau, \tag{16.13}$$

where $\delta\Psi = \delta\mathcal{T} + \delta\Phi$ is the "total pulsation energy," $\delta\mathcal{T}$ denoting the total pulsation kinetic energy (see eq. [16.6]). Integrating the right side of eq. (16.13) by parts, then integrating over a complete period, assuming the system to return precisely to its initial state at the end of the period, and dividing by the period Π, we finally obtain for the average (over a period) rate of change of $\delta\Psi$,

$$\left\langle\frac{d\delta\Psi}{dt}\right\rangle = \frac{1}{\Pi}\int_0^\Pi dt \int_V (\Gamma_3 - 1)\left(\frac{\delta\rho}{\rho}\right)\delta\left(\epsilon - \frac{1}{\rho}\nabla\cdot\mathbf{F}\right)\rho d\tau. \tag{16.14}$$

Here we have used eq. (7.10), valid for thermodynamically reversible processes.

To summarize, we see that the result established in Chapter 9 for purely radial oscillations (see eq. [9.3]) is precisely valid for nonradial oscillations as well. Equation (16.14) provides a physical basis for discussions of the stability of nonradial oscillations (damped or overstable) (see §16.2), just as in the case of radial oscillations.

16.2. INTEGRAL EXPRESSIONS FOR THE EIGENVALUES

In this section we shall show that integral expressions, exactly analogous to those derived in §9.3 for radial oscillations, exist for the eigenvalues of linear, nonradial oscillations. These integral expressions permit one to derive an expression for the stability coefficient ($= |e$-folding time for pulsation amplitude$|^{-1}$) for such oscillations. In view of the energy equation derived in §16.1 for nonradial oscillations, such a conclusion is perhaps not surprising. Again, we adopt the three assumptions made in §15.2.

In view of the assumption that the velocity is zero in the unperturbed state of the system, there is no difference between d/dt and $\partial/\partial t$, and the linearized equation of motion may be written in the form

$$\frac{\partial^2 \delta \mathbf{r}}{\partial t^2} = -\mathcal{L}(\delta \mathbf{r}) - \frac{1}{\rho}\nabla\left(\delta P - \frac{\Gamma_1 P}{\rho}\delta\rho\right), \tag{16.15}$$

where $\mathcal{L}$ is the linear operator introduced in Chapter 15. Here we have simply added on the nonadiabatic contribution (last term in parentheses in eq. [16.15]) to the Lagrangian pressure variation.

We now take $d/dt = \partial/\partial t$ of eq. (16.15), making use of the fact that the quantities in the unperturbed state of the system do not depend on the time. The time derivative of the nonadiabatic contribution is obtained from the linearized energy equation in the form of eq. (5.34a).

We now assume that

$$\delta \mathbf{r}(\mathbf{r},t) = \mathbf{u}(\mathbf{r})e^{i\omega t}, \tag{16.16}$$

with similar time dependences for the other pulsation variables. The time derivative of eq. (16.15) then becomes

$$-i\omega^3 \mathbf{u} + i\omega\mathcal{L}(\mathbf{u}) = -\frac{1}{\rho}\nabla\left\{\rho(\Gamma_3 - 1)\left[\delta\left(\epsilon - \frac{1}{\rho}\nabla\cdot\mathbf{F}\right)\right]_{\mathrm{sp}}\right\}, \tag{16.17}$$

where subscript "*sp*" means "space part." We now form the scalar product of $\mathbf{u}^*\rho d\tau$ with eq. (16.17) and integrate over the entire volume V of the configuration. We perform a parts integration on the right side of the equation and make use of the divergence theorem so as to obtain a surface integral. If, however, the density is zero at the surface of the configuration, then this surface integral will vanish. We also use the linearized mass equation,

$$\nabla\cdot\mathbf{u}^* = -(\delta\rho/\rho)^*_{\mathrm{sp}}, \tag{16.18}$$

in the other integral resulting from the parts integration. We now define the quantities

$$J \equiv \int_V \mathbf{u}^* \cdot \mathbf{u} \rho d\tau, \tag{16.19}$$

$$J\Sigma^2 \equiv \int_V \mathbf{u}^* \cdot \mathcal{L}(\mathbf{u}) \rho d\tau, \tag{16.20}$$

$$C \equiv \int_V (\Gamma_3 - 1) \left(\frac{\delta\rho}{\rho}\right)^*_{\text{sp}} \left[\delta\left(\epsilon - \frac{1}{\rho}\nabla \cdot \mathbf{F}\right)\right]_{\text{sp}} \rho d\tau. \tag{16.21}$$

We then obtain

$$i\omega(\omega^2 - \Sigma^2) = C/J, \tag{16.22}$$

which is formally exactly the same cubic equation for the eigenvalues ω as for the case of purely radial oscillations (see eq. [9.43]). Hence, all the discussion there (explicit formulae for ω, and so forth) may be applied directly to the case of nonradial oscillations. However, one should heed the warning remark near the bottom of page 522 in Ledoux and Walraven (1958).

In particular, for the case where $|C/J|$ is small compared to any of the individual terms on the left side of eq. (16.22), the solution for the oscillation amplitude is of the form

$$\delta\mathbf{r}(\mathbf{r},t) = \mathbf{u}(\mathbf{r})e^{\pm i\sigma t} \cdot e^{-\kappa t}, \tag{16.23}$$

where σ (the real part of ω) $\approx \Sigma = 2\pi/\text{period}$ is the oscillation angular frequency and κ is the stability coefficient, given approximately by eq. (9.13), except that C is here given by eq. (16.21). The quantities C_r (real part of C) and J are often evaluated by using the adiabatic eigenfunctions in the integrands. This procedure is called the *quasi-adiabatic* approximation, and frequently gives reasonably accurate results.

17

Linear, Adiabatic, Nonradial Oscillations of Spherical Stars

In this chapter we shall consider small, nonradial oscillations of self-gravitating, spherical stars about their "equilibrium" states. We shall assume, moreover, everywhere except in §§17.1–17.4 and in §17.14 that the oscillations are also *adiabatic* (described by eqs. [5.36]). Even in this approximation the problem is, as we shall see, mathematically quite complicated. The more difficult problem of small, *nonadiabatic,* nonradial oscillations will be considered briefly in Chapter 18.

For the equilibrium states about which the oscillations are supposed to take place, we shall, unless stated otherwise, adopt the three assumptions discussed in Chapter 15. These equilibrium states are also assumed to be in thermal equilibrium ($\epsilon_0 - dL_{r,0}/dm = 0$), and we neglect any changes in chemical composition that may result from nuclear transmutations.

Many of the general remarks made at the beginning of Chapter 8, on small, *radial,* adiabatic oscillations of stars, also apply to the present more general case of nonradial stellar oscillations.

In §§17.1–17.4 we shall be concerned primarily with general physical and mathematical preliminaries useful in the study of nonradial oscillations. In these sections we do not specifically assume adiabatic oscillations, although this assumption is made in §§17.5–17.13.

Procedures for calculating adiabatic eigenvalues and eigenfunctions for general stellar models are outlined in §17.5, and boundary conditions are discussed in §17.6. The homogeneous, compressible model, one of the few cases that can be treated analytically, is discussed briefly in §17.7. Some general properties of nonradial, adiabatic oscillations of stars are described, and some numerical results given, in §17.8. Certain oscillatory properties are qualitatively different from the above in sufficiently complicated stellar models, and some of these differences are discussed in §§17.10 and 17.11. An important approximation, in which the perturbations of the gravitational potential are neglected (the Cowling approximation), is described and discussed in §17.9. Section 17.12 contains a discussion of nonradial oscillations of very high-order modes of stars. A recent application of Epstein-type weight functions to nonradial stellar oscillations is reviewed in §17.13. Finally, some methods for computing damping times of nonradial oscillations, and some results, are described in §17.14.

Other excellent reviews of nonradial stellar oscillations are given by Ledoux (1974, 1978); see also J. P. Cox (1976a). For some nice insights and approaches, see Wolff (1979), Shibahashi (1979), and Christensen-Dalsgaard (1979). An excellent monograph dealing with nonradial stellar oscillations has recently been published by Unno, Osaki, Ando, and Shibahashi (1979).

17.1. GENERAL EQUATIONS

The appropriate linearized equations are the mass, momentum, energy, energy generation and transfer, and Poisson's equations. The first three of these are, respectively, eqs. (5.29a), (5.31) with $\mathbf{v}' = \delta\mathbf{v} = d(\delta\mathbf{r})/dt$, and (5.35a). It can easily be shown, using the relation (eq. [5.16]) between Eulerian and Lagrangian variations, that the momentum equation is valid also if the Eulerian variations therein are everywhere replaced by Lagrangian variations, for models which are spherical and in hydrostatic equilibrium in their unperturbed states. As usual, primes denote Eulerian variations, and we omit zero subscripts from unperturbed quantities.

The energy generation and transfer equation is

$$T\frac{d\delta s}{dt} = \delta\left(\epsilon - \frac{1}{\rho}\nabla \cdot \mathbf{F}\right), \tag{17.1}$$

where ϵ and $\mathbf{F}$ denote, respectively, the rate per unit mass of thermonuclear energy generation and the net vector heat flux including, in principle, all mechanisms that may be contributing to the heat transfer (we have assumed here thermodynamically reversible processes), and δs denotes the Lagrangian variation of the specific entropy s. The right side of eq. (17.1) in general depends, among other things, on $\delta\mathbf{r}$, ρ', T' and $\nabla T'$, and involves specific mechanisms of energy generation and heat transfer. Explicit expressions for $\delta([\nabla \cdot \mathbf{F}]/\rho)$ for radiative transfer in stars will be given in §17.14.

For some purposes it is more convenient to regard Poisson's equation as two first-order differential equations rather than as one second-order differential equation. The gravitational potential ψ is defined by the statement that the gravitational force per unit mass $\mathbf{f}$ is given by

$$\mathbf{f} = -\nabla\psi, \tag{17.2}$$

and Poisson's equation just states that

$$\nabla \cdot \mathbf{f} = -\nabla^2\psi = -4\pi G\rho. \tag{17.3}$$

The Eulerian variations of eqs. (17.2) and (17.3) are obtained simply by attaching primes to **f**, ψ, and ρ.

We shall first write the momentum equation in a particular form (see Ledoux and Walraven 1958, §74), which will be useful for our subsequent work. After some manipulation and after combining this equation with the linearized energy equation, we obtain for a spherical star

$$\frac{d^2\delta\mathbf{r}}{dt^2} = -\nabla\left(\frac{P'}{\rho} + \psi'\right) + \mathbf{A}\frac{\Gamma_1 P}{\rho}\nabla\cdot\delta\mathbf{r} - (\Gamma_3 - 1)T\,\delta s\,\frac{\nabla\rho}{\rho}, \quad (17.4)$$

where

$$\mathbf{A} \equiv \frac{1}{\rho}\nabla\rho - \frac{1}{\Gamma_1 P}\nabla P. \quad (17.5)$$

The physical interpretation of the important quantity A (radial component of **A**, the only component if the star is spherically symmetric) will be discussed in the next section (§17.2).

For adiabatic oscillations, $\delta s = 0$, and we simply drop the last term in eq. (17.4). We shall usually assume adiabatic oscillations in this chapter unless we specify otherwise.

It is interesting to note that if the star in its unperturbed state is in hydrostatic equilibrium (and therefore spherical), the quantity $(P'/\rho + \psi')$ can be replaced by $(\delta P/\rho + \delta\psi)$.

17.2. DISCUSSION OF A

In the case of a spherical star, the only component of **A** is the radial one, A:

$$A = \frac{1}{\rho}\frac{d\rho}{dr} - \frac{1}{\Gamma_1 P}\frac{dP}{dr}, \quad (17.6)$$

where all quantities in this equation denote equilibrium (or unperturbed) values, and P, ρ, and Γ_1 are functions only of radial distance r. The physical significance of A can easily be seen from the following considerations, based on the excellent and lucid discussions of Tolstoy (1963, 1973).

Consider a stratified fluid, with P, ρ, and Γ_1 depending only on r, and imagine a fluid element which was initially at r_0 that has been displaced to the new position $r_0 + \delta r$ (see Fig. 17.1). We assume that the element always retains its identity and does not mix with its surroundings. We denote by $\Delta\rho(r)$ the difference, at any level r, between the density $\rho_{\rm el}(r)$ of the element and the density $\rho_{\rm surr}(r)$ of its immediate surroundings, both at that level:

$$\Delta\rho(r) \equiv \rho_{el}(r) - \rho_{surr}(r). \tag{17.7}$$

Now, $\Delta\rho$ will in general be made up of two contributions. One, say $\Delta_1\rho$, merely results from the fact that the density in the unperturbed fluid varies from point to point. This contribution is given, for small displacements δr, by

$$\Delta_1\rho = -\delta r\,\frac{d\rho}{dr}, \tag{17.8}$$

where the derivative applies to the surrounding fluid and is to be evaluated at the level r_0. The negative sign appears because this contribution arises entirely from the variation of $\rho_{surr}(r)$ with r; in other words, $\rho_{el}(r)$ is regarded as constant here. The other contribution arises from the possible compressibility of the fluid, and takes any variation of $\rho_{el}(r)$ with r into account. If the pressure within the element at the new location differs from that at the old location, the volume occupied by the element at the new location may differ from its volume at the old location. If we suppose that the element has moved to its new location sufficiently slowly, then the element will always be in pressure equilibrium with its surroundings; that is, the pressure within the element will always be equal to the surrounding pressure. If, moreover, the motion has occurred without heat exchanges between the element and its surroundings (this situation could be essentially realized by sufficiently rapid movement of the element), then it would be moving adiabatically. In this case a density difference between the element and its surroundings might develop simply by virtue of the facts that the pressure varies from point to point in the unperturbed fluid

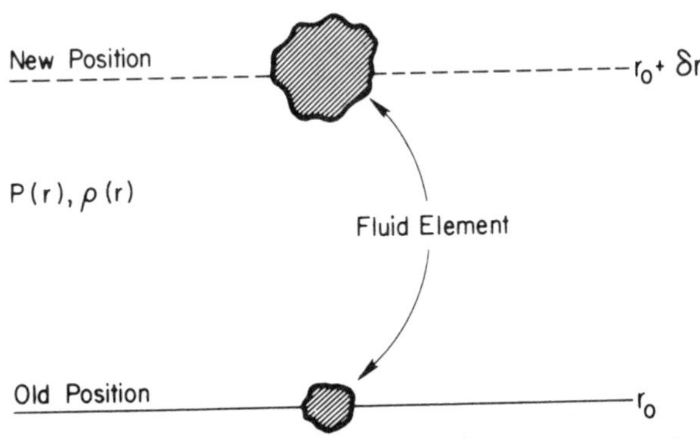

Figure 17.1. A displaced fluid element in a stratified fluid. The initial level is r_0 and the new level is $r_0 + \delta r$.

and that the fluid may be compressible. This difference would be, to first order in small quantities,

$$\Delta_2\rho = \delta r \left(\frac{d\rho}{dP}\right)_{\rm ad} \frac{dP}{dr}, \tag{17.9}$$

where $(d\rho/dP)_{\rm ad} = (\rho/P)\Gamma_1^{-1}$ is the derivative of ρ with respect to P along an adiabat (see §4.2.c), and dP/dr, which applies to the surrounding fluid, is to be evaluated at r_0. The case of an incompressible fluid could be treated simply by taking $\Gamma_1 = \infty$, in which case $\Delta_2\rho = 0$ (however, see below).

The total density excess over the density of its surroundings of an element so displaced by a distance δr in a stratified fluid is then

$$\begin{aligned}\Delta\rho &= \Delta_1\rho + \Delta_2\rho \\ &= -\rho\left(\frac{1}{\rho}\frac{d\rho}{dr} - \frac{1}{\Gamma_1 P}\frac{dP}{dr}\right)\delta r \\ &= -\rho A\,\delta r.\end{aligned} \tag{17.10}$$

(It will be recognized that $\Delta\rho$ is, in the linear approximation, just the Eulerian variation of ρ, denoted by ρ' in Chapter 5, and that $\Delta_2\rho$ is, also in the linear approximation, the Lagrangian variation $\delta\rho$; the first equality above is, in this approximation, just the relation [5.16] between the two types of variations.) The *buoyant force* per unit volume, f_B, acting on the element is then

$$f_B = -g\Delta\rho = \rho g A\,\delta r \tag{17.11}$$

to first order in small quantities, where $g \equiv Gm(r)/r^2$ is the local gravitational acceleration (G = constant of gravitation, $m[r]$ = mass interior to radial distance r). If, now, f_B and δr are of opposite signs, then the buoyant force will tend to return the element to its original level, and the fluid will be *convectively stable*. The condition for stability against convection is then, from eq. (17.11),

$$A < 0, \tag{17.12}$$

whereas $A > 0$ would represent *instability* against convection. The quantity A is therefore seen to be a local convective stability criterion.

In the case of stability against convection, and in the absence of heat exchanges, friction, and the like, our element would simply execute simple harmonic motion about its original level. The angular frequency N of this oscillatory motion would be given by writing $f_B = -\rho N^2\delta r$, whence

$$N^2 = -Ag; \tag{17.13}$$

N is usually referred to as the *Brunt-Väisälä* frequency.

The result (17.13) follows more formally from the work of §17.1 very easily. If pressure equilibrium between the element and its surroundings is always maintained, then $P' = 0$. Also, ψ' is obviously being neglected in the above simple picture. We neglect δs also, write $\delta P = \delta\mathbf{r} \cdot \nabla P$, use the adiabatic relation (eq. [5.36a]) between δP and $\delta\rho$, and remember that $(\nabla P)/\rho = \mathbf{g}$ (gravitational acceleration) since the unperturbed fluid is assumed to be in hydrostatic equilibrium. We obtain

$$\frac{d^2\delta\mathbf{r}}{dt^2} = -\mathbf{A}\,\delta\mathbf{r} \cdot \mathbf{g} = \mathbf{A}\,\delta r\, g, \tag{17.14}$$

where $\mathbf{g} = -g\,\mathbf{e}_r$ if $\mathbf{g}$ is directed straight downward, and δr is the radial component of $\delta\mathbf{r}$. Writing $\delta r \propto e^{iNt}$, we immediately obtain the result (17.13). In the case of an *incompressible* fluid, for which $\delta\rho = 0$, a more straightforward procedure is to go back to eq. (5.31), again neglecting P' and ψ' therein, and write $\rho' = -\delta\mathbf{r} \cdot \nabla\rho$. Equation (17.13) then follows exactly as above, except that $\mathbf{A}$ is replaced by $(\nabla\rho)/\rho$, the value of $\mathbf{A}$ for $\Gamma_1 = \infty$. Further discussion bearing on this matter will be found in §17.12.

(It is interesting to note two properties of eq. [17.13] with A replaced by $[d\rho/dr]/\rho$—the case of an incompressible fluid. First, N^2 is positive, corresponding to oscillatory behavior, only if $d\rho/dr < 0$, that is, only if the density *increases downward* [we are always assuming that $\mathbf{g}$ is nonzero and directed downward]. In the case of a density which *increases upward,* as would be the case with a heavier fluid lying on top of a lighter fluid, N^2 would be negative, corresponding to a dynamical instability. Thus, such a stratification would be unstable. This instability is sometimes called the Rayleigh-Taylor instability [see, e.g., Chandrasekhar 1961, Chap. X]. It is clear that such a Rayleigh-Taylor instability would exist for $d\rho/dr > 0$ also if the fluid were compressible, because the second term in the second equality in eq. (17.10) always gives a positive contribution to A for a system in hydrostatic equilibrium. Second, N would be *zero* [infinite period] if $d\rho/dr = 0$, that is, for a *homogeneous* [uniform density] incompressible fluid. Such a fluid would then be *neutrally stable* against the kind of perturbations considered here, and displaced fluid elements would not oscillate. This conclusion is essentially the reason why g modes [see §17.7] do not exist in a homogeneous incompressible sphere [see Ledoux 1974; J. P. Cox 1976a].)

The quantity A is closely related to the usual K. Schwarzschild convective stability criterion (Schwarzschild 1906). In fact, it can easily be shown (see, e.g., Cox and Giuli 1968, Chap. 13) that in the case of a continuously varying (in space) chemical composition,

$$A = \frac{1}{\lambda_P}\frac{\chi_T}{\chi_\rho}\left[\nabla - \nabla_{\mathrm{ad}} + \frac{\chi_\mu}{\chi_T}\left(\frac{d\ln\mu}{d\ln P}\right)\right], \tag{17.15}$$

where μ denotes mean molecular weight; $\lambda_P \equiv -(d \ln P/dr)^{-1}$ is the pressure scale height; χ_ρ, χ_T, were defined in §4.2.c;

$$\chi_\mu \equiv \left(\frac{\partial \ln P}{\partial \ln \mu}\right)_{\rho,T}; \tag{17.16}$$

and

$$\nabla \equiv \left(\frac{d \ln T}{d \ln P}\right) \tag{17.17}$$

is the actual logarithmic temperature gradient (with respect to pressure) in the star.

Finally, it may be useful to note that A is also an indicator of whether the specific entropy in a star is, locally, increasing or decreasing outward. We simply let $d \ln P$, for example, be the change in $\ln P$ associated with an outward increment dr in radial distance in the star. We therefore have, making use of a well-known thermodynamic identity,

$$A = -\frac{\Gamma_2 - 1}{\Gamma_2}\frac{\rho T}{P}\frac{ds}{dr}, \tag{17.18}$$

which shows that $A < 0$ in regions where the specific entropy s is increasing outward (stability against convection), and vice versa.

17.3. EXPRESSION OF PERTURBATION VARIABLES IN TERMS OF SPHERICAL HARMONICS

First, let us write the momentum equation (17.4) in terms of the components of the vector $\delta\mathbf{r}$ in a spherical polar coordinate system of the kind shown in Figure 6.1:

$$\delta\mathbf{r} = \mathbf{e}_r\delta r + \mathbf{e}_\theta\delta t_\theta + \mathbf{e}_\phi\delta t_\phi, \tag{17.19}$$

where the $\mathbf{e}$'s are the (dimensionless) unit vectors shown there; δr is the radial component, and δt_θ and δt_ϕ are the transverse components, all of $\delta\mathbf{r}$:

$$\delta t_\theta = r\delta\theta, \quad \delta t_\phi = r \sin\theta\, \delta\phi, \tag{17.20}$$

$\delta\theta$ and $\delta\phi$ being the corresponding increments of the polar and azimuthal angles θ and ϕ. Since the unperturbed system is assumed to be static, $d/dt = \partial/\partial t$. Also, we assume that all perturbation variables contain the time-dependent factor $e^{i\sigma t}$, so that $\partial^2/\partial t^2 = -\sigma^2$, where σ is the angular oscillation frequency. Remembering that $\mathbf{A}$ is purely radial for a spherical star ($\mathbf{A} = A\mathbf{e}_r$), we see that the three components of the momentum equation (17.4) for adiabatic motion are

$$\sigma^2 \delta r = \frac{\partial \chi}{\partial r} - A \frac{\Gamma_1 P}{\rho} \alpha, \tag{17.21a}$$

$$\sigma^2 \delta t_\theta = \frac{1}{r}\frac{\partial \chi}{\partial \theta} = \frac{\partial}{\partial \theta}\left(\frac{\chi}{r}\right), \tag{17.21b}$$

$$\sigma^2 \delta t_\phi = \frac{1}{r \sin\theta}\frac{\partial \chi}{\partial \phi} = \frac{1}{\sin\theta}\frac{\partial}{\partial \phi}\left(\frac{\chi}{r}\right), \tag{17.21c}$$

where

$$\alpha \equiv \nabla \cdot \delta\mathbf{r}(= -\delta\rho/\rho), \qquad \chi \equiv \frac{P'}{\rho} + \psi'. \tag{17.22, 23}$$

We shall also frequently have use of the expression for div $\delta\mathbf{r}$. We have, using eqs. (17.21),

$$\nabla \cdot \delta\mathbf{r} = \frac{1}{r^2}\frac{\partial(r^2 \delta r)}{\partial r} - \frac{1}{\sigma^2 r^2} L^2 \chi, \tag{17.24}$$

where the operator L^2 is sometimes called the *Legendrian:*

$$L^2 \equiv -\frac{1}{\sin\theta}\frac{\partial}{\partial\theta}\left(\sin\theta \frac{\partial}{\partial\theta}\right) - \frac{1}{\sin^2\theta}\frac{\partial^2}{\partial\phi^2}. \tag{17.25}$$

It is customary to assume that each perturbation variable is a manifestation of (owes its existence to) a "normal mode" of oscillation, and to treat each such normal mode as if only that one mode existed. In most applications each such mode is assumed to consist of the product of a function of r alone and a function only of the polar and azimuthal angles θ and ϕ. This last function is assumed to be proportional to a spherical harmonic $Y_l^m(\theta,\phi)$. (Good summaries of properties of spherical harmonics are in Condon and Shortley 1935, pp. 52–53; and Rojansky 1938, p. 414; for more background information, see, e.g., Jackson 1962, Chaps. 2, 3.) The order of the spherical harmonic is l, which can have only positive integral values, including zero: $l = 0, 1, 2, \ldots$. For each l, there are $2l + 1$ possible values of m: $m = -l, -l + 1, \ldots, l - 1, l$; m is sometimes called the azimuthal spherical harmonic index.[1]

The purpose of introducing these remarks about spherical harmonics, in the present context, is to point out that they are *eigenfunctions* of the

[1]This use of the symbol m should not be confused with use of the same symbol elsewhere in this book to denote interior mass. The meaning of m will always be made clear whenever it is used, or will be clear from the context.

operator L^2, with *eigenvalues* $l(l+1)$:

$$L^2 Y_l^m(\theta,\phi) = l(l+1) Y_l^m(\theta,\phi) \tag{17.26}$$

(see, e.g., Goertzel and Tralli 1960). Also, noting that χ is assumed to be proportional to Y_l^m, eq. (17.24) becomes

$$\alpha = \nabla \cdot \delta\mathbf{r} = -\delta\rho/\rho = \frac{1}{r^2}\frac{\partial(r^2 \delta r)}{\partial r} - \frac{l(l+1)}{\sigma^2 r^2}\chi \; (l = 0, 1, 2, \ldots). \tag{17.27}$$

This expression for div $\delta\mathbf{r}$ is valid for nonadiabatic as well as for adiabatic oscillations.

It may be noted that the case $l = 0$ represents purely radial oscillations.

We also note, from the second equalities in each of eqs. (17.21b) and (17.21c), that no new functions are involved in the two transverse components δt_θ and δt_ϕ of $\delta\mathbf{r}$ with this procedure. This result is just a consequence of the momentum equation (17.4), which gives the vector acceleration of a mass element as the gradient of a scalar function, χ, plus a purely radial vector (for a spherical star).

It will also be convenient for our later work to consider the expression for the divergence of a general Eulerian vector perturbation variable, say $\mathbf{f}'$, where in terms of components in a spherical coordinate system

$$\mathbf{f}' = f'_r \mathbf{e}_r + f'_\theta \mathbf{e}_\theta + f'_\phi \mathbf{e}_\phi \tag{17.28}$$

(similar considerations also apply to a Lagrangian vector perturbation variable $\delta\mathbf{f}$). However, for vectors that are equal to the gradient of a scalar function plus a purely radial vector, the two transverse components are evidently given by angular derivatives of only one function, which is often called the *tangential component,* and will be denoted here by f'_t (an example is given by eqs. [17.21b] and [17.21c]). Thus $f'_\theta = \partial f'_t/\partial\theta$, $f'_\phi = (\sin\theta)^{-1}\,\partial f'_t/\partial\phi$. If f'_t consists of the product of a function of r alone and Y_l^m, then we may also write

$$f'_\theta = f'_t \frac{1}{Y_l^m}\frac{\partial Y_l^m}{\partial\theta}, \qquad f'_\phi = f'_t \frac{1}{Y_l^m}\frac{\partial Y_l^m}{\sin\theta\,\partial\phi}. \tag{17.29}$$

Hence, we may express the vector $\mathbf{f}'$ as

$$\mathbf{f}' = f'_r \mathbf{e}_r + f'_t \mathbf{t}, \tag{17.30}$$

where $\mathbf{t}$ is a vector lying entirely in the tangential plane (perpendicular to a radius vector):

$$\mathbf{t} \equiv \mathbf{e}_\theta \frac{1}{Y_l^m}\frac{\partial Y_l^m}{\partial\theta} + \mathbf{e}_\phi \frac{1}{Y_l^m}\frac{\partial Y_l^m}{\sin\theta\,\partial\phi}. \tag{17.31}$$

However, **t** is not a *unit* vector. We therefore have

$$\nabla \cdot \mathbf{f}' = \frac{1}{r^2}\frac{\partial(r^2 f'_r)}{\partial r} - \frac{l(l+1)}{r} f'_t. \tag{17.32}$$

By comparison of eq. (17.32) with eq. (17.27), it is seen that the tangential component, say δt, of $\delta\mathbf{r}$ is

$$\delta t = \chi/(\sigma^2 r), \tag{17.33}$$

and the theta and phi components are given by eqs. (17.29), where χ was defined in eq. (17.23).

Note that only *scalar* perturbation variables are written as products of a function of r alone and a spherical harmonic.

In the special but important case where $\mathbf{f}'$ is equal to the gradient of a scalar function H which in turn is equal to the product of a function of r alone and a spherical harmonic, say

$$\mathbf{f}' = \nabla H \tag{17.34}$$

$$= \mathbf{e}_r f'_r + \mathbf{t} f'_t, \tag{17.35}$$

where $\mathbf{e}_r$ and **t** are the two vectors defined above, we have

$$f'_r = \frac{\partial H}{\partial r}, \qquad f'_t = \frac{H}{r}. \tag{17.36}$$

Equation (17.34) is actually a special case of *Helmholtz's theorem*, which states that any vector (subject to certain rather mild mathematical conditions, see, e.g., Phillips 1933, p. 158) can be written as the sum of the gradient of a scalar potential and the curl of a vector potential.

It should be noted that proportionality to a spherical harmonic implies the existence of an azimuthal running wave, whose phase velocity is

$$\frac{d\phi}{dt} = -\frac{\sigma}{m}. \tag{17.36'}$$

It is also useful to note here the effect of the Laplacian $\nabla^2 = \nabla \cdot \nabla$ operating on a function $y(r,\theta,\phi)$ which consists of the product of a function of r only and Y_l^m. We have then

$$\nabla^2 y = \frac{1}{r^2}\frac{\partial}{\partial r}\left(r^2 \frac{\partial y}{\partial r}\right) - \frac{l(l+1)}{r^2} y = \frac{1}{r}\frac{\partial^2(ry)}{\partial r^2} - \frac{l(l+1)}{r^2} y. \tag{17.37}$$

The separation of variables discussed above, on which practically all discussions of nonradial oscillations of stars are based, defines the class of spheroidal modes. The vector $\delta\mathbf{r}$ representing the displacement of a mass element satisfies in this class of modes

$$(\nabla \times \delta \mathbf{r})_r \equiv 0 \tag{17.38}$$

for $\sigma^2 \neq 0$, where subscript r signifies the radial component.

A second class of modes, the toroidal modes, can be found by using the separation

$$\delta r = 0, \tag{17.39a}$$

$$\delta t_\theta = \frac{T_{l,m}(r)}{r \sin \theta} \frac{\partial Y_l^m}{\partial \phi} e^{i\sigma t}, \tag{17.39b}$$

$$\delta t_\phi = -\frac{T_{l,m}(r)}{r} \frac{\partial Y_l^m}{\partial \theta} e^{i\sigma t}, \tag{17.39c}$$

where $T_{l,m}(r)$ is any arbitrary function of r alone which must satisfy the boundary conditions. These toroidal modes are time independent ($\sigma = 0$) in a static, spherical star and are characterized by identically vanishing Lagrangian and Eulerian variations of pressure, density, and gravitational potential. (In fact, it has been shown by Aizenman and Smeyers 1977, who give an excellent discussion of these modes, that the toroidal modes do not exist for such a star for $\sigma \neq 0$.) These modes represent a kind of slow twisting of the system. Their existence was first noted by Perdang (1968) and emphasized by Simon (1969); see also Chandrasekhar (1961, App. III). The author is indebted to P. Smeyers (private communication, 1975) for calling his attention to this class of modes. The radial component of $\nabla \times \boldsymbol{\delta r}$ is not zero at all points for these modes:

$$(\nabla \times \delta \mathbf{r})_r = l(l+1) \frac{T_{l,m}(r)}{r^2} Y_l^m e^{i\sigma t}. \tag{17.40}$$

The toroidal modes associated with $l = 1$ can be represented by (Aizenman and Smeyers 1977)

$$\delta \mathbf{r} = \mathbf{\Omega} \times \mathbf{r} e^{i\sigma t}, \tag{17.41}$$

where $\mathbf{\Omega}$ is in general a vector function of r. When $\mathbf{\Omega}$ is constant, these modes correspond to uniform rigid-body rotations.

It is usually assumed that the spheroidal and toroidal modes, together, form a *complete* set (see, e.g., R. Simon 1969; Kaniel and Kovetz 1967; Eisenfeld 1969). This assumption implies that an arbitrary function of position can be represented as an infinite sum of such modes.

We shall not consider the toroidal modes in detail in this book, and, unless explicitly indicated otherwise, we shall always assume that we are dealing with the spheroidal modes.

The toroidal modes have non-vanishing frequencies in the case of a rotating star. In fact, these modes become the "r modes" of Papaloizou and Pringle (1978) in this case. The toroidal modes also have non-vanishing frequencies in the case of a star part of whose interior can sustain shear, as in a partially crystalline white dwarf (see Hansen and Van Horn 1979).

17.4. DISCUSSION OF ORDERS, ETC.

What is the spatial order of the system of differential equations for small, nonradial, *nonadiabatic* oscillations? To answer this question, let us summarize the orders involved in the four basic differential equations and enumerate the various unknowns. We have the following equations:

Mass: This is a first-order scalar differential equation involving ρ' and the three components of $\delta\mathbf{r}$ (see, e.g., eq. [5.29a]).

Momentum: This is a first-order vector differential equation involving $d^2\delta\mathbf{r}/dt^2$, P', ρ' and ψ', that is, three first-order scalar differential equations (see, e.g., eq. [5.31]).

Energy: The specific entropy variation δs can be eliminated from this equation by taking the time derivative d/dt of, say, eq. (5.35a) and combining the result with eq. (17.1). The resulting equation is then seen to be a first-order (in space) scalar differential equation involving P', ρ' ϵ' and the three components of, respectively, $\delta\mathbf{r}$ and $\mathbf{F}'$. The relation among ϵ', ρ' and T' is ordinarily only an algebraic equation obtained from the energy-generation rate $\epsilon = \epsilon(\rho,T)$ (see, e.g., §4.3).

Flux: This equation has not yet been written down explicitly, since its most general form (including convective transfer) is not known quantitatively. Its form is, however, known in the case of radiative transfer in the diffusion approximation, including thermal conduction (see, e.g., eq. [4.41]). In this form, at least, one sees (see, e.g., §17.14) that the equation for $\mathbf{F}'$ is a first-order vector differential equation involving ρ', T' and the three components of, respectively, $\delta\mathbf{r}$ and $\mathbf{F}'$, that is, three first-order scalar differential equations. The relation among P', ρ', and T' is normally only an algebraic equation which can be obtained from the equation of state $P = P(\rho,T)$ of the material (see, e.g., §4.2).

Poisson's: For our present purposes this equation is more conveniently regarded as two first-order differential equations: one vector equation, given by the Eulerian variation of eq. (17.2), and one scalar equation, given by the linearized form of eq. (17.3). Poisson's equation therefore provides four first-order scalar differential equations in ψ' and in the three components of $\mathbf{f}'$, the Eulerian variation of the gravitational force per unit mass.

The dependent variables are therefore seen to be, for example, P', ρ', ψ', and the three components of each of $\delta\mathbf{r}$, $\mathbf{F}'$ and $\mathbf{f}'$, or twelve scalar variables in all. (Of course, Lagrangian variations can always be used in place of Eulerian variations as the dependent variables, in view of the relation, eq. [5.16], relating these two types of variations.) There are seen to be twelve scalar first-order (in space) differential equations in all. The system is therefore of the twelfth order in space. However, as we have seen (see, e.g., Chapter 9), in nonadiabatic oscillations each dependent variable must be complex, so that the spatial order is actually 24 in real variables.

But the assumption that each (scalar) variable is proportional to a spherical harmonic $Y_l^m(\theta,\phi)$ eliminates six variables, two components each of $\delta\mathbf{r}$, $\mathbf{F}'$ and $\mathbf{f}'$; and six equations: two equations of momentum, two of flux, and two of gravitational force per unit mass (the two tangential components of each vector are expressible in terms of algebraic equations in the scalar variables, for example, see eqs. [17.21b] and [17.21c]). The system is then of the sixth spatial order in complex variables, or of the twelfth order in real variables.

Another viewpoint is provided in §18.1.

In the *adiabatic* approximation the energy equation is not used, and accordingly neither is the flux equation. The order of the system is then only four in complex variables. However, in the case of adiabatic oscillations each dependent variable may be considered purely real (see, e.g., Chapter 8), so that in this approximation the system is actually only of the fourth order in real variables. The reason that the corresponding problem for *radial* oscillations is only of the second order is that for such oscillations Poisson's equation admits of a trivial solution; see Chapter 6. In all of the following we shall assume adiabatic oscillations, unless we state otherwise.

It is important to note that, since the azimuthal spherical harmonic index m does not appear in any of the equations (either adiabatic or nonadiabatic) of the problem, then the eigenvalues of small, nonradial oscillations of static, spherical stars are degenerate with respect to m; that is, for each value of l, there are $2l + 1$ values of m. And all of these $2l + 1$ eigenvalues for given l are exactly the same, in the absence of perturbations such as rotation and magnetic fields (see Chap. 19).

General methods of solution will be outlined in the next section (§17.5).

17.5. PROCEDURE FOR GENERAL MODELS

In this section we shall write some of the equations for small, adiabatic, nonradial oscillations of spherical stars (incorporating the assumptions stated in Chapter 15), in the forms that have actually been used to discuss such oscillations of general stellar models. The first form (§17.5a) is that described in Ledoux and Walraven (1958, §79); and used, for example, by Smeyers (1967). The second form (§17.5b) is that originated by Dziembowski (1971) and used, for example, by Osaki and Hansen (1973a).

17.5a. FIRST FORM

To obtain this form, we combine the energy and mass equations into one equation, and also use the radial component of the momentum equation. We have, in addition, Poisson's equation. Introducing the new variables,

$$u \equiv r^2\delta r, \qquad y \equiv P'/\rho, \tag{17.42,43}$$

we see that our three equations—the momentum, mass, and Poisson's equations, respectively—become, after some manipulation,

$$\frac{dy}{dr} = \frac{\sigma^2 + Ag}{r^2} u - Ay - \frac{d\psi'}{dr}, \tag{17.44}$$

$$\frac{du}{dr} = \frac{\rho g}{\Gamma_1 P} u + \left[\frac{l(l+1)}{\sigma^2} - \frac{r^2\rho}{\Gamma_1 P}\right] y + \frac{l(l+1)}{\sigma^2}\psi' \tag{17.45}$$

$$\frac{1}{r^2}\frac{d}{dr}\left(r^2\frac{d\psi'}{dr}\right) = -\frac{4\pi G\rho A}{r^2} u + \frac{4\pi G\rho^2}{\Gamma_1 P} y + \frac{l(l+1)}{r^2}\psi'. \tag{17.46}$$

In these equations the dependent variables u, y, and ψ' are regarded as functions only of r, that is, the common factors $Y_l^m(\theta,\phi)$ have been cancelled out. These differential equations obviously form a fourth-order system.

Since the above equations presumably apply to any value of l, they must describe, as a special case, adiabatic, purely *radial* oscillations, characterized by $l = 0$, of the kind discussed at length in Chapter 8. This conclusion has in fact been verified explicitly by Smeyers (1967). We shall therefore restrict most of our remarks in the following to the case $l > 0$.

17.5b. SECOND FORM

To obtain the second form of the differential equations (17.44)–(17.46), we define the four dimensionless dependent variables (Dziembowski 1971)

$$y_1 \equiv \frac{\delta r}{r}, \quad y_2 \equiv \frac{1}{gr}\left(\frac{P'}{\rho} + \psi'\right), \quad y_3 \equiv \frac{1}{gr}\psi', \quad y_4 \equiv \frac{1}{g}\frac{d\psi'}{dr}, \tag{17.47}$$

where g is the local gravitational acceleration. We also use the dimensionless angular frequency Ω (see eq. [8.20]) and define the dimensionless function

$$c_1 \equiv \left(\frac{r}{R}\right)^3 \frac{M}{m}, \tag{17.48}$$

where m is the mass interior to a sphere of radius r. The dimensionless variables of stellar structure theory (see, e.g., Chandrasekhar 1939),

$$U \equiv \frac{d \ln m}{d \ln r}, \qquad V \equiv -\frac{d \ln P}{d \ln r} = \frac{gr\rho}{P}, \tag{17.49}$$

where the second equality in V above applies to spherical stars in hydrostatic equilibrium, are also used. The above differential equations then become the following four first-order differential equations, where it has again been assumed that the common factors $Y_l^m(\theta,\phi)$ have been cancelled out:

$$r\frac{dy_1}{dr} = \left(\frac{V}{\Gamma_1} - 3\right)y_1 + \left[\frac{l(l+1)}{c_1\Omega^2} - \frac{V}{\Gamma_1}\right]y_2 + \frac{V}{\Gamma_1}y_3, \tag{17.50}$$

$$r\frac{dy_2}{dr} = (c_1\Omega^2 + Ar)y_1 + (1 - U - Ar)\,y_2 + Ary_3, \tag{17.51}$$

$$r\frac{dy_3}{dr} = (1 - U)y_3 + y_4, \tag{17.52}$$

$$r\frac{dy_4}{dr} = -UAry_1 + \frac{UV}{\Gamma_1}y_2 + \left[l(l+1) - \frac{UV}{\Gamma_1}\right]y_3 - Uy_4. \tag{17.53}$$

In the present formulation there are *four* first-order differential equations, instead of (as in the first formulation) two first-order ones, and one second-order differential equation, because Poisson's equation is really two first-order differential equations (see discussion near beginning of §17.1 above).

17.6. BOUNDARY CONDITIONS

In this section we shall consider the boundary conditions, both at center (§17.6a) and at surface (§17.6b) for small, adiabatic oscillations of spherical stars for which the three assumptions made in §15.2 apply. (However, for generality, we shall in this section not explicitly assume that the density vanishes at the surface of the system.) We shall find, not surprisingly, that these conditions are more involved than for purely radial oscillations of the kind considered in Chapter 8. We shall assume that $l > 0$ in most of the remainder of this section. We shall also work primarily in terms of the variables u, y, and ψ' introduced in §17.5a.

17.6a. CENTER

The appropriate equations are eqs. (17.44)–(17.46). We note, for reference, that A, g, $d\rho/dr$, and dP/dr all are proportional to r near $r = 0$ in a spherical star, whereas ρ, P, and Γ_1 all approach finite values as the center is approached. Hence Ag becomes small compared with σ^2 in eq. (17.44), and the second term in square brackets in eq. (17.45) becomes small compared with the first term, both as $r \longrightarrow 0$.

We now assume that, near $r = 0$:

$$y = r^a \sum_{\nu=0}^{\infty} Y_\nu r^\nu, \tag{17.54a}$$

$$u = r^b \sum_{\nu=0}^{\infty} U_\nu r^\nu, \tag{17.54b}$$

$$\psi' = r^c \sum_{\nu=0}^{\infty} \Psi_\nu r^\nu, \tag{17.54c}$$

where the Y_ν, U_ν, and Ψ_ν are constants to be determined, and these expansions are assumed to be convergent. We now assume that at least some two of our three dependent variables, say δr and P', remain finite at $r = 0$. Then the differential equations require that the third variable, in this case ψ', must also remain finite at $r = 0$, as is physically required. Since the values of a, b, and c in eqs. (17.54) determine how our dependent variables behave near $r = 0$, we must accordingly have $a \geq 0$, $b \geq 2$, and $c \geq 0$, in order that these central boundary conditions be satisfied. A further consequence of these considerations is that $\delta P \propto r^a$ and, for adiabatic oscillations, $\rho' \propto r^a$, both as $r \longrightarrow 0$.

However, from the fact that our dependent variables must obey the differential equations, the quantities a, b, and c must be related to one another. In fact, it can be shown that, for $l \neq 0$,

$$c = b - 1 = a. \tag{17.55}$$

We now insert the expansions (17.54) into the differential equations (17.44)–(17.46), use eq. (17.55), treat ρ, P, and Γ_1 as constants near $r = 0$, and equate coefficients of the lowest powers of r. We obtain

$$aY_0 = \sigma^2 U_0 - a\Psi_0, \tag{17.56}$$

$$(a + 1)U_0 = \frac{l(l + 1)}{\sigma^2}(Y_0 + \Psi_0), \tag{17.57}$$

$$a(a + 1)\Psi_0 = l(l + 1)\Psi_0. \tag{17.58}$$

This last condition gives, for $l \neq 0$,

$$a = l, \quad -l - 1 \tag{17.59}$$

if $\Psi_0 \neq 0$. Equations (17.56) and (17.57) also yield this same result, if $U_0 \neq 0$. The choice $a = -l - 1$ must be discarded in order to avoid singularities at $r = 0$. The expansions (17.54) must then be, in order to obey the boundary conditions at $r = 0$, for $l \neq 0$:

$$\delta r = r^{l-1} \sum_{\nu=0}^{\infty} U_\nu r^\nu, \tag{17.60a}$$

$$P'/\rho = r^l \sum_{\nu=0}^{\infty} Y_\nu r^\nu, \tag{17.60b}$$

$$\psi' = r^l \sum_{\nu=0}^{\infty} \Psi_\nu r^\nu. \tag{17.60c}$$

We note, also, that we have the condition (see eq. [17.56])

$$\sigma^2 U_0 = l(Y_0 + \Psi_0). \tag{17.61}$$

Hence, for given σ^2 and l, the central boundary conditions leave two of the three constants (U_0, Y_0, Ψ_0) undetermined. As we shall see in §17.6b, the values of these two constants must be chosen so that the *surface* boundary conditions are satisfied.

The tangential component δt of the Lagrangian displacement $\delta\mathbf{r}$ bears an interesting relation to the radial component δr at the stellar center. Using eq. (17.33), we have $\delta t = (P'/\rho + \psi')/(\sigma^2 r)$. Substituting the expansions (17.60) into this, evaluating at $r = 0$, and using the condition (17.61), we obtain at $r = 0$,

$$\delta t = (\delta r)/l. \tag{17.62}$$

Equation (17.62) is actually only a special case of a more general result that applies to any vector which is proportional to the gradient of a scalar that is, in turn, proportional to a spherical harmonic. If F_t and F_r are, respectively, the tangential and radial components of such a vector, say $\mathbf{F}$ (see eq. [17.30]), then the more general result, obtained from the required finiteness of $\nabla \cdot \mathbf{F}$ at $r = 0$ (see below), is

$$(F_r/F_t)_{r=0} = l. \tag{17.62'}$$

The above rather mathematical treatment of the central boundary conditions can also be looked at from a more physical standpoint. We note that in the present case of linear, nonradial, adiabatic oscillations only two vectors are involved, the Lagrangian displacement $\delta\mathbf{r}$ of a mass element and the force per unit mass, say $\mathbf{f}'$, due to the Eulerian variation ψ' of the gravitational potential: $\mathbf{f}' = -\nabla\psi'$. Moreover, we note that two divergences are involved: $\nabla \cdot \delta\mathbf{r}$ appears in the mass equation, and $\nabla \cdot \mathbf{f}'$ appears in Poisson's equation. The requirements that $\delta\rho$ and ρ' remain finite at $r = 0$ then lead to the requirement that both of the above divergences remain finite here. It can be shown that the behavior of the variables near $r = 0$, as given by expansions (17.60), are necessary and sufficient conditions for satisfaction of this last requirement (see also eq. [17.62']).

We note that nonradial oscillations ($l > 0$) differ from radial oscillations ($l = 0$) in that P' and δP must both vanish at the center ($r = 0$) in the former case but not in the latter. Also, $\delta r = 0$ at $r = 0$ for nonradial oscillations only for $l > 1$. The "dipole" mode, $l = 1$, for which $\delta r \neq 0$ at $r = 0$, corresponds to a displacement of the *geometrical center* of the configuration. For an incompressible fluid, this case would correspond to a bodily translation of the entire configuration. For a compressible fluid, however, this case may also correspond to a displacement which leaves the *center of mass* of the system unaltered, as was first shown by Smeyers (1966) (see also §17.8). A simple example of such oscillations, for $m = 0$, would be a person on ice skates holding a heavy dumbbell in each hand and swinging the dumbbells in unison back and forth along straight lines parallel to the direction he is facing. In this case the center of mass of the *man* is oscillating, but the center of mass of the *system* is stationary. Another example would be provided by a person pacing back and forth on a barge floating in the water. For $m = \pm 1$, a simple example of the case $l = 1$ would be a person on ice skates holding dumbbells at a fixed distance out in front of him while he spins about a vertical axis. Another example of this case is the earth-moon system: the center of the earth is displaced away from the moon some 2800 miles from the center of mass of this system, while the line joining the earth and the moon rotates with the lunar sidereal period about the center of mass of the system.

In order to obtain recursion relations from which values of the coefficients in the expansions (17.60) for $\nu > 0$ can be obtained, it is necessary to expand the coefficients appearing in the differential equations (17.44)–(17.46) in series about $r = 0$. The values of these coefficients of course depend on the properties of the unperturbed model. Assuming spherical symmetry for this model, we may write, near $r = 0$,

$$\frac{A}{r} = \sum_{\nu=0}^{\infty} A_{2\nu} r^{2\nu}, \tag{17.63a}$$

$$\frac{Ag}{r^2} = \sum_{\nu=0}^{\infty} B_{2\nu} r^{2\nu}, \tag{17.63b}$$

$$\frac{\rho}{\Gamma_1 P} = \sum_{\nu=0}^{\infty} C_{2\nu} r^{2\nu}, \tag{17.63c}$$

$$\rho = \sum_{\nu=0}^{\infty} D_{2\nu} r^{2\nu}, \tag{17.63d}$$

$$\frac{\rho g}{r\Gamma_1 P} = \sum_{\nu=0}^{\infty} E_{2\nu} r^{2\nu}, \tag{17.63e}$$

where the values of the coefficients are assumed known from the properties of the unperturbed model. Substituting these expansions and the expansions (17.60) into eqs. (17.44)–(17.46), we obtain, respectively, for $l \neq 0$,

$$r^{l-1} \sum_{\nu=0}^{\infty} \left[(l+\nu)(Y_\nu + \Psi_\nu) - \sigma^2 U_\nu \right] r^\nu = r^{l+1} \sum_{\mu,\nu=0}^{\infty} (B_{2\mu} U_\nu - A_{2\mu} Y_\nu) r^{2\mu+\nu} \tag{17.64}$$

$$r^{l} \sum_{\nu=0}^{\infty} \left[(l+1+\nu) U_\nu - \frac{l(l+1)}{\sigma^2} (Y_\nu + \Psi_\nu) \right] r^\nu = r^{l+2} \sum_{\mu,\nu=0}^{\infty} (E_{2\mu} U_\nu - C_{2\mu} Y_\nu) r^{2\mu+\nu} \tag{17.65}$$

$$r^{l-2} \sum_{\nu=0}^{\infty} \left[\nu(2l+1+\nu) \right] \Psi_\nu r^\nu = -4\pi G r^l \sum_{\mu,\nu,\xi=0}^{\infty} [D_{2\mu}(A_{2\nu} U_\xi - C_{2\nu} Y_\xi)] r^{2\mu+2\nu+\xi}. \tag{17.66}$$

We note that the expansions on the left sides of these equations start out with two powers of r less than those on the right side. Hence the coefficients of r^ν with $\nu = 0$ and $\nu = 1$ on the left sides must vanish. The appropriate coefficient with $\nu = 0$ in eq. (17.64) set equal to zero just gives us back eq. (17.61). A similar statement applies to eq. (17.65); eq. (17.66) just yields a trivial identity. Equating to zero the appropriate coefficients on the left sides of eqs. (17.64)–(17.66) for $\nu = 1$ yields $U_1 = 0$, $Y_1 = 0$, and $\Psi_1 = 0$. These last results are a consequence of the assumed spherical symmetry of the unperturbed model. This assumption is embodied implicitly in the expansions (17.63) used for the variables in the unperturbed model.

Equating to zero the coefficients of the next higher powers of r in eqs. (17.64)–(17.66) yields our first real recursion relations:

$$(l+2)(Y_2 + \Psi_2) - \sigma^2 U_2 = B_0 U_0 - A_0 Y_0, \tag{17.67a}$$

$$(l+3)U_2 - \frac{l(l+1)}{\sigma^2}(Y_2 + \Psi_2) = E_0 U_0 - C_0 Y_0, \qquad (17.67b)$$

$$(2l+3)\Psi_2 = -2\pi G D_0 (A_0 U_0 - C_0 Y_0). \qquad (17.67c)$$

From these equations the values of U_2, Y_2, and Ψ_2 can be calculated, for given l and σ^2, in terms of, for example, U_0 and Y_0 (see eqs. [17.56]–[17.58]) and the appropriate coefficients with zero subscripts characterizing the unperturbed model. A similar procedure for the next higher powers of r yields $U_3 = 0$, $Y_3 = 0$, and $\Psi_3 = 0$. In fact, all the U_ν, Y_ν, and Ψ_ν having odd values of ν would be found to vanish. By continuing in this manner, it is clear that as many coefficients in the expansions (17.60) as are needed could be calculated. One would then have power series expansions, with known coefficients, representing the solution of eqs. (17.44)–(17.46) near $r = 0$ for given l and σ^2.

17.6b. SURFACE

At the *surface* of a spherical star we have $r = R$ (stellar radius), the equilibrium value (sometimes denoted in general by zero subscripts), which is being assumed constant here. As was discussed ealier (see Chap. 15 and the beginning of the present chapter), the assumption of vanishing surface pressure, $P = 0$, is usually adequate and is almost universally made. This assumption requires that the Lagrangian pressure variation also vanish at the stellar surface:

$$\delta P = 0. \qquad (17.68)$$

Since ∇P_0 in a spherical star contains only the radial component dP_0/dr, and since this vanishes at the surface if the density vanishes here, then also $P' = 0$ here in this case. More general surface boundary conditions are discussed in Chapter 3 of Unno, Osaki, Ando, and Shibahashi (1979).

The surface pressure boundary condition (17.68) is sometimes written more explicitly. If the pressure P_0 in the unperturbed model vanishes at the surface, then for regular solutions (the only kind, generally, of physical interest) the quantity $\delta P/P$ must remain finite everywhere, in particular at the stellar surface (more precisely, it is the surface value of the quantity P/ρ [pressure/density] that is usually assumed small or vanishing in the unperturbed model). In complete analogy to the case of purely radial oscillations considered in Chapter 8, the surface value of $\delta P/P$, consistent with the boundary condition (17.68), can be calculated in the case of nonradial oscillations from the linearized momentum equation. As was the case with purely radial oscillations, the appropriate value of $\delta P/P$ at the stellar surface can be calculated by two methods, as has been emphasized by Dziembowski (1971). (a) One can consider the adiabatic, nonradial oscillations of an isothermal atmosphere, in analogy to the calculation presented in §8.4. The condition that the acoustic energy per unit volume remain finite as r (radial distance) $\rightarrow \infty$, then determines the appropriate value of $\delta P/P$. (This condition is sometimes called the reflection condition; see, for example, Ando and Osaki 1975.) (b) One can use the linearized momentum equation itself to obtain directly the appropriate value of $\delta P/P$ at the surface, somewhat analogously to the method used in §8.3 in the case of purely radial oscillations. The results of method (b) apply also to nonadiabatic oscillations (see Chapter 18), and are therefore more general than those of method (a). Moreover, method (b) is much easier computationally than method (a). Method (a) has been used by Ando and Osaki (1975) and by Hill, Rosenwald, and Caudell (1977) (see also Hill, Caudell, and Rosenwald 1977), to derive the reflection boundary condition. Here we shall present the results of method (b).

It is probably best to start with the linearized momentum equation in the form of eq. (5.32), with the force per unit mass $\mathbf{f} = -\nabla\psi$, where ψ is the gravitational potential. We consider only the case of a star which in its unperturbed state is purely static and spherically symmetric. We can therefore replace d/dt by $\partial/\partial t$, and we assume that all variations are proportional to $\exp(i\sigma t)$. After some manipulation and with use of the mass equation in the form of eqs. (17.22) and (17.27), and of the momentum equation for the unperturbed model ($\rho^{-1}\nabla P = \mathbf{g}$), the linearized momentum equation can be solved for $\nabla(\delta P/P)$. The right side of the resulting equation can be expressed as $1/\lambda_P$ times a quantity which contains, among other things, $\delta P/P$, where $\lambda_P = P/(\rho g)$ is the pressure scale height, g being the local gravitational acceleration (see eq. [17.11]). (This equation is analogous to eq. [8.10].) Since $|\nabla(\delta P/P)|$ must remain finite near the surface, and since λ_P will be small or zero here, then it follows that the quantity by which $1/\lambda_P$ is multiplied must likewise be small or zero. Taking the radial component of the above quantity, setting it equal to zero, and solving for $\delta P/P$, we obtain

$$\frac{\delta P}{P} = -\left(\frac{\sigma^2 r^3}{Gm} + 4 - \frac{4\pi\rho r^3}{m}\right)\frac{\delta r}{r} + \frac{l(l+1)}{\sigma^2 r^2}\left(\frac{P'}{\rho} + \psi'\right) + \frac{1}{g}\frac{\partial\psi'}{\partial r}; \quad (17.69)$$

or at least that the right side of eq. (17.69) must be multiplied by the factor $(1 + \mathrm{O}[\lambda_P/R])$, where $\mathrm{O}(\lambda_P/R)$ may represent a power series in λ_P/R whose leading term is of order λ_P/R (Equation [17.69] is analogous to eq. [8.11].) The *surface* value of $\delta P/P$ is of course obtained by inserting on the right side the surface values of all relevant quantities. In particular, one can eliminate the surface value of $\partial\psi'/\partial r$ by combining it with the surface boundary condition (17.71) presented below.

The equation which results from this elimination of $\partial\psi'/\partial r$ is interesting and has been discussed in an enlightening manner by Buta and Smith (1979). Making the assumption that $\delta P = 0$ at the surface and rearranging, we obtain

$$\frac{\delta P}{P} = \left[\frac{l(l+1)}{\sigma^2 r^3} Gm - \frac{\sigma^2 r^3}{Gm} - 4\right]\frac{\delta r}{r} + \left[\frac{l(l+1)}{\sigma^2 r^3} Gm - l - 1\right]\left(\frac{\psi'}{gr}\right). \tag{17.69'}$$

In the Cowling approximation (see §17.9) only the term containing $\delta r/r$ survives.

For purely radial oscillations, $l = 0$, and it is easy to show that

$$\frac{\partial\psi'}{\partial r} = -4\pi G\rho\,\delta r. \tag{17.70}$$

Hence, for this case eq. (17.69) properly reduces to eq. (8.11).

Equations (17.68) or (17.69) (or [17.69']) provide one surface boundary condition (this is the only one required for the purely radial case; see Chapter 8). The other surface boundary condition concerns the perturbation ψ' in the gravitational potential. It is shown in Ledoux and Walraven (1958, §75) that this boundary condition is

$$\frac{\partial\psi'}{\partial r} + \frac{l+1}{R}\psi' = -4\pi G\rho\,\delta r. \tag{17.71}$$

Equations (17.68) or (17.69) (or [17.69']) and (17.71) are the two boundary conditions which must be obeyed at the surface of a spherically symmetric configuration for small, adiabatic nonradial oscillations with vanishing surface pressure. Physically, this last condition expresses the requirements that the Eulerian variation of the gravitational potential, and the gravitational force per unit mass, both be continuous across the (perturbed) stellar surface. Satisfaction of these conditions will determine the values, for given l and σ^2, of the two parameters which were not determined by the central boundary conditions discussed in the preceding subsection.

It will be recalled that there were also two central boundary conditions involved, namely that some *two* of the dependent variables shall remain finite at the center. These two conditions, plus the above two surface boundary conditions, make *four* boundary conditions, in all, that must be satisfied for such oscillations. However, since our fourth-order system of differential equations is homogeneous, then one of the four constants of integration must be left arbitrary, since the normalization is clearly arbitrary. Only for certain values of σ^2, then, can all four boundary conditions be satisfied. Thus, we again have an eigenvalue problem, with σ^2 the eigenvalue. For nonradial oscillations, however, the equations do not form a Sturm-Liouville system except in certain limits (see §17.9), so that, in general, there exists no simple and unique method of ordering the eigenvalues.

Finally, it is interesting to examine the value of the ratio $\delta t/\delta r$ at the stellar surface, where δt and δr are, respectively, the tangential and radial components of the Lagrangian displacement $\delta\mathbf{r}$ of a mass element. It can easily be shown that

$$\left(\frac{\delta t}{\delta r}\right)_{r=R} = \frac{1}{\Omega^2}\left[1 + \frac{\psi'}{(P'/\rho)}\right]_{r=R}, \tag{17.72}$$

where Ω^2 is the dimensionless angular frequency defined in Chapter 8. Since $|\psi'/(P'/\rho)|$ is of order unity (moreover, ψ' is often neglected entirely, see §17.9), it follows that $(\delta t/\delta r)_{r=R} \sim \Omega^{-2}$.

17.7. THE HOMOGENEOUS COMPRESSIBLE MODEL

The homogeneous compressible model (density ρ not a function of position) is of course highly unrealistic. It is nevertheless interesting, in that for this model, the differential equations describing small, adiabatic, nonradial oscillations can be reduced to a system of only the second order in space. Because of the relatively low order of the differential equations for the nonradial oscillations of this model, a fairly simple power-series solution of these equations can be obtained. Note that the assumption stated in Chapter 15, that the surface density vanishes, is not valid for the homogeneous model. The remaining assumptions, however, are still assumed for this model.

Pekeris (1938) was the first to discuss this model, and a good synopsis of this work may be found in Ledoux and Walraven (1958, §76). Pekeris (1938) combined the linearized mass, momentum, and Poisson's equations with the linearized energy equation for adiabatic oscillations, for a general spherically symmetric stellar model in hydrostatic equilibrium, into two simultaneous, second-order differential equations in terms of δr and $\alpha \equiv \nabla \cdot \delta\mathbf{r} (= -\delta\rho/\rho)$ (see §17.1). These two equations become decoupled for the homogeneous model, and one of them becomes a second-order differential equation in α alone. This equation was solved by Pekeris (1938) by a power-series solution. The imposition of boundary conditions then yields fairly simple analytic formulae for the eigenvalues for this model. These analytic formulae provide an introduction to the so-called p- and g-modes of nonradial oscillations, which will be discussed later in this section, in later sections of this chapter, and in later chapters of this book.

For the *homogeneous* model we have, for $r \leq R$,

$$\rho(r) = \text{constant} \equiv \rho_0, \tag{17.73}$$

$$P(r) = \tfrac{2}{3}\pi G\rho_0{}^2 R^2(1 - r^2/R^2) \tag{17.74}$$

(R = equilibrium stellar radius), and

$$A = -\frac{1}{\Gamma_1 P}\frac{dP}{dr}\,(\geq 0). \tag{17.75}$$

The second-order differential equation in α mentioned above may be written, for constant Γ_1, as in eq. (76.6) of Ledoux and Walraven (1958, §76). The appropriate power series solution of this equation shows that $\alpha \propto r^l$ for r near zero. The condition that this power series shall be only a polynomial (Pekeris 1938 showed that this series does not converge for $r = R$) then determines the eigenfrequencies of this model. This condition is that

$$\Omega^2 - \frac{l(l+1)}{\Omega^2} = -4 + \Gamma_1[n(2l + 2n + 5) + 2l + 3]$$
$$\equiv 2D_n \; (n = 0,1,2,\ldots), \tag{17.76}$$

which defines D_n, where Ω is the dimensionless frequency defined in Chapter 8. It is interesting to note that, for $l = 0$ (radial oscillations) eq. (17.76) yields the appropriate eigenfrequencies for this case (see Chap. 8).

Equation (17.76) is a quadratic in Ω^2, with solutions

$$\Omega_{l,n}{}^2 = D_n \pm [D_n{}^2 + l(l+1)]^{1/2}. \tag{17.77}$$

We note that, for given l, n, one of the two roots $\Omega_{l,n}{}^2$ is always negative; this situation corresponds to *dynamical instability*. We shall see below (§17.12) that this behavior is related to the fact that the homogeneous model is convectively unstable (that is, $A > 0$) throughout.

We now consider fixed l and large n. The two sets of roots of eq. (17.77) are then, for $n \gg 1$

$$\Omega_{l,n}{}^2 \approx 2\Gamma_1 n^2 + \frac{l(l+1)}{2\Gamma_1 n^2} \quad (p \text{ modes}), \tag{17.78a}$$

$$\Omega_{l,n}{}^2 \approx -\frac{l(l+1)}{2\Gamma_1 n^2} \quad (g \text{ modes}). \tag{17.78b}$$

For given l, then, we see that there are two spectra, one characterized by eigenvalues (Ω^2) tending toward infinity as $n \to \infty$, and dynamically stable; the other characterized by eigenvalues tending toward zero in absolute value as $n \to \infty$, and dynamically unstable. The first set of eigenvalues and eigenfunctions is called p modes, the second set g modes. (We shall see below that g modes are not always dynamically unstable).

The terminology, originated by T. G. Cowling (1941), stems from the following facts, which are not all applicable to the present homogeneous compressible model (see below). The p modes are predominantly radial and are characterized by relatively large Eulerian pressure and density variations during the oscillations; these modes are sometimes called acoustic modes. The g modes are predominantly transverse and are characterized by relatively small Eulerian pressure and density variations during the oscillations; these modes are sometimes called gravity modes. These gravity modes have been compared to a fisherman's bob floating in the water by Hill (1978). That the motions are mainly transverse for the g modes can easily be seen, for example, from eq. (17.14), in which P' and ψ' have both been neglected. According to this equation, the acceleration of a mass element is proportional to the component δr of $\delta\mathbf{r}$ along the gravitational acceleration $\mathbf{g}$, which is directed radially for a spherical star in hydrostatic equilibrium. The smallness of this acceleration (proportional to $|\sigma^2|$ for the high-order g modes) demands that the motion be characterized by small radial, mainly transverse, displacements. In the words of Cowling (1941), "... the motion [corresponding to g modes] is chiefly horizontal and is due to the action of gravity in attempting to smooth out the density differences on any sphere concentric with the star." We may note that the adiabatic approximation may actually not be very good for very long-period g modes. Other arguments for the facts that p modes are predominantly radial and that g modes are predominantly transverse, will be presented in §17.12.

On the other hand, for $l = n$ and $n \gg 1$, we have for g modes

$$\Omega_{n,n}{}^2 \approx -\frac{1}{4\Gamma_1}, \tag{17.79}$$

which is independent of n. Physically, this case of $l = n$ and $n \gg 1$ corresponds to "blobs" of material having roughly the same size in all dimensions, where this size is much smaller than the stellar radius. As we shall see, g modes of large l and n can be considered to describe, at least in some sense, stellar convection. The independence of $\Omega_{n,n}{}^2$ on n in this limit corresponds to the fact that the Brunt-Väisälä frequency (see §17.2) is independent of the dimensions of a fluid element.

For the homogeneous model, each pair of p and g modes for given l,n has the *same* eigenfunction $\alpha_{l,n}$ ($= -\delta\rho/\rho_0 = -\rho'/\rho_0$ for the homogeneous model), as can be seen from the equations presented in Ledoux and Walraven (1958, §76). The same statement also applies to $\psi'_{l,n}$ and $\delta P_{l,n}$, but not to $\delta r_{l,n}$.

A discussion of this case by Sauvenier-Goffin (1951) (summarized in Ledoux and Walraven 1958, §76) shows the following.

For the p modes, δr has one more node than does α. For the g modes, δr and α have the same number of nodes. For the lowest mode, $n = 0$, we have the following results for p and g modes having $l > 0$ (we do not count any nodes at the stellar center): For the p mode having $n = 0$, α, ψ', δP, and $\delta \rho$ have *no* nodes, whereas δr has *one* node. For the g mode having $n = 0$, α, ψ', δP, $\delta \rho$, and δr have *no* nodes (that is, each quantity has the same sign for all r in $0 \leq r \leq R$). It may be noted that the $n = 0$ p mode is normally regarded as the closest nonradial analogue to the fundamental mode of purely radial oscillations of this model. More generally, this statement applies to the lowest p mode of any stellar model (see below).

Another pure mode, a particular case of a class of Kelvin modes, is possible for the homogeneous compressible model, as was first pointed out by Chandrasekhar (1964d). This mode is characterized by having, for this model, purely solenoidal (divergence-free) displacements, that is, $\nabla \cdot \delta \mathbf{r} = 0$. It then follows that $\delta P = 0$, $\delta \rho = 0$, just as for an incompressible fluid, where the first of these relations is valid for adiabatic oscillations. The eigenvalues for this mode are (Chandrasekhar 1964d, Smeyers 1966)

$$\Omega_f^2 = \frac{2l(l-1)}{2l+1}. \tag{17.80}$$

This expression also applies to the homogeneous incompressible sphere, first investigated by Kelvin (Thomson 1863). This mode, in fact, is the only possible mode of oscillation for this model, and the term "Kelvin" derives from this case. This is the f mode of Cowling (1941), and it is not, for general models, divergence-free, as was shown by Robe (1965). For this reason, the f mode for general models is sometimes called the pseudo-Kelvin mode.

We may note that (1) $\Omega_f^2 = 0$ for $l = 0$ or $l = 1$; (2) Ω_f^2 is independent of Γ_1 (this is a result of the facts that $\delta P = 0$ and $\delta \rho = 0$); and (3) Ω_f^2 is independent of n, that is, only the lowest mode ($n = 0$) exists.

By applying the entire fourth-order system of equations to the homogeneous model (as opposed to reducing these equations to a second-order system), Smeyers (1966) showed that the f mode was naturally obtained as part of his results (see also Hurley, Roberts, and Wright 1966 and Deupree 1974). Smeyers' (1966) analysis showed that for the solenoidal Kelvin modes of the homogeneous compressible model, the radial component δr of

TABLE 17.1

Dimensionless Eigenvalues Ω^2 for Small, Adiabatic Oscillations of the Homogeneous Compressible Model ($\Gamma_1 = 5/3, \Omega^2 = 3\sigma^2/4\pi G\rho_0 = \sigma^2 R^3/GM$)

l	n = 0		1		2	
0 (radial)	1.000		12.677		31.000	
	4.754	(p)	19.44	(p)	41.05	(p)
1	—	(f)	—		—	
	−0.4207	(g)	−0.1029	(g)	−0.04872	(g)
	8.382	(p)	26.23	(p)	51.12	(p)
2	0.8000	(f)	—		—	
	−0.7158	(g)	−0.2288	(g)	−0.1174	(g)
3	12.00	(p)	33.03	(p)	61.20	(p)
	1.714	(f)	—		—	
	−1.000	(g)	−0.3633	(g)	−0.1961	(g)
	15.61	(p)	39.84	(p)	71.28	(p)
4	2.667	(f)	—		—	
	−1.281	(g)	−0.5021	(g)	−0.2806	(g)

the vector displacement $\delta\mathbf{r}$ of a mass element was given by $\delta r = \text{const.}\ r^{l-1}$. The tangential component of $\delta\mathbf{r}$ was $\delta t = (\delta r)/l$.

Some numerical results for the small, adiabatic oscillations of the homogeneous, compressible model with $\Gamma_1 = 5/3$ are given in Table 17.1, based in part on results given in Table 15 of Ledoux and Walraven (1958).

17.8. GENERAL DESCRIPTION OF p, g, AND f MODES

In this section we shall present a general description of certain features of nonradial stellar oscillations; here we do not consider the class of toroidal modes (see §17.3). This general description is based on the linear, adiabatic theory, but the assumption is usually made that such effects as nonlinearity or nonadiabicity (especially if small) do not significantly modify the qualitative features to be described here. Some nonadiabatic effects will be discussed briefly in Chapter 18. Also, most of the remarks in this section are subject to the important qualification that they are restricted to sufficiently simple stellar models. By this is meant models which are chemically homogeneous, do not have several spatially alternating radiative and convective zones, and are not too highly centrally condensed. Such more complicated models reveal qualitatively different

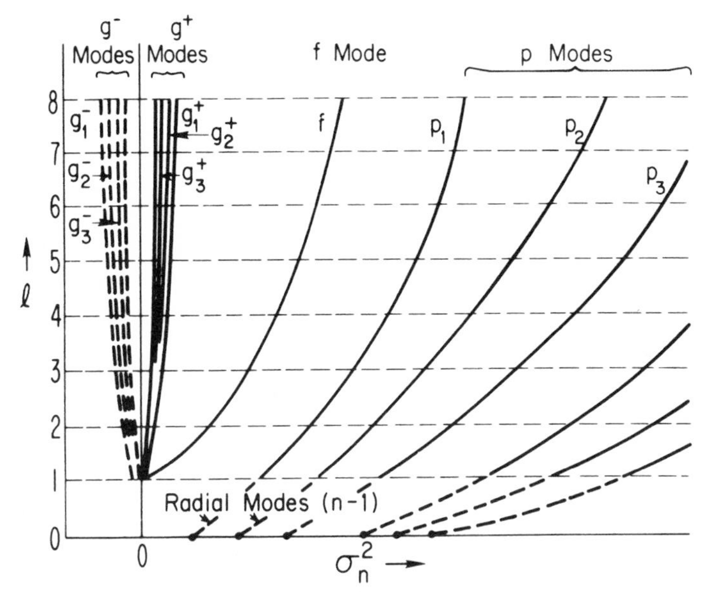

Figure 17.2. Eigenvalues $\sigma_n{}^2 = (2\pi/\text{Period})^2$ of linear, adiabatic, nonradial oscillations for various modes (n) versus the angular spherical harmonic index l (schematic). Shown are the four types of nonradial oscillations (p, f, g^+, and g^-) of the spheroidal modes as originally classified by Cowling (1941). The extensions (shown by dashed lines) of the p modes to the horizontal axis give the corresponding eigenvalues for $n - 1$ for the radial ($l = 0$) modes.

oscillatory characteristics, as compared with simpler models, and will be discussed in §17.10. The classification of nonradial oscillations into *p*, *g*, and *f* modes for the homogeneous compressible model, following the conventional Cowling (1941) classification scheme, was introduced in §17.7. Throughout this section and in the remainder of this book our designations "*p*," "*g*," and "*f*," will be in accordance with the above conventional scheme.

The spectra of the linear, adiabatic, nonradial oscillations of normal stars are illustrated schematically in Figure 17.2, which is qualitatively similar to a diagram used by Smeyers (1967) and others. In this diagram the squares of the angular oscillation frequencies $\sigma_n{}^2$ increase along the horizontal axis, where n denotes the order of the mode ($n = 1,2, \ldots$)[2]; and l, the spherical harmonic index, increases along the vertical axis. The purely radial eigenfrequencies (corresponding to $l = 0$) are indicated by points along the horizontal axis.

[2]Note that for *general* models it is conventional to let the lowest mode of nonradial oscillations correspond to $n = 1$, whereas the lowest mode of the *homogeneous compressible* model corresponded to $n = 0$ (see §17.7).

The diagram shows that the nonradial eigenfrequencies are normally clearly separated into the four distinct classes of modes: the p modes, the single f mode (for $l \geq 2$), the g^+ modes, and the g^- modes. In general, the p modes are characterized by relatively large Eulerian pressure variations. These variations, in fact, are responsible for most of the restoring forces which act during the oscillations. Note that the σ_n^2 increase both with increasing n and increasing l.

The g modes are characterized by relatively small Eulerian pressure variations for a given (vector) displacement of a mass element (see §17.12 below). For these modes most of the restoring force is due to the (relatively weak) force of gravity (see below and §17.12). In this case, for given $l, |\sigma_n^2| \rightarrow 0$ as $n \rightarrow \infty$. On the other hand, it often happens that, for given n, $|\sigma_n^2|$ becomes independent of l for sufficiently large l (see §17.12); this, however, is not true for the homogeneous model (see §17.7). Finally, if $n \approx l$ as $n \rightarrow \infty$, then $|\sigma_n^2|$ becomes independent of n (see §17.10). The g^+ modes are dynamically stable ($\sigma_n^2 > 0$), while the g^- modes are dynamically unstable ($\sigma_n^2 < 0$). (The difference in sign of σ_n^2 is the only general defining difference between g^+ and g^- modes.) The criterion that determines which of these two cases prevails is the quantity A, defined and discussed in §17.2. Further discussion of the relation between g^+ and g^- modes and the sign of A will be provided below.

The unique f, or Kelvin, mode exists in stars only for the lowest mode ($n = 0$ in the homogeneous model, $n = 1$ in general models) and for $l \geq 2$. Its eigenvalue (σ^2) is, for sufficiently simple models, intermediate between the eigenvalues for the p_1 and g_1 modes, for all $l \geq 2$ (we are here taking $n = 1$ for the lowest mode).

We note from Figure 17.2 that the f and g modes do not exist for $l = 0$. We note also that when the curves in Figure 17.2 for the p_n modes for $l > 0$ are extended down to $l = 0$, the points of intersection with the horizontal line for $l = 0$ give the eigenvalues (σ_n^2) for the corresponding purely radial modes (see remarks at end of paragraph). For this reason the p_n mode is usually regarded as the nonradial analogue of the purely radial mode for $n - 1$. There are no radial analogues for the f and g modes. Although the g^+ modes may appear from Figure 17.2 to have $\sigma^2 = 0$ for $l = 1$, actually they have $\sigma^2 > 0$ for this value of l. Therefore the line for the f mode (for which $\sigma^2 = 0$ for $l = 1$) would appear to have to cross all of the lines for the g^+ modes as it approaches $\sigma^2 = 0$. These crossings actually take place in a series of "bumpings" or "avoided crossings" as discussed by Aizenman, Smeyers, and Weigert (1977). They also discuss the transition of the p_n nonradial modes to the corresponding radial modes for $n - 1$ as $l \rightarrow 0$.

The dipole mode, $l = 1$, was for some time following the work of Pekeris (1938), believed to be unphysical because it can correspond to a displacement of the center of mass of the system, as must in fact be the case with an incompressible sphere. However, Smeyers (1966) showed that $l = 1$ *was* physically significant for a compressible sphere, at least for adiabatic oscillations. In this case there is no displacement of the center of mass of the system, but there *is* a displacement of the geometrical center of the object (see §17.6a). For all other values of l, the center of mass and the geometrical center always coincide, and their common displacement is zero. The proof that the center of mass does not change for $l = 1$ in the more general case of nonadiabatic oscillations has been effected by Christensen-Dalsgaard (1976).

We consider now the *eigenfunctions* of small, adiabatic, nonradial oscillations. We base much of the following discussion on Smeyers (1967), and many of the following remarks are based on the accumulated experience derived from examination of numerous numerical results. The validity of many of these remarks may therefore not necessarily be obvious from the equations. Also, the generality of many of these remarks is somewhat uncertain. General proofs have never been presented; indeed, such general proofs probably do not exist. These remarks are almost certainly valid for simpler stellar models, but exceptions to their validity are not excluded for more complicated models (see also §17.10 below). Also, for orientation we have summarized in Table 17.2 certain information regarding nonradial (and radial, for comparison), adiabatic oscillations of polytropes with $0 \leq \nu \leq 4$, based on the work of Robe (1968), Hurley, Roberts and Wright (1966), and Schwank (1975, 1976). Here (and here alone) ν denotes the polytropic index. Note from eq. (17.33) that P' is proportional to δt, the tangential component of $\delta \mathbf{r}$, when the perturbation ψ' in gravitational potential is neglected (the Cowling approximation, see §17.9). In Table 17.2 the values of the pulsation constant Q (see, e.g., eq. [2.4]), with units in days (denoted by subscript "d" in the table), are given. Further numerical results are given in Ledoux (1974) and in J. P. Cox (1976a); for numerical results concerning upper main sequence stars, see Aizenman, Hansen, and Ross (1975).

For the p modes, $\delta r/r$ has a node closer to the center of the star than does $\delta t/r$, and vice versa for the g modes; this property seems to persist even for the highly centrally concentrated models (see §17.10 below). For the f mode, $\delta r/r$ and $\delta t/r$ have *no* nodes, except in the case of polytropes with large polytropic index (≥ 3.25) or of more complicated models. Also, at least for the simpler models, δr and ρ' have the same sign for this mode at every radial distance. From this fact and the relation, eq. (5.16), between Eulerian and Lagrangian variations, it follows that $\delta\rho$ is relatively small everywhere for this mode. For the homogeneous (compressible or incompressible) model, as we have seen, $\delta\rho \equiv 0$ everywhere for this mode. For all

TABLE 17.2a

Results for Polytropes ($\Gamma_1 = 5/3, l = 2$)*
$\nu = 0$

Mode	$\sigma^2 R^3/GM$	Q_d	Number of nodes in δr	P'	ρ'
p_1	8.38	4.00	1	1	1
f	0.800	12.96	0	0	0
g_1	−0.72	—	0	1	1
g_2	−0.22	—	1	2	2
g_3	−0.117	—	2	3	3
F	1	11.59	0	—	0
$1H$	12.67	3.256	1	—	1
$2H$	31.00	2.082	2	—	2

TABLE 17.2b

Results for Polytropes ($\Gamma_1 = 5/3, l = 2$)*
$\nu = 1$

Mode	$\sigma^2 R^3/GM$	$Q_d \times 10^2$	Number of nodes in δr	P'	ρ'
p_{10}	312	0.656	10	10	10
p_9	260	0.719	9	9	9
p_8	211.6	0.7968	8	8	8
p_7	168.4	0.8931	7	7	7
p_6	129.9	1.017	6	6	6
p_5	96.15	1.182	5	5	5
p_4	67.08	1.415	4	4	4
p_3	42.83	1.771	3	3	3
p_2	23.49	2.391	2	2	2
p_1	9.308	3.799	1	1	1
f	1.498	9.470	0	0	0
g_1	−0.3029	—	0	1	1
g_2	−0.1383	—	1	2	2
g_3	−0.08025	—	2	3	3
g_4	−0.05272	—	3	4	4
g_5	−0.03742	—	4	5	5
g_6	−0.02798	—	5	6	6
g_7	−0.02174	—	6	7	7
g_8	−0.01739	—	7	8	8
g_9	−0.01424	—	8	9	9
g_{10}	−0.01187	—	9	10	10
F	1.892	8.426	0	—	—
$1H$	12.09	3.333	1	—	—
$2H$	27.08	2.227	2	—	—

*For comparison, we have included in the last three rows the results for adiabatic radial oscillations ($l = 0$), from Hurley, Roberts, and Wright (1966). Here "F", "$1H$," and "$2H$" mean, respectively, "fundamental," "first overtone," and "second overtone" (see footnote in §8.12c).

TABLE 17.2c

Results for Polytropes ($\Gamma_1 = 5/3, l = 2$)*
$\nu = 1.5$

Mode	$\sigma^2 R^3/GM$	$Q_d \times 10^2$	Number of nodes in δr	P'	ρ'
p_{10}	—	—	—	—	—
p_9	—	—	—	—	—
p_8	—	—	—	—	—
p_7	—	—	—	—	—
p_6	—	—	—	—	—
p_5	—	—	—	—	—
p_4	—	—	—	—	—
p_3	41.31	1.803	—	—	—
p_2	23.52	2.390	—	—	—
p_1	10.29	3.613	—	—	—
f	2.119	7.962	—	—	—
g_1	0	∞	—	—	—
g_2	0	∞	—	—	—
g_3	0	∞	—	—	—
g_4	0	∞	—	—	—
g_5	0	∞	—	—	—
g_6	0	∞	—	—	—
g_7	0	∞	—	—	—
g_8	0	∞	—	—	—
g_9	0	∞	—	—	—
g_{10}	0	∞	—	—	—
F	2.706	7.046	0	—	—
$1H$	12.54	3.273	1	—	—
$2H$	26.58	2.248	2	—	—

TABLE 17.2d

Results for Polytropes ($\Gamma_1 = 5/3, l = 2$)*
$\nu = 2$

Mode	$\sigma^2 R^3/GM$	$Q_d \times 10^2$	Number of nodes in δr	P'	ρ'
p_{10}	264	0.713	10	10	10
p_9	221	0.780	9	9	9
p_8	182	0.859	8	8	8
p_7	145.7	0.9602	7	7	7
p_6	113.7	1.087	6	6	6
p_5	85.50	1.253	5	5	5
p_4	61.13	1.482	4	4	4
p_3	40.63	1.818	3	3	3
p_2	24.07	2.362	2	2	2
p_1	11.56	3.409	1	1	1
f	3.113	6.569	0	0	0
g_1	0.5633	15.44	1	1	1
g_2	0.2968	21.27	2	2	2
g_3	0.1839	27.03	3	3	3
g_4	0.1254	32.73	4	4	4
g_5	0.09112	38.40	5	5	5
g_6	0.06928	44.03	6	6	6
g_7	0.05449	49.65	7	7	7
g_8	0.04401	55.25	8	8	8
g_9	0.03630	60.83	9	9	9
g_{10}	0.03046	66.41	10	10	10
F	4.001	5.794	0	—	—
$1H$	13.34	3.173	1	—	—
$2H$	26.58	2.248	2	—	—

*For comparison, we have included in the last three rows the results for adiabatic radial oscillations ($l = 0$), from Hurley, Roberts, and Wright (1966). Here "F", "$1H$," and "$2H$" mean, respectively, "fundamental," "first overtone," and "second overtone" (see footnote in §8.12c).

TABLE 17.2e

Results for Polytropes ($\Gamma_1 = 5/3, l = 2$)*
$\nu = 3$

Mode	$\sigma^2 R^3/GM$	$Q_d \times 10^2$	Number of nodes in δr	P'	ρ'
p_{10}	234	0.758	10	10	10
p_9	196	0.828	9	9	9
p_8	163	0.908	8	8	8
p_7	132.4	1.007	7	7	7
p_6	104.8	1.132	6	6	6
p_5	80.55	1.291	5	5	5
p_4	59.42	1.504	4	4	4
p_3	41.47	1.800	3	3	3
p_2	26.72	2.242	2	2	2
p_1	15.26	2.967	1	1	1
f	8.175	4.054	0	0	0
g_1	4.915	5.228	1	1	1
g_2	2.828	6.892	2	2	2
g_3	1.822	8.586	3	3	3
g_4	1.270	10.28	4	4	4
g_5	0.9360	11.98	5	5	5
g_6	0.7188	13.67	6	6	6
g_7	0.5691	15.36	7	7	7
g_8	0.462	17.1	8	8	8
g_9	0.382	18.8	9	9	9
g_{10}	0.322	20.4	10	10	10
F	9.255	3.810	0	—	—
$1H$	16.98	2.813	1	—	—
$2H$	28.48	2.172	2	—	—

TABLE 17.2f

Results for Polytropes ($\Gamma_1 = 5/3, l = 2$)*
$\nu = 3.25$

Mode	$\sigma^2 R^3/GM$	$Q_d \times 10^2$	Number of nodes in δr	P'	ρ'
p_{10}	227	0.769	10	10	10
p_9	192	0.836	9	9	9
p_8	160	0.916	8	8	8
p_7	130	1.02	7	7	7
p_6	103.6	1.139	6	6	6
p_5	80.18	1.294	5	5	5
p_4	59.66	1.500	4	4	4
p_3	42.22	1.784	3	3	3
p_2	27.88	2.195	2	2	2
p_1	16.96	2.814	1	1	1
f	11.24	3.457	0	2	0
g_1	7.972	4.105	1	1	1
g_2	4.857	5.259	2	2	2
g_3	3.172	6.508	3	3	3
g_4	2.224	7.772	4	4	4
g_5	1.644	9.039	5	5	5
g_6	1.264	10.31	6	6	6
g_7	1.003	11.57	7	7	7
g_8	0.818	12.8	8	8	8
g_9	0.674	14.1	9	9	9
g_{10}	0.569	15.4	10	10	10
F	11.03	3.490	0	—	—
$1H$	18.89	2.667	1	—	—
$2H$	29.87	2.121	2	—	—

*For comparison, we have included in the last three rows the results for adiabatic radial oscillations ($l = 0$), from Hurley, Roberts, and Wright (1966). Here "F", "$1H$," and "$2H$" mean, respectively, "fundamental," "first overtone," and "second overtone" (see footnote in §8.12c).

TABLE 17.2g

Results for Polytropes ($\Gamma_1 = 5/3, l = 2$)
$\nu = 3.50$

Mode	$\sigma^2 R^3/GM$	$Q_d \times 10^2$	Number of nodes in		
			δr	P'	ρ'
p_{10}	223	0.776	10	10	10
p_9	189	0.843	9	9	9
p_8	158	0.922	8	8	8
p_7	128	1.02	7	7	7
p_6	102.9	1.142	6	6	6
p_5	80.10	1.295	5	5	5
p_4	60.29	1.493	4	4	4
p_3	43.41	1.759	3	3	3
p_2	29.76	2.124	2	2	2
p_1	20.93	2.533	1	1	1
f	16.16	2.883	2	2	2
g_1	12.10	3.332	1	3	1
g_2	8.542	3.966	2	2	2
g_3	5.741	4.837	3	3	3
g_4	4.063	5.750	4	4	4
g_5	3.017	6.673	5	5	5
g_6	2.324	7.603	6	6	6
g_7	1.848	8.526	7	7	7
g_8	1.50	9.46	8	8	8
g_9	1.24	10.4	9	9	9
g_{10}	1.05	11.3	10	10	10
F	12.64	3.260	0	—	—
$1H$	21.21	2.516	1	—	—
$2H$	32.08	2.046	2	—	—

TABLE 17.2h

Results for Polytropes ($\Gamma_1 = 5/3, l = 2$)*
$\nu = 4$

Mode	$\sigma^2 R^3/GM$	$Q_d \times 10^2$	Number of nodes in		
			δr	P'	ρ'
p_{10}	216	0.789	10	10	10
p_9	184	0.854	9	9	9
p_8	154	0.934	8	8	8
p_7	128	1.02	7	7	7
p_6	105.2	1.130	6	6	6
p_5	87.45	1.239	5	5	5
p_4	76.65	1.324	4	6	6
p_3	62.86	1.462	5	5	5
p_2	50.81	1.626	4	4	4
p_1	42.14	1.785	3	5	5
f	34.33	1.978	4	4	4
g_1	27.59	2.206	3	5	5
g_2	23.00	2.417	4	4	4
g_3	17.99	2.732	5	5	5
g_4	15.36	2.957	4	4	4
g_5	12.76	3.245	5	5	5
g_6	10.08	3.650	6	6	6
g_7	8.085	4.076	7	7	7
g_8	6.62	4.50	8	8	8
g_9	5.50	4.94	9	9	9
g_{10}	4.65	5.37	10	10	10
F	15.15	2.978	0	—	—
$1H$	24.94	2.321	1	—	—
$2H$	37.07	1.904	2	—	—

*For comparison, we have included in the last three rows the results for adiabatic radial oscillations ($l = 0$), from Hurley, Roberts, and Wright (1966). Here "F", "$1H$," and "$2H$" mean, respectively, "fundamental," "first overtone," and "second overtone" (see footnote in §8.12c).

nonradial modes, $\delta r/r$ and $\delta t/r$ have the same number of nodes (Smeyers 1967); however, this property may not pertain to more complicated models.

For the p modes, δr always has at least one more node for nonradial oscillations of a given mode (say n) than does δr for the corresponding mode ($n - 1$) of purely radial oscillations. Also, the amplitude of $\delta r/r$ for the p modes is generally small in most of the star, becoming relatively large near the surface. On the other hand, $\delta r/r$ for the g modes can be relatively large in the deep stellar interior (see §17.13 below). (However, $\delta r/r$ is relatively small throughout most of the interior for g modes in white dwarf stars, for the reasons given near the end of §18.3.)

Note also that for sufficiently simple stellar models there are simple relations among the various pulsation quantities. For example, for such models the number of nodes (zeros) in the radial component δr of $\delta \mathbf{r}$ is equal to the order of the mode (not counting the node at $r = 0$ for $l \neq 1$) for the p and g^+ modes, and to zero for the f mode.

Certain properties of the g-mode eigenfunctions for massive stellar models ($\sim 30\ M_\odot - 200\ M_\odot$, $M_\odot$ = solar mass) consisting of a convective core and a radiative exterior are shown in several figures (which will not be reproduced here) in Smeyers (1967). These figures show that there is no node in $\delta\rho$ (and hence none also in δP, for adiabatic oscillations) corresponding to the innermost node in δr for these g^+ modes. In contrast to the case of radial oscillations, a node in δr does not necessarily imply the existence of a node in $\delta\rho$ nearby. This conclusion follows because of the presence of the additional term on the right side of eq. (17.27) for div $\delta\mathbf{r}$ ($\propto |\delta\rho|$) for nonradial oscillations: horizontal displacements can also compress or expand the material. It is also true from these results that, except possibly near the stellar surface, the signs of δr and of $\delta\rho$ are opposite from each other for these g^+ modes. This fact implies, in general, from eq. (5.16) between Eulerian and Lagrangian variations, that for such modes in such a star the term ρ' is smaller in value than the term $\delta r\, d\rho/dr$ (we are assuming that $d\rho/dr < 0$); the extreme case would be $\rho' = 0$. Similarly, P' ought to be relatively small for these modes. In a general way, then, we see that for such oscillations the restoring force per unit mass, $|-\rho^{-1}\nabla P' + \rho'\rho^{-2}\nabla P - \nabla\psi'|$, ought to be relatively small. The oscillation frequency could accordingly also be quite small, as is indeed the case with g^+ modes. More physically, we may say that each mass element that moves into a given location has almost the same physical characteristics as the surrounding material (which is, by assumption, in hydrostatic equilibrium), and is itself not far from hydrostatic equilibrium. The oscillation frequencies may therefore be very small. On the other hand, with p modes (or with ordinary sound waves) ρ' and P' may be relatively large, strong restoring forces may be involved, and oscillation frequencies may be very high.

Christensen-Dalsgaard (1979) has shown that the angular frequency σ for g modes must always be smaller than the highest maximum of the Brunt-Väisälä frequency N.

The spatially oscillatory character of dynamically stable (g^+, $\sigma^2 > 0$) and unstable (g^-, $\sigma^2 < 0$) g modes in radiative ($A < 0$) or convective ($A > 0$) regions of stellar models has been discussed by Ledoux and Smeyers (1966) and by Scuflaire (1974a). Both studies were based on the Cowling approximation (see §17.9), in which the perturbation ψ' of the gravitational potential is neglected. Ledoux and Smeyers (1966) considered only the asymptotic limit in which $|\sigma^2|$ is small, whereas no restrictions were imposed on the size of $|\sigma^2|$ by Scuflaire (1974a). The conclusion of these studies is that the eigenfunctions of g^- modes can be oscillatory only in convectively unstable regions ($A > 0$). On the other hand, the eigenfunctions of g^+ modes can be oscillatory only in radiative regions ($A < 0$), unless these regions are in the outer parts of a star. Here the oscillatory character of these eigenfunctions is independent of the sign of A. In regions where the eigenfunctions are not required to be oscillatory, they generally decrease exponentially with increasing distance from the boundaries of the oscillatory region(s). The situation is illustrated schematically in Figure 17.3 for a stellar model consisting of a convective core, a convective envelope, and a radiative intermediate zone.

Finally, the connection of the g modes with convective instability (sign of A, see §17.2) is interesting. The following remarks are based on the rigorous demonstrations of Lebovitz (1965a,b; 1966) for small, adiabatic, nonradial oscillations of spherical stars. See also the illuminating discussion of Ledoux and Smeyers (1966), based on the neglect of the perturbations of the gravitational potential ψ' (the Cowling approximation, see §17.9). (These remarks do not apply to the toroidal modes mentioned in §17.3 and called trivial modes by Lebovitz 1965b.) If $A < 0$ (convective stability) everywhere in a star, then $\sigma_n^2 > 0$ for all n. In other words, g^- modes do not exist in this case. If $A = 0$ in any finite subinterval of $0 \leq r \leq R$, then there exist neutral modes ($\sigma_n^2 = 0$). In particular, if $A = 0$ everywhere in a star, then $\sigma_n^2 = 0$ (all n) for the g modes; that is, such modes do not exist. The vanishing of A in some finite subinterval of $0 \leq r \leq R$ has, in fact, been shown to be necessary and sufficient that $\sigma_n^2 = 0$ for some modes by Lebovitz (1965b). On the other hand, it has been shown by Lebovitz (1966) that the condition $A > 0$ in some finite subinterval of $0 \leq r \leq R$ is a necessary and sufficient condition that dynamically unstable modes ($\sigma_n^2 < 0$) exist. In particular, g^+ modes do not exist if $A > 0$

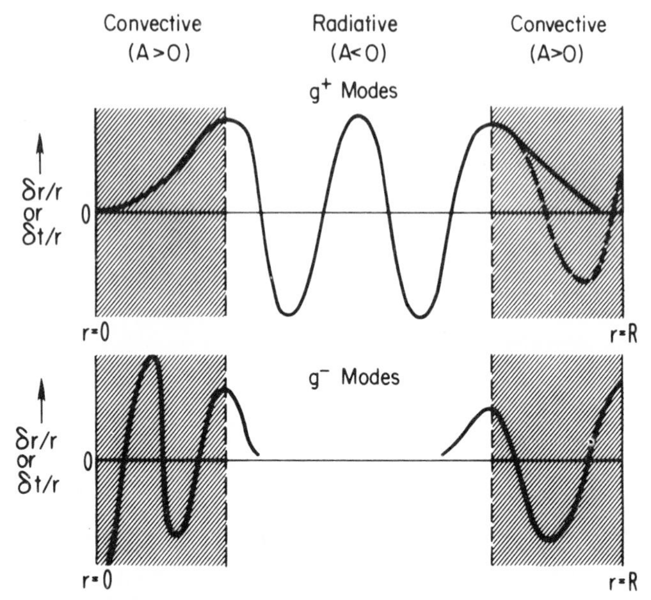

Figure 17.3. Eigenfunctions (schematic) of g^+ modes (top) and g^- modes (bottom) in a stellar model consisting of a convective core ($A > 0$), a convective envelope ($A > 0$), and a radiative intermediate zone ($A < 0$). Here δr and δt denote, respectively, the radial and transverse components of the Lagrangian displacement of a mass element, and r denotes equilibrium radial distance. In the top figure, the solid and dashed curves show two possibilities for the behavior of the eigenfunction of g^+ modes in the outer convective envelope.

everywhere. Finally, if $A < 0$ in some parts of the star and $A > 0$ in other parts, then both g^+ and g^- modes exist. In fact, it follows from the above conclusions that the presence of both g^+ and g^- modes *requires* that A have one sign in some part(s) of the star, and the opposite sign in some other part(s) of the star. It is also useful to note that, for sufficiently small $|A|$,

$$\sigma_n^2 \propto \langle -A \rangle \tag{17.81}$$

for the g modes, where angular brackets denote an appropriate average over the star (see Ledoux and Walraven 1958, §79).

The connection between g modes and convective instability has also been discussed by Sobouti (1977).

17.9. THE COWLING APPROXIMATION

In one of the earliest investigations of small, adiabatic, nonradial oscillations in stars, in which the terms "p, f, and g modes" were first introduced, Cowling (1941) neglected the (Eulerian) perturbations ψ' in gravitational potential altogether. The qualitative justification for this neglect is that ψ' is more or less an average, over the whole star, of the effects of the gravitational potential at each point in the star. This averaging effect should tend to smooth out local fluctuations in the gravitational potential. This expectation is certainly intuitively plausible, and has in fact been shown to be the case, especially for high-order modes (large n and/or l) and for strongly centrally concentrated stellar models, by Cowling (1941), Sauvenier-Goffin (1951), Kopal (1949), Owen (1957), Robe (1968), and others. This approximation of neglecting ψ' is usually called "Cowling's approximation" or "the Cowling approximation," and it is usually assumed not to alter the basic mathematical nature of the problem (see, e.g., Ledoux and Walraven 1958, §79). This approximation is nearly always made in calculations by meteorologists, oceanographers, and the like (see, e.g., Tolstoy 1963, 1973; Eckart 1960).

That $|\psi'|$ should be small for the higher-order modes may be seen, for example, from Poisson's equation. If we replace d/dr by ik, where k is the radial wave number (see remarks near the beginning of §17.12 below), then we can easily derive the result

$$|\psi'| \propto \left[k^2 + \frac{l(l+1)}{r^2}\right]^{-1}. \tag{17.82}$$

This relation shows that, indeed, $|\psi'| \to 0$ as k or $l \to \infty$, that is, for "high-order" modes.

If ψ' is dropped, the remaining equations clearly form only a second-order system which has been discussed at some length mostly from a mathematical standpoint by Ledoux and Walraven (1958, §79); Ledoux (1974, 1978); Scuflaire (1974a,b); Osaki (1975); Wolff (1979); Shibahashi (1979); Christensen-Dalsgaard (1979); and others. Some of this discussion may be summarized here by introducing two new variables which have been used extensively in most of the above discussions:

$$v \equiv uP^{1/\Gamma_1} = r^2 \delta r P^{1/\Gamma_1}, \tag{17.83a}$$

$$w \equiv y\rho P^{-1/\Gamma_1} = P'/P^{1/\Gamma_1}. \tag{17.83b}$$

In the Cowling approximation, ψ' is neglected; if we also assume that Γ_1 is constant (as we shall, unless indicated otherwise, in the remainder of this section), the basic equations become

$$\frac{dv}{dr} = \left[\frac{l(l+1)}{\sigma^2} - \frac{r^2\rho}{\Gamma_1 P}\right]\frac{P^{2/\Gamma_1}}{\rho}w, \tag{17.84a}$$

$$\frac{dw}{dr} = \frac{\sigma^2 + Ag}{r^2}\frac{\rho}{P^{2/\Gamma_1}}v. \tag{17.84b}$$

(For a generalization to the case where Γ_1 is not constant, see Osaki 1975.)

We may note that in the Cowling approximation the variable y is proportional to the transverse, or tangential ("horizontal"), component δt of the Lagrangian displacement of a mass element. Hence $w \propto \delta t$. Note also that $v \propto \delta r$.

The equivalent second-order systems in only one variable may be obtained by eliminating v or w from eqs. (17.84). We obtain

$$\frac{d}{dr}\left\{\frac{\rho}{P^{2/\Gamma_1}} \frac{1}{\left[\dfrac{l(l+1)}{\sigma^2} - \dfrac{r^2\rho}{\Gamma_1 P}\right]} \frac{dv}{dr}\right\} - \frac{\sigma^2 + Ag}{r^2} \frac{\rho}{P^{2/\Gamma_1}} v = 0, \tag{17.85a}$$

$$\frac{d}{dr}\left\{\frac{P^{2/\Gamma_1}}{\rho} \frac{r^2}{\sigma^2 + Ag} \frac{dw}{dr}\right\} - \left[\frac{l(l+1)}{\sigma^2} - \frac{r^2\rho}{\Gamma_1 P}\right] w = 0. \tag{17.85b}$$

To these equations we may add the boundary conditions on v and w:

$$\begin{aligned} v(0) &= 0, \qquad v(R) = 0; \\ w(0) &= 0, \qquad w(R) = 0. \end{aligned} \tag{17.86}$$

This last boundary condition on $w(R)$ can be shown to be valid only if the effective polytropic index $\nu_e > 1/(\Gamma_1 - 1)$, assuming that P becomes very small or is zero at the surface. Otherwise, $w(R)$ could be finite or infinite at the surface. Since the above condition is also the condition for stability of the radiative gradient against convection (e.g., Cox and Giuli 1968, Chap. 13), this last boundary condition in eq. (17.86) is valid only if the stellar model is radiative at the surface (which is nearly always the case in actual stellar models; see, e.g., Cox and Giuli 1968, Chap. 20).

From eqs. (17.85) it is now clear, as was first pointed out by Cowling (1941) (see also Ledoux and Walraven 1958, §79), that for $|\sigma^2|$ either very large or very small, these equations assume that Sturm-Liouville form, and the full theory of Sturm-Liouville systems (see, e.g., Ince 1944, Chaps. 9–11) may then be applied to these equations. For example, for σ^2 very large, eq. (17.85a) becomes a Sturm-Liouville equation with σ^2 as the eigenvalue. There should then exist an infinite, discrete set of eigenvalues tending toward infinity for very high-order modes; these modes were called p modes by Cowling (1941). Moreover, this asymptotic form of eq. (17.85a) shows that σ^2 is always positive (see also below). For σ^2 very small, eq. (17.85b) becomes a Sturm-Liouville equation with $1/\sigma^2$ as the eigenvalue. There should then exist an infinite, discrete spectrum of eigenvalues with σ^2 tending toward zero for very high-order modes; these modes were called g modes by Cowling (1941). Moreover, as was first pointed out by Ledoux and Smeyers (1966), this asymptotic form of eq.

(17.85b) shows the following. If $A < 0$ (radiative) everywhere, then $\sigma^2 > 0$; if $A > 0$ (convective) everywhere, then $\sigma^2 < 0$; finally, if $A < 0$ in some parts of the star and $A > 0$ in other parts, then both signs of σ^2 must exist, that is, both g^+ and g^- modes must be present.

Integral expressions for σ^2 in these two limits have been derived from equations analogous to eqs. (17.85) by Ledoux and Walraven (1958, §79); we refer the interested reader to this discussion.

Some calculations summarized in Ledoux and Walraven (1958, §79) (see also Robe 1968) suggest that the Cowling approximation is almost never more than about 20% in error in the case of σ^2, and is often much more accurate than this.

17.10. NONRADIAL MODES FOR COMPLICATED STELLAR MODELS

By "complicated" stellar models we mean, somewhat loosely speaking, those with large central mass concentrations, alternating radiative and convective regions, and/or chemically inhomogeneous models such as those representing evolved stars. The classic examples of such models are the relatively centrally condensed polytropes, that is, those having effective polytropic index (see, e.g., Cox and Giuli 1968, Chap. 12) $\nu_e \geq 3.25$. The nonradial oscillation properties of these models were given in Table 17.2, and were first discussed by Robe (1968). For example, the f mode for $l = 2$ for polytropes having $\nu_e \leq 3.25$ has *no nodes* in δr. As ν_e increases above this value, however, this mode acquires two extra nodes (for $l = 2$, this acquisition occurs at $\nu_e = 3.42$, corresponding to a ratio of central to mean densities of 127.5, according to Scuflaire 1975b), and then four extra nodes, and so on. The adjoining modes behave in the same qualitative way: as the central mass concentration increases, these modes acquire additional nodes in δr, two at a time. At the same time, these intermediate-frequency modes take on the characteristics of gravity modes in the central regions of the models, and of acoustic modes in the outer stellar layers. The causes of these qualitative features will be discussed later in this section and in §17.11. Other examples of such complicated models have been provided by Osaki (1975, 1976) (representing slightly evolved, immediately post-main-sequence stars); Aizenman, Cox, and Lesh (1975) (representing somewhat more evolved post-main-sequence stars); Goossens and Smeyers (1974) (composite polytropes); Dziembowski (1971, 1975, 1977b); Osaki (1977) (representing highly evolved, strongly centrally concentrated stars in the core helium burning phases); and Noels, Boury, Scuflaire, and Gabrial (1974) (representing evolved low-mass post-main-sequence stars having an external convective zone). It is well known that the simple relations mentioned above and in §17.8 regarding modes and number of nodes in δr, and also a number of other generalities mentioned there, break down for these more complicated models. Such a breakdown should actually, from a mathematical standpoint, not be surprising, for the equations of nonradial, adiabatic stellar pulsation, even in the Cowling approximation, do not in general form a Sturm-Liouville system.

Let us write eqs. (17.84) (which are valid only in the Cowling approximation and for Γ_1 constant) as

$$\frac{dv}{dr} = \alpha(r)w, \tag{17.87a}$$

$$\frac{dw}{dr} = \beta(r)v, \tag{17.87b}$$

where $v(\propto \delta r)$ and $w(\propto \delta t)$ were defined in eqs. (17.83), and where

$$\alpha(r) \equiv \left(\frac{S_l^2}{\sigma^2} - 1\right)\left(\frac{\rho}{\Gamma_1 P}\right)\frac{r^2 P^{2/\Gamma_1}}{\rho}, \tag{17.88a}$$

$$\beta(r) \equiv \left(\frac{\sigma^2}{N^2} - 1\right)(-Ag)\frac{\rho}{r^2 P^{2/\Gamma_1}}. \tag{17.88b}$$

Here S_l is a *critical acoustic frequency,* defined by the relation

$$S_l^2 \equiv \frac{l(l+1)\Gamma_1 P}{r^2\rho}; \tag{17.89}$$

moreover, $N = (-Ag)^{1/2}$ is the Brunt-Väisälä frequency (see §17.2). Physically, S_l^{-1} is the time which would be required for a sound wave to traverse one horizontal wavelength $2\pi r/[l(l+1)]^{1/2}$ of the disturbance (see §17.12 below) along the circumference of a circle of radius r about the center of the star; S_l is also sometimes called the *Lamb frequency.*

On the basis of eqs. (17.87), we may discern two qualitatively different types of behavior of the solutions v and w. We consider for the time being only a region (r_1,r_2) of radial distance r within which each of the functions $\alpha(r)$ and $\beta(r)$ retains the same sign. We are implicitly assuming here that σ^2 is somehow known (at least roughly), so that the functions $\alpha(r)$ and $\beta(r)$ can be evaluated (at least so that their signs can be determined) at each radial distance. According to one of the Sturm's comparison theorems (see, e.g., Ince 1944, §10.32), the functions v and w may be oscillatory with r within the interval (r_1,r_2) only if α and β have opposite signs in this interval. These facts were pointed out by Unno (1975) and were further elaborated upon by Osaki (1975) and Scuflaire (1974b); see also Wolff (1979), Shibahashi (1979), and Christensen-Dalsgaard (1979).

Opposite signs of α and β in (r_1,r_2) can be realized in two ways, as inspection of eqs. (17.88) for α and β will show:

(1) σ^2 very large; more specifically, $\sigma^2 > S_l^2, N^2$. In this case $\alpha < 0$ and $\beta > 0$. (We are here assuming that $-A > 0$ [radiative], and that $\sigma^2 \geq 0$. These assumptions will always be made unless we specify otherwise. For cases in which these assumptions are not valid, see the discussion near the end of §17.8; also, see below). For the reasons to be given below and in §17.11 and to comply with Osaki's (1975) and Scuflaire's (1974b) notations, we shall refer to this case of large σ^2 as the acoustic case.

(2) σ^2 very small; more specifically, $\sigma^2 < S_l^2, N^2$. In this case $\alpha > 0$ and $\beta < 0$–at least if $-A > 0$ (radiative). In regions were $A > 0$ (convective), it is seen that $\beta > 0$ always for $\sigma^2 > 0$, since $N^2 < 0$ in this case. If $\sigma^2 > S_l^2$, we will have $\alpha < 0$ and $\beta > 0$ in such regions. Such convective regions must, as pointed out by Scuflaire (1974a), occupy the outermost stellar layers; this eventuality should actually be considered as belonging to case (1) above. For sufficiently small σ^2, however (in particular, $\sigma^2 < S_l^2$), both α and β are positive everywhere in such a convective region, and so the corresponding behavior of v and w must be as described for case (3) below. We shall refer to this case of σ^2 small as the gravity case.

(3) σ^2 is intermediate in value; more specifically, σ^2 lies between S_l^2 and N^2. In this case α and β have the same signs; both are positive if $N^2 < \sigma^2 < S_l^2$ and both are negative if $S_l^2 < \sigma^2 < N^2$. We shall refer to this case as a mixed case. The terms "acoustic" and "gravity" are related to certain properties of, respectively, p modes and g modes (see §7.8).

In cases (1) and (2) v and w may be oscillatory (in space), and in these cases the corresponding time-dependent solutions are said to represent propagating waves. In the cases (3), v and w cannot be oscillatory. The corresponding time-dependent solutions are said to be evanescent (Osaki 1975).

To get a rough idea of the kinds of solutions of eqs. (17.87) that we may expect in the above three cases, let us neglect for the time being the dependence of α and β on r and treat them as constants.[3] Then we may easily combine eqs. (17.87a,b) into two second-order differential equations each in only one variable. For example, we have

$$\frac{d^2 v}{dr^2} = \alpha\beta v, \tag{17.90}$$

with an identical equation for w. If we assume that each of v and w is proportional to the factor exp (ikr), where k is the local radial wave

[3]The more general case where α and β are not constants can easily be treated by a standard transformation of variables. Thus, a general second-order, linear differential equation such as eq. (17.85a), for example, can always be written in the "normal" form, $d^2f/dr^2 + \gamma^2(r)f = 0$ where $\gamma^2(r)$ is a function of r, by means of such a transformation; see, for example, Forsyth (1929), §59.

number, we obtain.

$$k^2 = -\alpha\beta$$

$$= -\left(\frac{S_l^2}{\sigma^2} - 1\right)\left(\frac{\rho}{\Gamma_1 P}\right)\left(\frac{\sigma^2}{N^2} - 1\right)(-Ag). \qquad (17.91)$$

We see that, in case (1) (σ^2 larger than both S_l^2 or N^2), $k^2 > 0$, so that $v(\propto \delta r)$ and $w(\propto \delta t)$ are oscillatory in space, and the corresponding time-dependent solutions represent propagating acoustic waves. In fact, for $\sigma^2 \longrightarrow \infty$, it follows from eq. (17.91) that $k^2 \longrightarrow \sigma^2(\rho/\Gamma_1 P)$, which is the relation to be expected for ordinary sound waves. In case (2) (σ^2 smaller than both S_l^2 and N^2), again $k^2 > 0$, so we again have oscillatory solutions which represent propagating gravity waves. For $\sigma^2 \longrightarrow 0$, k must become very large, and this case therefore corresponds to disturbances with very short radial wavelengths. However, the horizontal wavelengths are much larger (see eq. [17.96] below). In the cases (3), however (σ^2 intermediate in value between S_l^2 and N^2), $k^2 < 0$, so that v and w are nonoscillatory and represent nonpropagating (evanescent) waves. In these evanescent regions the wave amplitude may become exponentially small, as may be seen from eq. (17.91).

How do the frequencies S_l and N vary with radial distance r in a spherical star? Clearly, $S_l \longrightarrow \infty$ as $r \longrightarrow 0$ and $S_l \propto P/\rho$ near the stellar surface, $r \approx R$. Ordinarily, P/ρ vanishes or becomes small near the surface. Calculations show that, typically, S_l decreases more or less monotonically from infinity at $r = 0$ to zero or a very small value at $r = R$. On the other hand, A may be written as the product of a spatially slowly-varying function (which is constant in the case of polytropes) of r times the reciprocal of the pressure scale height, that is, times $\rho g/P$ in the case of a star in hydrostatic equilibrium, where g is the local gravitational acceleration. Since $g \propto r$ as $r \longrightarrow 0$, we have that $|A| \propto r$ and $N^2 = -Ag \propto r^2$ as $r \longrightarrow 0$. Near the surface, $N^2 \propto \rho/P$, and thus N^2 becomes large or infinite there. Thus N^2 increases from zero at $r = 0$ to a large or infinite value at $r = R$. Calculations show that, for the more simple stellar models, this increase of N^2 from center to surface is more-or-less monotonic, and the curves for S_l^2 and N^2 cross at some intermediate point in the star. For more realistic models, the curve of N^2 versus r may exhibit wiggles and irregularities, because A involves the difference between two derivatives.

The curves of S_l^2 and N^2 versus r are shown schematically in Figure 17.4 for a simple stellar model. In this case the acoustic (A) and gravity (G) regions are clearly separated in frequency. The f mode for such models would be represented by a horizontal line in Figure 17.4 passing near the crossing point of the S_l^2 and N^2 curves. The p modes (or g modes) would all

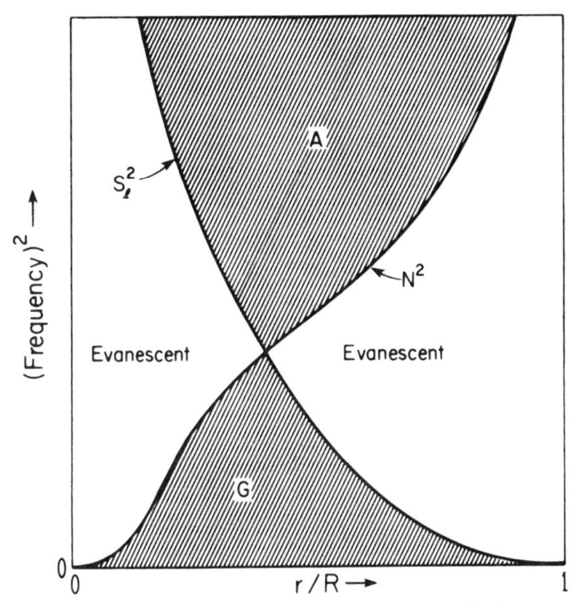

Figure 17.4. Schematic curve of the (critical acoustic frequency)^2 S_l^2 (eq. [17.89]) and of the (Brunt-Väisälä frequency)^2 N^2 (eq. [17.13]) versus fractional radial distance r/R in a "simple" stellar model. The symbols "A" and "G" stand for, respectively, "acoustic" and "gravity." Cross-hatched areas are the regions where the solutions of eqs. (17.87) are oscillatory in space, and correspond to propagating waves.

correspond to horizontal lines in this figure. The horizontal lines for the p modes would lie above this crossing point, whereas these lines for the g modes would lie below this crossing point. For such simple models, then, the relations between modes and number of nodes discussed in §17.8, as well as most of the general remarks made there, apply. Diagrams such as that in Figure 17.4 have been called "propagation diagrams" by Osaki (1975) and are widely used in geophysics to study wave propagation in the earth's oceans and atmosphere (see, e.g., Eckart 1960, Tolstoy 1963, 1973).

For example, S_l^2 and N^2 are plotted by Scuflaire (1974b) for the polytrope of index 3 for $l = 2$ and for $\Gamma_1 = 5/3$, and this plot qualitatively resembles Fig. 17.4. This model is evidently sufficiently simple that the above remarks are borne out.

For more complicated models, however, of which the polytrope of index 4 may be considered an example, we obtain the situation depicted in Figure 17.5 for $l = 2$ and $\Gamma_1 = 5/3$, from Scuflaire (1974b). In this case the A and G regions are no longer clearly separated (vertically): There is a considerable range of intermediate frequencies, embracing the f mode and

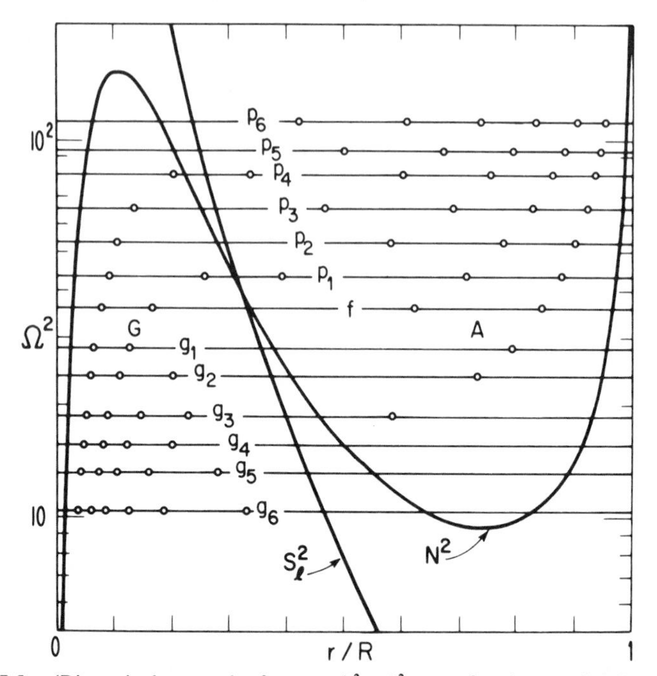

Figure 17.5. (Dimensionless angular frequency)$^2 = \Omega^2$ versus fractional radial distance r/R for a polytrope of index 4 (from Scuflaire 1974b). The solid curves show the squares of the two frequencies S_l for $l = 2$ (eq. [17.89]) and N for $\Gamma_1 = 5/3$ (eq. [17.13]). In this case the A ("acoustic") and G ("gravity") regions are not clearly separated (vertically); there exists a considerable range of "intermediate" frequencies for which the outer parts of the star lie predominantly in the A region, whereas the inner parts lie predominantly in the G region. Horizontal straight lines show the (eigenfrequencies),^{2} appropriately labelled, and small open circles show the positions of the nodes of δr.

several of the adjoining lower p and g modes, in which the inner parts of the star lie predominantly in the G region, and the outer parts lie predominantly in the A region. These intermediate frequency modes, then, possess a mixed character: they behave like gravity waves in the inner parts of the star and like acoustic waves in the outer parts. Note, as shown in Figure 17.5, that nodes in δr may occur in both the G and the A regions. The numerous nodes and the mixed character of these intermediate-frequency modes, as described earlier in this section for the highly centrally condensed polytropes, are accounted for in terms of these A and G regions, as was pointed out by Scuflaire (1974b). Robe's (1968) explanation of these features was essentially the same as that described here, although Robe's explanation was expressed in different terms. Other examples of propagation diagrams may be found in, for example, Ando and Osaki (1975); Ando (1976); Shibahashi and Osaki (1976a,b); Osaki (1977); and Unno, Osaki, Ando, and Shibahashi (1979).

The reason that N^2 possesses two well-defined extrema for these more centrally concentrated models is probably mostly the result of the strong influence of the local gravitational acceleration g. For such models g possesses a strong maximum roughly at the radial distance interior to which most of the interior mass is contained. Physically, the gravitational force is so strong here that the Brunt-Väisälä frequency is locally relatively large. Exterior to this point g varies approximately as r^{-2}; whereas it should be recalled that $g \propto r$ near the stellar center. This behavior of g, together with the earlier remarks in this section about A, probably account mostly for the qualitative features of N^2 as shown in Figure 17.5.

Note that, for frequencies above or below these two extrema in N^2, the waves are confined to be entirely within an A or G region, respectively. Note also that, as is pointed out by Ledoux (1978), the waves in an A region are more and more nearly confined to the surface region, the larger l is. The reason for this is that S_l is proportional to $l(l + 1)$ (see eq. [17.89]). It was pointed out by Osaki (1975) that the evolution of moderately massive main sequence models (~10 solar masses), in which the convective core shrinks in mass as hydrogen is consumed, can result in a relatively thin region in which the chemical composition varies continuously in space, immediately surrounding the convective core. In this case the mean molecular weight μ would increase inward in this region, $-A$ would be enhanced here (see eq. [17.13]) (the material would be even further stabilized against convection), and so the Brunt-Väisälä frequency $N = (-Ag)^{1/2}$ would be relatively large here. The situation is depicted in several propagation diagrams in Osaki (1975, 1976), each corresponding to a different evolutionary stage; see also Figure 17.6. These diagrams show a highly peaked, trapezoid-shaped extension of the G region, in the zone of continuously spatially varying chemical composition, becoming higher and wider as the evolution proceeds (an example is shown in Fig. 17.6). As was explained in the preceding paragraph, the large value of N in this region causes the eigenfunctions to be oscillatory in space here, as was first pointed out by Osaki ($\sigma^2 < S_l^2, N^2$). Osaki (1975) has referred to this oscillatory behavior of the eigenfunctions in these isolated regions of continuously varying chemical composition as a "trapping" of gravity modes. He has also compared these localized regions to the potential wells often encountered in quantum mechanics.

A relatively new phenomenon (at least in the context of pulsating stars) has emerged in the work of Osaki (1975). This phenomenon manifests itself as apparent, abrupt, stepwise changes in the dimensionless frequencies of the nonradial modes during the evolution, as shown in Figure 17.7. According to Osaki, these abrupt changes are related to the appearance of new nodes in the zone of continuously varying chemical composition in the models, as they evolve. A qualitatively similar phenomenon has been

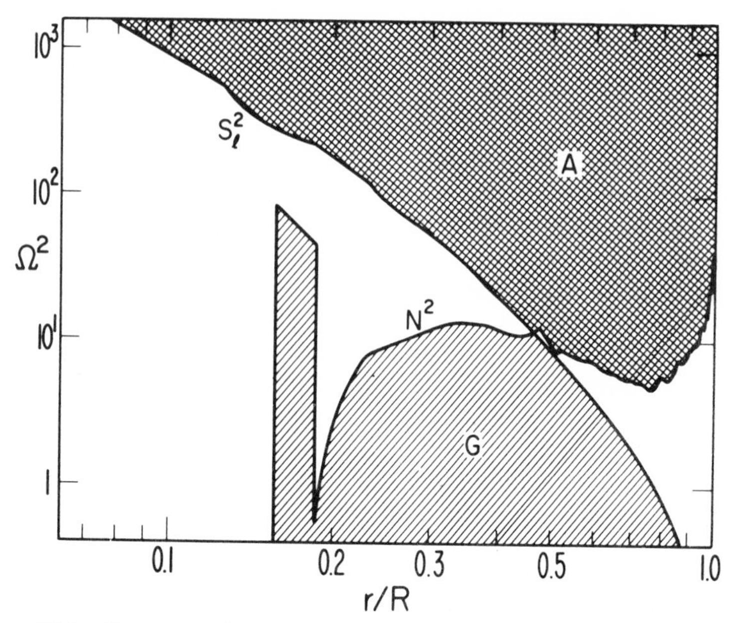

Figure 17.6. The square of the critical acoustic frequency S_l (eq. [17.89]), and of the Brunt-Väisälä frequency N (eq. [17.13]), where all frequencies are dimensionless (that is, in the units GM/R^3), versus fractional radius r/R for $l = 2$ for a model star of 10 solar masses in which the relative mass fraction of hydrogen in the central convective core has been reduced to 0.48 (from Osaki 1975). Note the peaked, trapezoid-shaped extension of the G region; this extension occurs in the zone surrounding the convective core. In this surrounding region the chemical composition varies continuously in space, with mean molecular weight μ increasing inward. This inward increase in μ causes $-A$ to be large here (see eq. [17.13]).

encountered in the work of Hansen, Aizenman, and Ross (1976) on uniformly rotating, isothermal gas cylinders. These last authors have referred to the phenomenon as a "bumping" of modes.

This "bumping" phenomenon has been discussed in an important paper by Aizenman, Smeyers, and Weigert (1977) in terms of coupling of "g-like" modes with "p-like" modes. See also Shibahashi (1979) and Christensen-Dalsgaard (1979).

17.11. PHASE DIAGRAMS

Plots of v versus w (or of δr versus δt), with radial distance r as the parameter, have been called "phase diagrams" by Osaki (1975) and Scuflaire (1974b). More specifically, v is plotted along the x axis, w along the y axis. As was first pointed out by these authors, such phase diagrams can give a very vivid pictorial representation of some of the qualitative features of adiabatic, nonradial oscillations as were discussed in §17.10.

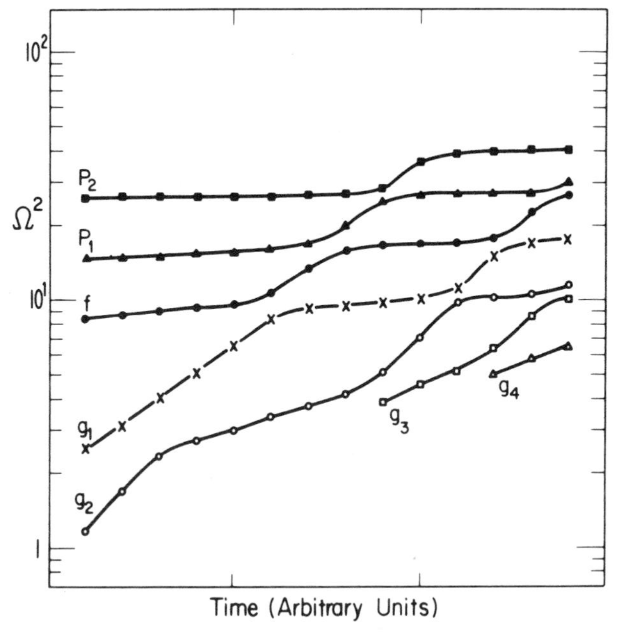

Figure 17.7 Dimensionless eigenfrequencies for the nonradial modes having $l = 2$, versus time in arbitrary units for an evolving 10 – solar mass model (from Osaki 1975).

One can obtain a qualitative idea as to how v and w vary on such a phase diagram by treating $\alpha(r)$ and $\beta(r)$ constants. Then, as suggested by Osaki (1975), we can multiply eq. (17.87a) by βv, eq. (17.87b) by αw, and then subtract one equation from the other. After integrating, we obtain

$$\beta v^2 - \alpha w^2 = \text{const.} \tag{17.92}$$

If α and β have the same signs, as would be the case in an evanescent region (see §17.10), then eq. (17.92) is the equation of a hyperbola. If α and β are of opposite signs, as would be the case in acoustic (A) or gravity (G) regions (see §17.10), then eq. (17.92) is the equation of an ellipse. In the more general case where $\alpha(r)$ and $\beta(r)$ are functions of r but have the above sign relationships, then the curves of w versus v would no longer be simple geometric figures such as hyperbolas and ellipses, but their general, qualitative features would not be different from those described above.

As radial distance r is varied, the representative point on a vw plot, or on a phase diagram, will move in accordance with eqs. (17.87). For acoustic waves, or in an A region, for which $\alpha < 0$ and $\beta > 0$, this point will clearly move *counterclockwise* as r increases. For gravity waves, or in a G region for which $\alpha > 0$ and $\beta < 0$, this point will move *clockwise* with increasing r. At the stellar center ($r = 0$), the central boundary condition $\delta t = (\delta r)/l$

(see eq. [17.62]) requires that v and w have the same sign here. This sign is usually taken as positive, since the normalization is arbitrary. Thus, according to this normalization, the point on the phase diagram representing the stellar center would lie in the *first* quadrant, where our numbering of quadrants is conventional.

Recall that in our discussion in this section we have assumed, merely for simplicity, the Cowling approximation (§17.9); however, this approximation is not supposed to alter the mathematical nature of the problem (see Ledoux and Walraven 1958, §79).

At the stellar surface ($r = R$), the boundary condition in the Cowling approximation may be taken as $\delta t = (\delta r)/\Omega^2$ (see eq. [17.72]). Thus, as long as $\Omega^2 > 0$ (as will essentially always be true in cases of interest), then, again v and w must have the same signs. Hence, the representative point on a phase diagram for the stellar surface would have to lie in the first or third quadrants.

Thus, as r increases from 0 to R, the representative point on a phase diagram would have to start out in the first quadrant, spiral counterclockwise or clockwise according to whether the region under consideration is, respectively, an A or a G region, and finally end up in either the first or third quadrant. For the simple stellar models, as has been pointed out by Osaki (1975) and Scuflaire (1974b), the number of times that the representative point crosses with w axis (that is, that δr vanishes) is equal to the order of the mode.

In this scheme the f mode is assigned the number 0, the p_n mode the number $+n$, and the g_n mode the number $-n$; $+n$ is equal to the number of *counterclockwise* crossings of the w (or δt) axis, $-n$ to the number of *clockwise* crossings of this axis, both as r increases. Note that, in agreement with the remarks made in §17.8, this scheme requires that δr have a node closer to the stellar center than δt for p modes, and conversely for g modes.

These points are illustrated in the phase diagrams of Figures 17.8a,b,c for a polytrope of index 3, for $l = 2$ and $\Gamma_1 = 5/3$, for, respectively, the f, p_3 and g_3 modes, from Scuflaire (1974b). The variables plotted in these figures are actually ξ and ζ, defined by

$$\xi \equiv \pm\log_{10}\left(1 + \left|\frac{\delta r}{R}\right|\right), \tag{17.93a}$$

$$\zeta \equiv \pm\log_{10}\left(1 + \left|\frac{RP'}{GM\rho}\right|\right), \tag{17.93b}$$

where the signs are the same as the signs of δr and of P' (all symbols have their usual meanings). Note that ξ is analogous to v or δr, and ζ to w or δt.

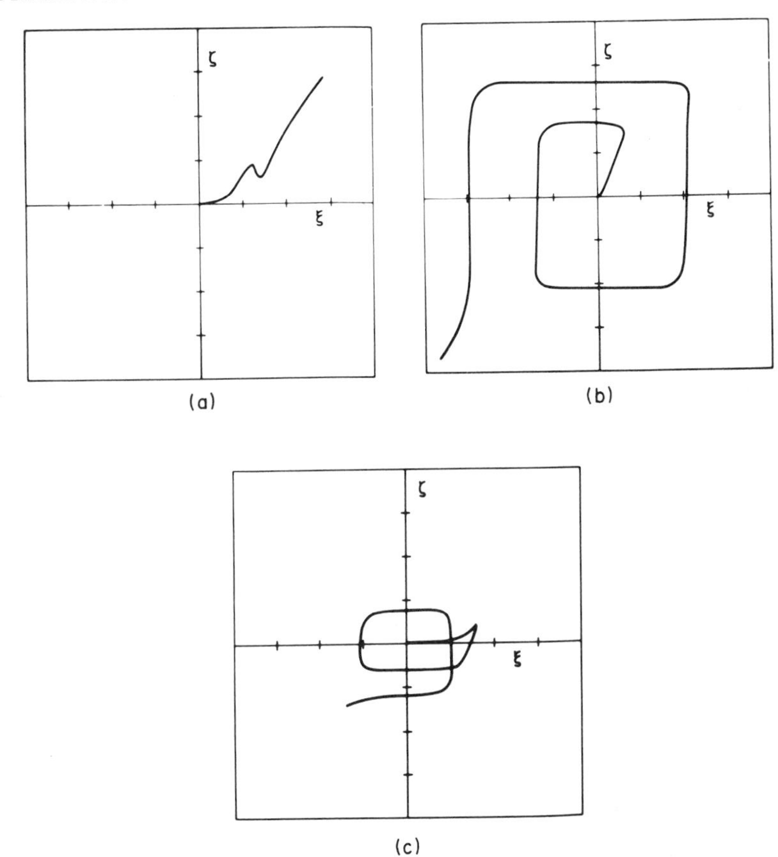

Figure 17.8a,b,c: Phase diagrams for the f (a), p_3(b), and g_3(c) modes for $l = 2$ and $\Gamma_1 = 5/3$ for the polytrope of index 3. What is plotted are ξ and ζ, given by eqs. (17.93), equivalent, respectively, to v and w. Each division of the axes represents one unit of ξ or ζ. Note that the stellar center ($r = 0$) corresponds to $\xi = 0$, $\zeta = 0$, and that r increases as one follows the curve away from the origin (see text for further explanation (from Scuflaire 1974).

For the more complicated models, however, the path of a representative point on a phase diagram is, probably not surprisingly, more convoluted than described above. For example, Figures 17.9a,b,c show such phase diagrams for the polytrope of index 4, for $l = 2$ and $\Gamma_1 = 5/3$ (from Scuflaire 1974b). Note that, in accordance with the above remarks, this point spirals clockwise in G regions and counterclockwise in A regions. It has been suggested by Osaki (1975) and Scuflaire (1974b) that the conventional mode classification scheme as described in the preceding paragraph for simple stellar models can be retained, at least for $l = 2$, for the more complicated models by subtracting, for each mode, the number of clockwise crossings of the w (or ζ, or δt) axis from the number of

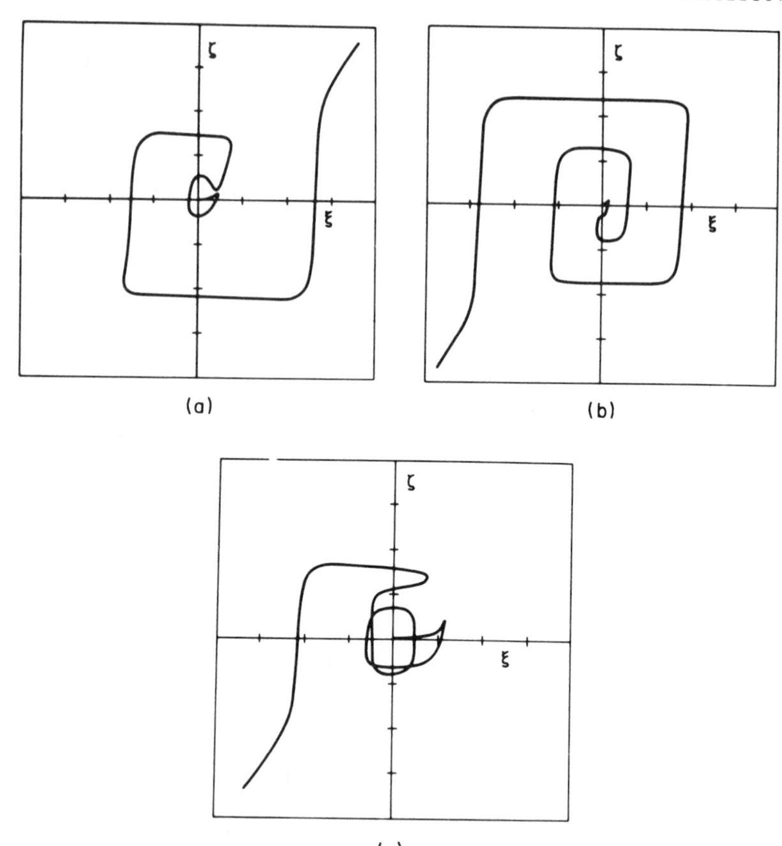

Figure 17.9a,b,c: Same as Figure 17.8a,b,c except for the polytrope of index 4 ($l = 2$, $\Gamma_1 = 5/3$), from Scuflaire (1974b). Shown in (a), (b), and (c) are, respectively, the f, p_3, and g_3 modes.

counterclockwise crossings of this axis, as r increases. For example, for the p_3 mode of the polytrope of index 4, the number of counterclockwise crossings of the ζ axis as r increases is 4, the number of clockwise crossings of this axis is 1, and $4 - 1 = +3$.

An alternative mode classification scheme was introduced by Shibahashi and Osaki (1976a), in which the "trapping" dichotomy of nonradial oscillations (see §18.3) was taken into account. Still another classification scheme, somewhat similar to that of Shibahashi and Osaki (1976a), but based to a large extent on the properties of high-order p and g modes, was suggested by Wolff (1979).

Phase diagrams are presented for the ZAMS model of 10 solar masses for $l = 2$ for the g_2, g_1, f, p_1, and p_2 modes and for the g_1, f, and p_1 modes in the later evolutionary state of the same model in Osaki (1975), but will not be shown here. Other examples of phase diagrams may be found in Unno, Osaki, Ando, and Shibahashi (1979).

17.12. NONRADIAL OSCILLATIONS FOR VERY HIGH-ORDER MODES

In this section we shall present some approximate results for nonradial oscillations of modes of very high orders, that is, those for which $l \gg 1$ and/or the radial wave number $k = 2\pi/(\text{radial "wavelength"})$ is very large, say large compared with the reciprocal of the radial distance r or of the pressure scale height $\lambda_P = P/(\rho g)$ for a star in hydrostatic equilibrium. In this case, and also because our treatment in this section is only approximate, we shall treat the critical acoustic frequency S_l (defined in eq. [17.89]), the Brunt-Väisälä frequency N (defined in eq. [17.13]), and the Laplacian sound speed $(\Gamma_1 P/\rho)^{1/2}$ all as constants, and shall assign them values in each application appropriate to a certain region in the star. These various regions will be discussed further below and in §17.13 and also in §18.2. We shall also, in this short wavelength limit, adopt the Cowling approximation (§17.9) and assume that the radial variation of each of the dependent variables (say v or w, see §17.9) is proportional to $\exp(ikr)$, at least in the first parts of this discussion. This assumption regarding the spatial dependence of the variations is justified because in this limit of very short wavelengths all coefficients in the differential equations are almost spatially constant over a distance of the order of a wavelength. Therefore, these coefficients appear practically constant on this length scale, and it is well known that, in the case of linear differential equations with constant coefficients, the solution is proportional to an exponential. The differential equations then yield eq. (17.91), in which all quantities will be treated as constants. High-order nonradial oscillations have also been discussed by Andrew (1967), Shibahashi (1979), and Christensen-Dalsgaard (1979).

We now assume that the angular oscillation frequency σ is determined primarily by physical conditions in only one region of the star, and this region will be chosen to be that appropriate to the values used for S_l, N, and $\Gamma_1 P/\rho$ in eq. (17.91). If all quantities in this equation are regarded as constants, then it is a quadratic in σ^2, whose solution is

$$\sigma^2 = \frac{1}{2}\left[S_l^2 + N^2 + k^2\frac{\Gamma_1 P}{\rho}\right] \left\{1 \pm \left(1 - \frac{4S_l^2 N^2}{\left[S_l^2 + N^2 + k^2\dfrac{\Gamma_1 P}{\rho}\right]^2}\right)^{1/2}\right\}, \quad (17.94)$$

where we have assumed that the quantity in square brackets is nonvanishing.

At least for the high-order modes presently under consideration, the plus sign in eq. (17.94) gives σ^2 for the p modes, the minus sign for the g modes. We assume that $|N^2|$ is small compared with either of the other two terms in the square brackets in eq. (17.94). This assumption can be shown to be sufficient to guarantee that the second, or last term under the radical sign in eq. (17.94), is small. Neglecting the aforementioned term, we then obtain for high-order p modes:

$$\sigma_p^2 \approx \left[\frac{l(l+1)}{r^2} + k^2\right]\frac{\Gamma_1 P}{\rho}. \tag{17.95}$$

The quantity $l(l + 1)/r^2$ is sometimes referred to as the square of the "horizontal" wave number for the disturbance.

This equation shows, not surprisingly, that σ_p increases both with increasing k and with increasing l. The equation also reveals the physically plausible result that σ_p is almost independent of l for $l(l + 1)/r^2$ considerably less than k^2; that is, for p waves whose "vertical wavelength" is small compared with their "horizontal wavelength," the value of σ_p is almost independent of l, at least in this limit. In particular, we have that $\sigma_p^2 \approx k^2\Gamma_1 P/\rho$ for $l = 0$ (purely radial oscillations), and this is the relation which applies to ordinary, high-order sound waves (see §§5.6 and 17.10).

For the high-order g modes, we may expand the radical to lowest order in the small term referred to in the paragraph before last. We obtain, to this order,

$$\sigma_g^2 \approx \frac{l(l+1)/r^2}{l(l+1)/r^2 + k^2} N^2. \tag{17.96}$$

We note, first, that σ_g^2 has the same sign as $-A$ in the region under consideration, which in this treatment is assumed to be the only region that determines σ_g. Thus, $\sigma_g^2 > 0$ (dynamical stability) if $A < 0$ (stability against convection). Also, $\sigma_g^2 < 0$ (dynamical instability) if $A > 0$ (instability against convection). As we shall see later in this section, in this last case $|\sigma_g^2|$ is of the order of the reciprocal of the square of the eddy turn-over time in a convection zone.

Second, for fixed l, and $k \to \infty$, $\sigma_g^2 \propto 1/k^2 \to 0$; this behavior is consistent with the asymptotic behavior of high-order g modes as described in §17.8. Similarly, $\sigma_g^2 = 0$ for $l = 0$ and any (finite) value of k. This result is just a manifestation of the fact that g modes cannot exist in purely radial motion. For fixed k, and $l \to \infty$, eq. (17.96) shows that $\sigma_g^2 \approx N^2$. In other words, in this limit of fixed "vertical wavelength" and arbitrarily short "horizontal wavelength," σ_g becomes independent of l and approximately equal to the Brunt-Väisälä frequency N.[4] However, l might have to be very large indeed before this independence is realized, depending on the value of k (see near the end of §18.3). Physically, this last result is just a manifestation of the fact that N is defined only with respect to *vertical* oscillations of a mass element in a stratified fluid. This limit of fixed k and $l \rightarrow \infty$ corresponds to many closely-spaced columns of rising and falling material. The conclusion that σ_g becomes independent of l for large l and fixed k does not, however, apply to the homogeneous (constant density), compressible model. The analytic solution obtained in §17.7 shows that, in this limit, $|\sigma_g| \propto \sqrt{l}$.[5] Finally, eq. (17.96) shows that $\sigma_g^2 \sim N^2$ when $k^2 \sim l(l+1)/r^2$, that is, when the vertical wave number k and the horizontal wave number $[l(l+1)]^{1/2}/r$ are comparable to each other. In this limit we obtain the same picture as in §17.2 of blobs of fluid with about the same dimensions in all directions.

Third and finally, it is interesting to consider the relation between the *phase velocity* $v_{\rm ph} = -\sigma/k$ and the group velocity $v_{\rm gr} = -d\sigma/dk$ for these high-order (large k) modes (see, e.g., Jenkins and White 1957 or Born and Wolf 1959). (The minus signs above arise from the time dependence assumed: $e^{+i\sigma t}$.) Clearly, $v_{\rm gr} = v_{\rm ph}$ if $v_{\rm ph}$ is independent of the vertical wavelength $2\pi/k$.

In the case of high-order p modes we obtain

$$|v_{\rm ph}| = \left[\frac{l(l+1)}{r^2k^2} + 1\right]^{1/2}\left(\frac{\Gamma_1 P}{\rho}\right)^{1/2} \tag{17.97}$$

which shows that $v_{\rm ph}$ becomes independent of k as $k \rightarrow \infty$ at fixed l. Thus $v_{gr} \rightarrow v_{\rm ph}$ in this case.

For high-order g modes we obtain

$$|v_{\rm ph}| = \frac{[l(l+1)]^{1/2}}{r}\frac{N}{k^2}\left[1 + \frac{l(l+1)}{r^2k^2}\right]^{-1/2}, \tag{17.98}$$

which shows that $v_{\rm ph} \propto 1/k^2$ as $k \rightarrow \infty$ at fixed l. In this limit, then, $v_{\rm gr} \rightarrow -v_{\rm ph}$; that is, for high-order g modes of fixed l, $v_{\rm gr}$ and $v_{\rm ph}$ have opposite signs.

Some of the characteristics of high-order p and g modes may be

[4] See footnote 5.

[5] Christensen-Dalsgaard (1979) has shown that this case of σ_g^2 being independent of l for large l applies only to polytropes having polytropic index $\nu > 2.191$; and then only to the case of what he calls "modes trapped near a local maximum of N^2." In fact, as discussed by Christensen-Dalsgaard (1979), the behavior of σ^2 for large l is considerably more complicated in complete stars than discussed here. For example, has he states, for any stellar model there is a class of modes trapped near the surface with σ approximately proportional to $l^{1/2}$.

obtained as follows. Our procedure will be to use the linearized equations of nonradial, nonadiabatic oscillations to obtain general expressions for $\delta t/\delta r$, where δt is the tangential, or transverse, component of $\delta \mathbf{r}$ (see §17.3), and for $\delta\rho/\rho$, the fractional Lagrangian density variation of the mass elements. These expressions will then be examined in the high frequency limit ($\sigma^2 \to \infty$, high-order p modes) and the low-frequency limit ($\sigma^2 \to 0$, high-order g modes). This procedure enables us to see more easily why the motions of mass elements become nearly "vertical," or radial, for high frequencies, and nearly "horizontal," or transverse, for low frequencies.

We start with the expression for δt, which was given by eq. (17.33), and which can also be written, for a spherical star in hydrostatic equilibrium, in terms of the Lagrangian variations δP and $\delta\psi$. We write $\delta\psi = \psi' - \delta\mathbf{r} \cdot \mathbf{g}$, where $\mathbf{g} = -\mathbf{e}_r Gm(r)/r^2$ is the local gravitational acceleration. Next, we express δP in terms of $\delta\rho$ and δs (Lagrangian variations of, respectively, density and specific entropy) by means of the thermodynamic identity, eq. (5.35a). The result is then an expression for δt in terms of $\delta\rho$, δs, $\delta\mathbf{r}$, ψ', and $\mathbf{g}$. We express $\delta\rho$ in terms of $\partial(r^2\delta r)/r^2\partial r$ and δt by means of the mass equation, eq. (17.27). Solving the resulting equation for δt and dividing through by δr, we finally obtain

$$\left[1 - \frac{\Gamma_1 P}{\rho}\frac{l(l+1)}{\sigma^2 r^2}\right]\frac{\delta t}{\delta r} = -\frac{1}{\sigma^2 r}\frac{\Gamma_1 P}{\rho}\frac{1}{\delta r}\frac{\partial(r^2\delta r)}{r^2\partial r} + \frac{(\Gamma_3 - 1)T\delta s}{\sigma^2 r\delta r} - \frac{1}{\sigma^2 r}\frac{\delta\mathbf{r}\cdot\mathbf{g}}{\delta r} + \frac{1}{\sigma^2 r}\frac{\psi'}{\delta r}. \tag{17.99}$$

In a somewhat similar manner we obtain

$$\left[1 - \frac{\Gamma_1 P}{\rho}\frac{l(l+1)}{\sigma^2 r^2}\right]\frac{\delta\rho}{\rho} = -\frac{\partial(r^2\delta r)}{r^2\partial r} + \frac{l(l+1)}{\sigma^2 r^2}\left[(\Gamma_3 - 1)T\delta s - \delta\mathbf{r}\cdot\mathbf{g} + \psi'\right]. \tag{17.100}$$

In most of the following we shall assume that l remains fixed in value. We first consider the high-frequency limit, $\sigma^2 \to \infty$ (high-order p modes). In this case the second terms in square brackets on the left sides of eqs. (17.99) and (17.100) become so small as to be negligible. Also, the last three terms on the right sides of each of these two equations become very small in this limit (this is obvious for all terms except for the first one on the right side of eq. [17.99]; but this procedure can be shown to be correct); δs and ψ' are in any case neglected for adiabatic oscillations in the Cowling approximation (cf. §17.9).

We first note that both $\delta t/\delta r$ and $\delta\rho/\rho$ become independent of l in this high-frequency limit. Next, we note that the expression for $\delta\rho/\rho$ is in this limit the same as for purely radial oscillations, an indication that such high-frequency oscillations are mostly radial. In particular, if we replace $\partial/\partial r$ by ik and treat r as approximately constant, we have

$$\frac{\delta\rho}{\rho} \approx -ik\,\delta r, \tag{17.101}$$

which shows that, for such acoustic waves, δr and $\delta\rho$ are approximately 90° out of phase.

For such high-order p modes we write $\sigma_p{}^2 \approx k^2\Gamma_1 P/\rho$, valid for $k^2 \gg l(l+1)/r^2$ and $\partial(r^2\delta r)/r^2\partial r \approx ik\,\delta r$, both in the first term on the right side of eq. (17.99). Neglecting the last three terms on the right side of eq. (17.99) (because σ^2 is large), we obtain

$$\frac{\delta t}{\delta r} \approx -i\frac{1}{rk}. \tag{17.102}$$

This relation says that for such modes, $|\delta t/\delta r|$ is of the order of the ratio of the vertical wavelength of the disturbance to the radial distance r; and this ratio has been assumed small in the present treatment. Hence, as was stated above, such high-order p modes are nearly vertical, or longitudinal. Note also that δt *lags behind* δr by just 90°, so that the particles are executing a kind of elliptical motion, with the long axis of the ellipse in the vertical direction.

We consider now the low-frequency limit, $\sigma^2 \to 0$ (high-order g modes). In this case the second term in brackets on the left side of eq. (17.99) becomes large in magnitude, and the factor σ^{-2} cancels out of the equation. Retaining only the first term on the right side, and writing $\partial(r^2\delta r)/r^2\partial r \approx ik\,\delta r$, we obtain

$$\frac{\delta t}{\delta r} \approx i\frac{rk}{l(l+1)}. \tag{17.103}$$

Thus, we see that for very high-order g modes (large k), the ratio $|\delta t/\delta r|$ is of the order of the number of vertical wavelengths within the distance r; and this ratio has been assumed large in the present treatment. Such very high-order g modes therefore represent predominantly horizontal, or transverse, motions, as was stated earlier. Note that the phase relation of δt to δr is just π different from that for very high-order p modes (cf. eq. [17.102]). Thus in the present case the particles move again on approximately elliptical paths, but in the opposite direction than that for high-order p modes, and with the long axis of the ellipse oriented horizontally.

It is of some interest to obtain an expression for $\delta\rho/\rho$ for high-order g modes. We obtain from eq. (17.100) (assuming that $P/[\rho\sigma^2 r^2]$ is much larger than one)

$$\frac{\delta\rho}{\rho} \approx -\frac{1}{\Gamma_1}\frac{\delta r}{\lambda_P} - \frac{1}{\Gamma_1 g\lambda_P}[(\Gamma_3 - 1)T\delta s + \psi'], \qquad (17.104)$$

where λ_P is the pressure scale height, and the remaining symbols have their usual significance. This expression shows that $|\delta\rho/\rho|$ does not increase as k increases (that is, $|\delta\rho/\rho|$ becomes independent of [vertical] wavelength), in contrast to the case with high-order p modes. Also, this expression shows that nonadiabatic effects (large $|\delta s|$) may have a significant effect on $|\delta\rho/\rho|$ for high-order g modes.

That the above conclusions regarding the vertical and horizontal characteristics of the motions of mass elements in high-order p and g modes, respectively, are also physically plausible, may be seen from the following considerations.

The net force, and hence the acceleration, of a given mass element may always be resolved into vertical and horizontal components. In nonradial oscillations, the magnitude of the horizontal component (relative to the magnitude of the radial component) depends at each instant only on the departure of the perturbed configuration from sphericity at that instant. If the perturbed quantities are assumed to be proportional to spherical harmonics, then the indices l and m are measures of this departure from sphericity. (However, since $|m| \leq l$, we actually only need to refer to l.) In general, l is roughly equal to the number of complete "waves" that will fit into the circumference of any circle lying in a plane perpendicular to the axis. And the magnitudes of the horizontal forces (produced by horizontal pressure gradients and gravity) relative to those of the radial forces at any given instant will increase as l increases. However, in the above arguments we have assumed that l remains fixed, and so the relative magnitude of the horizontal forces also remans fixed, that is, essentially independent of the frequency.

Now, for very high frequencies the total accelerations of mass elements must become relatively large. Since for the above reasons the horizontal accelerations of mass elements do not depend strongly on frequency, it follows that the total accelerations, and hence motions, of mass elements must be nearly vertical in this high-frequency limit.

At very low frequencies, however, the total accelerations of mass elements must be relatively small. Since, again, the horizontal components of the accelerations or forces do not depend strongly on frequency, the vertical components must become relatively small (a vector cannot be shorter than one of its perpendicular components). Hence the motion of mass elements must be mainly horizontal. A crude, terrestrial mechanical analogy would be a frictionless mass sliding down an inclined plane whose angle, say θ, to the horizontal is very small. In this case one can easily show that the ratio of the horizontal to the vertical distance traveled by the mass is $\sin(2\theta)/(2\sin^2\theta) \approx 1/\theta$ for θ very small.

An expression for $(P'/P)/(\delta r/r)$ also follows from the present considerations. We write, using eq. (17.33),

$$\frac{P'/P}{\delta r/r} = \sigma^2 r^2 \frac{\rho}{P}\frac{\delta t}{\delta r} - \frac{\psi'}{\delta r/r}\frac{\rho}{P}. \tag{17.105}$$

For high-order g modes we may use eq. (17.96) for σ^2, and the appropriate limiting form of eq. (17.99) for $\delta t/\delta r$. However, for uniform chemical composition $|A| \sim |\nabla - \nabla_{\text{ad}}|/\lambda_P$ (see §17.2), where λ_P is the pressure scale height. We also set $r/\lambda_P \sim 1$ and drop ψ' (as is consistent with the Cowling approximation); we obtain

$$\left|\frac{P'/P}{\delta r/r}\right| \sim \frac{|\nabla - \nabla_{\text{ad}}|}{\lambda_P} \cdot \frac{k}{k^2 + \dfrac{l(l+1)}{r^2}}. \tag{17.106}$$

According to this expression, $|(P'/P)/(\delta r/r)| \to 0$ as k or $l \to \infty$. In other words, for very high-order g modes the relative pressure variations at a given point (the *Eulerian* pressure variations) become negligibly small compared to the relative displacement of a fluid element at that point, a situation which sounds reminiscent of that usually assumed in conventional mixing-length theories of convection. The close connection between high-order g modes and convection will be discussed next.

For this purpose we use the formula $N^2 = -Ag$ and eq. (17.15) for A, in which we treat the mean molecular weight μ as spatially constant, in the approximate expression (17.96) for σ_g^2. We set the factor multiplying N^2 in this equation equal to unity (since we desire only an order-of-magnitude result); $\chi_T \sim \chi_\rho \sim 1$; $g \sim GM/R^2$; $\lambda_P \sim R$; and we obtain for these high-order g modes

$$|\sigma_g| \sim |\nabla - \nabla_{\text{ad}}|/t_{ff}, \tag{17.107}$$

where $t_{ff} \sim (R^3/GM)^{1/2}$ is the free-fall time (see Chap. 2). This expression for $|\sigma_g|$ is also the same, to order of magnitude, as the expression for the reciprocal of the eddy turn-over time in a stellar convective region, according to the mixing-length theory of convection (see, e.g., Cox and Giuli 1968, Chap. 14). The reason for this result is, of course, that this time is also, to order of magnitude, equal to $|N|^{-1}$. The connection between high-order g modes and convective instability has also been discussed by Sobouti (1977).

17.13. WEIGHT FUNCTIONS

The use of the integral expressions derived in Chapter 8 for determining which portions of a star are important for determining the pulsation period via "weight functions," as described in §8.13, proved extremely fruitful for radial oscillations (see Epstein 1950). This general method has now been extended to nonradial oscillations (see, e.g., Schwank 1976, Goossens and Smeyers 1974). This section is mainly concerned with the application of these techniques to nonradial oscillations. We follow mostly the discussion of Schwank (1976).

Schwank applied these methods only to polytropes having indices between 1 and 4, and based his weight functions on the integral expressions presented in Chapters 15 and 16 for adiabatic, nonradial oscillations.

We note that the denominator of the integral expression for σ^2 involves the integral (the oscillatory moment of inertia)

$$J = \int_V \boldsymbol{\xi}^* \cdot \boldsymbol{\xi} \, \rho r^2 \, dr \sin\theta \, d\theta d\phi, \tag{17.108}$$

where $\boldsymbol{\xi} \equiv \delta\mathbf{r}$, we have dropped zero subscripts from unperturbed quantities, and we have written the expression for the volume element $d\tau$ in spherical polar coordinates. The integral J is proportional to the total kinetic energy in the oscillations, for the mode considered. We write $\boldsymbol{\xi}$ in terms of its three components in an orthogonal coordinate system, as in eq. (17.19), and we assume that each component is proportional to a spherical harmonic (see §17.3) and to exp $(i\sigma t)$. Using the properties of spherical harmonics, we obtain after a little manipulation

$$J = \text{const.} \int_0^R \left[(\delta r)^2 + l(l+1) \frac{\chi^2}{\sigma^4 r^2} \right] 4\pi\rho r^2 \, dr, \tag{17.109}$$

where the value of the constant may depend on l and m. If we had been dealing with the more general case of nonadiabatic oscillations, the quantities $(\delta r)^2$, χ^2, and σ^4 would have to be replaced by, respectively, $|\delta r|^2$, $|\chi|^2$, and $|\sigma^2|^2$.

We note that, for $l = 0$, this result reduces to that appropriate for purely radial oscillations (see Chap. 8). The term in the integrand involving l represents the kinetic energy per unit mass contained in the transverse (or "horizontal") components of the motion of oscillating mass elements. Using the dimensionless variable $x \equiv r/R$, we see that the integral in eq. (17.109) becomes an integral between 0 and 1 over what Schwank (1976) called $g(x)$, which is essentially the integrand of this integral. Similarly, the integrand of the integral in the numerator of the integral expressions (the "weight function") was called $f(x)$ by Schwank (1976). Here we shall not go into the details involved in computing $f(x)$.

Schwank found, perhaps not surprisingly, that the frequencies of p modes were determined predominantly by conditions in the outer stellar layers. Conversely, the frequencies of g modes were determined mostly by conditions in the deeper, interior stellar layers. Finally, all regions in the star contributed roughly equally to the frequencies of f modes.

17.14. DAMPING TIMES

In this section we shall write down the expression, derived by two different methods in Chapter 16, for the stability coefficient explicitly for nonradial oscillations. As explained earlier in this chapter (see §17.3), we shall assume that each scalar perturbed quantity is proportional to a spherical harmonic.

As explained in Chapter 16, the time dependence of all perturbed quantities is as given in eq. (16.23). In particular, the nonperiodic time dependence is exp $(-\kappa t)$, where the real quantity κ, the stability coefficient, is given for the case of "small" nonadiabaticity by eq. (9.13), with C therein given by eq. (16.21). We may recall that $|\kappa|^{-1}$ is the e-folding time for growth or decay of the oscillation amplitude.

The quantity $\delta\rho/\rho$ appearing in C can be computed from the mass equation, eq. (17.27). We therefore consider the quantity $\delta[\epsilon - (\nabla \cdot \mathbf{F})/\rho]$. Using the relation (5.16) relating Eulerian and Lagrangian variations, we obtain

$$\delta[\epsilon - (\nabla \cdot \mathbf{F})/\rho] = \delta\epsilon + \mathscr{E}(\delta\rho/\rho) - (\mathscr{E}/\rho_0)(d\rho_0/dr) - (\nabla \cdot \mathbf{F}')/\rho_0 - \delta r\, d\mathscr{E}/dr, \quad (17.110)$$

where zero subscripts denote quantities in the unperturbed state of the star, assumed here to be spherically symmetric. The quantity $\mathscr{E}$ is defined by the relation

$$\mathscr{E} \equiv [(\nabla \cdot \mathbf{F})/\rho]_0 = (dL_r/dm)_0, \quad (17.111)$$

where $L_{r,0} = 4\pi r^2 F_0$ is the ordinary interior luminosity (the net rate of energy issuing from a sphere of radius r), F_0 being the radial (and only) component of $\mathbf{F}_0$, the vector energy flux; and m is the usual interior mass.

In order to compute $\nabla \cdot \mathbf{F}'$ explicitly, we assume that $\mathbf{F}$ is only the radiative flux, and we assume, furthermore, that it is given by a diffusion-type equation of the form

$$\mathbf{F} = -K\nabla T \quad (17.112)$$

where the radiative conductivity K may in general be a function of ρ and T (an explicit expression will be given below, see eq. [17.117]). By restricting ourselves to an expression of the form of eq. (17.112), we are thereby restricting ourselves to the deep stellar interior; for a more general treatment, see, for example, Ando and Osaki (1975). Taking the Eulerian variation of **F**, using eq. (17.32) for $\nabla \cdot \mathbf{F}'$, eq. (17.27) for $\delta\rho/\rho_0 (= -\nabla \cdot [\delta\mathbf{r}])$, and assembling the above results, we obtain

$$\begin{aligned}\delta[\epsilon - (\nabla \cdot \mathbf{F})/\rho] = {} & \delta\epsilon - \mathscr{E}[\partial(r^2\delta r)/(r^2\partial r)] \\ & + \mathscr{E}\{[l(l+1)]/(\sigma^2 r^2)\}(P'/\rho_0 + \psi') - (\mathscr{E}/\rho_0)\delta r(d\rho_0/dr) \\ & - [1/(4\pi r^2\rho_0)]\partial[4\pi r^2(\mathbf{F}')_r]/\partial r \\ & + \{[l(l+1)]/(4\pi\rho_0 r^3)\}[L_{r,0}/(d\ln T_0/dr)](T'/T_0) \\ & - \delta r\, d\mathscr{E}/dr. \end{aligned} \tag{17.113}$$

In order to simplify this expression, we make the formal assumption that the instantaneous interior luminosity L_r is given by the relation

$$L_r = 4\pi r^2(\mathbf{F})_r, \tag{17.114}$$

where $(\mathbf{F})_r$ is the radial component of the energy flux **F**. Now, because $(\mathbf{F})_r$ in general varies with the polar angles θ and ϕ at each instant of time, so does L_r. Thus, L_r may be interpreted as the instantaneous value of the interior luminosity at radial distance r, if, at given r, θ, and ϕ, a sphere of radius r were drawn about the stellar center, and $(\mathbf{F})_r$ were assumed to be the outward flux at all θ, ϕ.

We take the Lagrangian variation of L_r and use the relation (5.16) connecting Lagrangian and Eulerian variations. We get, after a small bit of manipulation, a relation from which $4\pi r^2(\mathbf{F}')_r$ can be obtained in terms of δL_r and δr. We therefore obtain

$$\begin{aligned}\frac{1}{4\pi r^2\rho_0}\frac{\partial[4\pi r^2(\mathbf{F}')_r]}{\partial r} = {} & \frac{L_{r,0}}{4\pi r^2\rho_0}\frac{\partial}{\partial r}\left(\frac{\delta L_r}{L_{r,0}}\right) + \mathscr{E}\frac{\delta L_r}{L_{r,0}} \\ & - \mathscr{E}\frac{\partial(r^2\delta r)}{r^2\partial r} - \frac{\mathscr{E}}{\rho_0}\frac{d\rho_0}{dr}\delta r - \delta r\frac{d\mathscr{E}}{dr}.\end{aligned} \tag{17.115}$$

Combining eq. (17.115) with eq. (17.113), we see that many terms cancel, and we finally obtain, dropping zero subscripts,

$$\begin{aligned}\delta[\epsilon - (\nabla \cdot \mathbf{F})/\rho] = {} & \delta\epsilon + \mathscr{E}\left[\frac{l(l+1)}{\sigma^2 r^2}\left(\frac{P'}{\rho} + \psi'\right) - \frac{\delta L_r}{L_r}\right] \\ & - \frac{L_r}{4\pi r^2\rho}\frac{\partial}{\partial r}\left(\frac{\delta L_r}{L_r}\right) + \frac{l(l+1)L_r}{4\pi\rho r^3}\frac{1}{(d\ln T/d\ln r)}\frac{T'}{T}.\end{aligned} \tag{17.116}$$

It is seen that, for $l = 0$ (purely radial oscillations), this equation reduces to the one appropriate to purely radial oscillations. The first term on the right side of eq. (17.116) involving l arises from $\delta\rho/\rho$, and the second (and last) term involving l arises from horizontal heat flows which may occur in nonradial oscillations.

Finally, we must obtain an explicit equation for $\delta L_r/L_r$ for nonradial oscillations. We take the Lagrangian variation of L_r from eq. (17.114), use eq. (17.112) for $\mathbf{F}$, and the following relation for the radiative conductivity K:

$$K = \frac{4ac}{3}\frac{T^3}{\kappa\rho}. \tag{17.117}$$

Taking the Lagrangian variation of K, and using eq. (17.27) for $\delta\rho/\rho_0$, we obtain after some manipulation, again dropping zero subscripts,

$$\frac{\delta L_r}{L_r} = 4\frac{\delta r}{r} + 4\frac{\delta T}{T} - \frac{\delta\kappa}{\kappa} + \frac{T_0}{(dT_0/dr)}\frac{\partial}{\partial r}\left(\frac{\delta T}{T_0}\right) - \frac{l(l+1)}{\sigma^2 r^2}\left(\frac{P'}{\rho} + \psi'\right). \tag{17.118}$$

For $l = 0$, this expression clearly reduces to the one (eq. [7.11a]) appropriate for purely radial oscillations in the case of radiative transfer in the diffusion approximation. We note that the term in eq. (17.118) involving l arises solely from $\delta\rho/\rho$, which in the case of nonradial oscillations may arise from horizontal motions of mass elements.

In the quasi-adiabatic approximation all quantities in eqs. (17.116) and (17.118) are computed using the adiabatic relations among $\delta P/P$, $\delta\rho/\rho$ and $\delta T/T$. The latter quantities are computed from the eigenfunctions of the linear, adiabatic, nonradial equations as described earlier in this chapter. However, eqs. (17.116) and (17.118), being merely identities, are valid also for linear, *nonadiabatic*, nonradial oscillations, to be considered in the next chapter.

Numerical values of these quasi-adiabatically computed stability coefficients can be found in the rather abundant astrophysical literature, for example, Osaki (1975, 1976); Shibahashi and Osaki (1976a,b); and Unno, Osaki, Ando, and Shibahashi (1979). Typically, these times may range all the way from a year or some fraction thereof up to some 10^5 to 10^7 years, depending on the oscillation mode and on the stellar model.

18

Linear, Nonadiabatic, Nonradial Oscillations of Spherical Stars

Just as in the case of radial oscillations, nonradial oscillations must in nature actually be nonadiabatic; that is, heat gains and losses by the mass elements must accompany the oscillations. In the case of nonradial oscillations these nonadiabatic effects could be even more important than for radial oscillations. The reason is that in nonradial oscillations there is the possibility that the oscillation frequency might be very small, as in the case of high-order *g* modes (see Chapter 17); or that the wavelength associated with a variation might be very small, as in certain regions of somewhat complicated stellar models (see §17.10). In both cases conditions are especially favorable for heat exchanges by the mass elements. The above statements are true in spite of the fact that nonadiabatic effects in nonradial oscillations are usually neglected—perhaps primarily because the mathematical problem is decidedly nontrivial, but partly because nonadiabatic effects are often relatively small throughout most of the stellar mass. As was pointed out in §17.4 and as will be emphasized further in §18.1, the problem of linear, nonradial, nonadiabatic oscillations in static spherical stars is equivalent, mathematically, to a system of differential equations of the sixth spatial order in complex variables, or of the twelfth order in real variables. By contrast, for small, *adiabatic,* nonradial oscillations, the system is only of the fourth order in real variables. And the adiabatic problem is already sufficiently complicated that it contains a wealth of information. Many of its features are only now beginning to be understood (see, e.g., Aizenman and Smeyers 1977; Wolff 1979; Shibahashi 1979; Christensen-Dalsgaard 1979), and it is extremely likely that new results in this field yet await discovery. Consequently, not very much is known about nonadiabatic effects in nonradial oscillations; so many of our remarks in this chapter will necessarily be either severely curtailed or not very secure, or possibly both.

In §18.1 we shall make some comments about orders, boundary conditions, and the like, for nonadiabatic oscillations, and shall point out that the main characteristics of *p* and *g* modes are not altered by nonadiabatic effects, unless these effects are enormous. We shall present a "local" (or short-wavelength) analysis in §18.2. Here it will be shown that the expressions presented in §17.12 for σ^2 for high-order *p* and *g* modes are not appreciably altered by nonadiabatic effects unless, again, these effects are enormous. This analysis will, moreover, yield some information regarding the stability of these modes. Finally, in §18.3 we shall survey some recent work in the field of linear, nonadiabatic, nonradial oscillation theory.

18.1. DISCUSSION OF ORDERS, BOUNDARY CONDITIONS

The fact that the problem of small, nonradial, nonadiabatic oscillations in spherical stars is, mathematically, of the sixth spatial order in complex variables, or of the twelfth order in real variables (see §17.4), may also be understood as follows. We note that for a spherical star the variables are separable for each mode, and the angular (θ,ϕ) dependence of each scalar perturbation variable can be expressed in terms of a spherical harmonic. This fact has the consequence that each vector differential equation is equivalent to only a scalar differential equation. Now, there are three first-order vector differential equations in the problem: the momentum equation, the flux equation, and one of Poisson's equations ($\mathbf{f}' = -\nabla\psi'$, where $\mathbf{f}$ is the gravitational force per unit mass, and a prime denotes, as usual, the Eulerian variation). In view of the above considerations, these are, for a spherical star, equivalent to three first-order *scalar* differential equations. The other three equations (each first-order in space) are already scalar differential equations. These are the mass equation, the energy equation, and the remaining one of Poisson's equations ($\nabla \cdot \mathbf{f}' = -4\pi G\rho'$). Thus, there are effectively six first-order scalar differential equations in, say, the complex variables P', ρ', δr, ψ', $d\psi'/dr$, and L_r'.

Note that, just as in the case of *adiabatic,* nonradial oscillations of a spherical star, *nonadiabatic* nonradial oscillations of such a star also do not depend on the azimuthal spherical harmonic index m; thus the nonadiabatic eigenfrequencies are also $(2l + 1)$-fold degenerate.

We consider now the boundary conditions. As we might expect, there are just six, three at the center and three at the surface. Four of these (two at the center and two at the surface) are exactly the same as for linear, nonradial, *adiabatic* oscillations discussed in §17.6. The remaining two come about as follows.

The extra central boundary condition comes from the fact that, in nonadiabatic oscillations, there is now an extra vector involved, and this is the Eulerian variation (say) of the vector net flux, $\mathbf{F}'$. Since its divergence enters into the energy equation, then, clearly, $\nabla \cdot \mathbf{F}'$ must remain finite at the center. Moreover, $\mathbf{F}'$ may be computed by taking the Eulerian variation of the diffusion equation, say eq. (17.112) (see Ando and Osaki 1975 for a more general treatment). It then follows that

$$T' \propto r^l \qquad (18.1)$$

near $r = 0$. It was shown in §17.6 that the central boundary conditions required (in order that $\nabla \cdot \delta\mathbf{r}$ and $\nabla^2\psi'$ remain finite) that $P' \propto r^l$ near $r = 0$. This, together with eq. (18.1) and the equation of state ($P' = \chi_\rho \rho' + \chi_T T'$), imply that, also, $\rho' \propto r^l$ near $r = 0$. For these reasons (as well as others [see §18.2]), the main characteristics of nonradial oscillations as inferred from the adiabatic approximation are not appreciably altered by nonadiabatic effects, at least when these latter effects are small.

The additional boundary condition at the surface involves (among, perhaps, other things) the relative luminosity, radius, and temperature variations. There are two possible cases, depending on the unperturbed model: one, either this model has a photosphere (however crude), in which case the surface temperature is nearly equal to the effective temperature T_e; or, two, this model is characterized by vanishing surface temperature. In either case we may write, at least formally, $L_r = 4\pi r^2(\mathbf{F})_r$ (see eq. [17.114]), where L_r is the interior luminosity at radial distance r, and $(\mathbf{F})_r$ is the radial component of the vector net flux $\mathbf{F}$. As was pointed out in §17.14, L_r does not have exactly the same interpretation for nonradial oscillations as for radial oscillations and, in fact, L_r has a somewhat fictitious character for nonradial oscillations. Nevertheless, we may formally define L_r as above and take its relative Lagrangian variation, to obtain a relation for $\delta L_r/L_r$ in terms of $\delta r/r$ and $(\delta\mathbf{F})_r/(\mathbf{F})_r$.

Now, in the first case (in which the model has a photosphere) we may assume that $(\mathbf{F})_r \propto B(T) \propto T^4$, where $B(T)$ is the integrated Planck function (T = temperature), so that

$$\frac{\delta L_r}{L_r} = 2\frac{\delta r}{r} + 4\frac{\delta T}{T}. \tag{18.2}$$

All quantities in this equation are to be evaluated, of course, at the surface of the model. Below the surface, $\delta L_r/L_r$ is to be computed from whatever formulae are being used (for example, those derived from the diffusion approximation, see below). The spatial derivative of $\delta L_r/L_r$ is used in the energy equation (see, e.g., eq. [17.116]).

Equation (18.2) also obtains if it is assumed that no radiation is incident on the star from the outside, as in Ando and Osaki (1975) or in Saio and Cox (1979b).

In the second case (in which the model has zero surface temperature), the diffusion approximation is assumed to hold all the way to the surface. That is, we may compute $\delta L_r/L_r$ below the surface as described in §17.14 (eq. [17.118]). *At* the surface, all we need do is to assume that $\partial(\delta T/T)/\partial r$ remains finite here (we are dropping zero subscripts from unperturbed quantities). Then we obtain from eq. (17.118)

$$\frac{\delta L_r}{L_r} = 4\frac{\delta r}{r} + 4\frac{\delta T}{T} - \frac{\delta \kappa}{\kappa} - \frac{l(l+1)}{\sigma^2 r^2}\left(\frac{P'}{\rho} + \psi'\right), \tag{18.3}$$

where all symbols have their usual meaning. This is the surface flux boundary condition to be used in the second case, assuming, of course, that the diffusion approximation, on which this formula is based, is satisfactory.

18.2. LOCAL ANALYSIS

Much useful information can often be obtained on the basis of a local analysis. In such an analysis one assumes that the wavelength of a disturbance is much smaller than essentially any other length of interest (the local pressure scale height, for example). The results of such a local analysis, particularly as regards stability, should, however, always be viewed with caution; for only a global analysis can really be trusted, in most cases, in questions of stability. Nevertheless, the results of a local analysis are often correct. Also, a local analysis, being mathematically simpler than the full problem, may yield some insight as to what is going on physically. Finally, a local analysis, while perhaps not yielding definitive results, will usually indicate trends which may be qualitatively correct.

A local analysis has been used to good advantage by Cowling (1951), Goldreich and Schubert (1967), Kato (1966), Defouw (1973), Dilke and Gough (1972), Osaki (1975), and Kruskal, Schwarzschild, and Härm (1977) (see also the excellent discussion by Ledoux 1974), as well as by others. In most cases the authors have been interested in the low-frequency limit, which applies to high-order *g* modes. Accordingly, some approximations are usually made (and have been made in most of the above works), as follows.

(a) The Cowling approximation is assumed, $\psi' = 0$ (see §17.9); as was pointed out in §17.9, such an approximation may be expected to be quite good for high-order modes.

(b) The fluid is assumed to behave as if it were incompressible, that is,

$$\delta\rho = 0. \tag{18.4}$$

This condition is likely to be a good approximation for the high-order *g* modes of interest (see Wolff 1979). Physically, for these modes the motion is sufficiently slow that any compression or expansion of the fluid would have time to be eliminated, that is, that eq. (18.4) would be satisfied. Also, as was shown in §17.12, $|\delta\rho|$ does not become excessively large for high-order g modes: it becomes independent of the wavelength for very short-wavelength g modes. Thus, setting $\delta\rho = 0$ may not be a bad approximation for the relatively fast changes (see next section) in the state of the fluid associated with high-order g modes. On the other hand, for the much slower changes associated with heat exchanges, $\delta\rho$ is not restricted to be zero. Thus, for example, the fluid may expand when heated. (This approximation of setting $\delta\rho = 0$ for relatively fast changes but $\delta\rho \neq 0$ for relatively slow changes is a consequence of the "Boussinesque approximation," see, e.g., Spiegel and Veronis 1960.)

(c) The Eulerian variation of the total pressure is ignored everywhere, $P' = 0$, except in the equation of motion. In this equation P' is not ignored because its gradient appears; and while $|P'|$ itself may be small, $|\nabla P'|$ may not be (see remarks below in connection with eq. [18.18]). As was pointed out in §17.2, ignoring P' (except as noted above) is usually an excellent approximation for high-order g modes.

Because of the "local" nature of the analysis, a coordinate system is usually used which is a locally Cartesian xyz system, with the z-axis pointing along an outward radius vector.

We shall first (§18.2a) present such an analysis specifically for high-order g modes, in which the assumptions (a), (b), and (c) are made. Next (§18.2b), we shall drop assumptions (b) and (c), so that we will then no longer be restricted to high-order g modes; that is, we may (and we shall) consider arbitrarily high frequencies, as would be associated with, for example, high-order p modes or sound waves. This more general case, of course, embraces the more restricted situation of high-order g modes as a special case.

As we are concerned only with very short wavelengths, we shall assume that all variations contain the factor $\exp(\mathscr{s}t + i\mathbf{k}\cdot\mathbf{r})$, where $\mathscr{s}$ is the complex angular frequency and $\mathbf{k}$ is the propagation vector for the wave under consideration. This spatial dependence is justified for the same reasons as were explained in §17.12; namely, on the scale of the very short wavelengths under consideration, all coefficients of the differential equations are nearly spatially constant, and an exponential spatial dependence of the variations is consistent with this near constancy of the coefficients. The imaginary part of $\mathscr{s}$ gives the oscillation angular frequency, while the real part of $\mathscr{s}$ contains information concerning the stability of the wave: a positive value of the real part of $\mathscr{s}$ denotes instability, or overstability; a negative value, stability. The radial component of $\mathbf{k}$ will be written as k_r (this component was written simply as k in Chap. 17), the tangential component as k_H:

$$k_H = (k_x^2 + k_y^2)^{1/2}, \tag{18.5}$$

where k_x and k_y are, respectively, the x and y components of $\mathbf{k}$. The total wave number will be denoted by k_T:

$$k_T = (k_r^2 + k_H^2)^{1/2}. \tag{18.6}$$

In accordance with the above assumed temporal and spatial dependences, whenever a derivative of a variation is taken, the derivative operation will be replaced be either $\mathscr{s}$ or $i\mathbf{k}$, as follows: $\partial/\partial t \rightarrow \mathscr{s}$, $\nabla \rightarrow i\mathbf{k}$ (except as explained below in connection with eq. [18.19]).

18.2a. LOW FREQUENCIES

We shall now write down the basic differential equations, under the above assumptions. The radial component of the equation of motion is, in the Cowling approximation of assumption (a),

$$\mathscr{s}^2\delta r = -ik_r\frac{P'}{\rho} - \frac{\rho'}{\rho}g, \tag{18.7}$$

where we have assumed that $dP/dr = -\rho g$.

The mass equation is

$$\frac{\delta\rho}{\rho} = -\nabla\cdot\delta\mathbf{r} = -\frac{\partial\delta x}{\partial x} - \frac{\partial\delta y}{\partial y} - \frac{\partial\delta z}{\partial z} = -ik_x\delta x - ik_y\delta y - ik_r\delta r, \tag{18.8}$$

where δx, δy, and $\delta z \equiv \delta r$ are, respectively, the x, y, and z components of $\delta\mathbf{r}$. However, the x and y components of the equation of motion yield

$$\mathscr{s}^2\delta x = -ik_xP'/\rho \tag{18.9}$$

and a similar equation involving δy and k_y. Inserting these results into the mass equation, we obtain

$$\frac{\delta\rho}{\rho} = 0 = -\frac{k_H^2}{\mathscr{s}^2}\frac{P'}{\rho} - ik_r\delta r, \tag{18.10}$$

where we have explicitly made assumption (b) (the fluid is effectively incompressible).

We may now use eq. (18.10) to eliminate P'/ρ from the radial component of the equation of motion (18.7). We obtain

$$\mathscr{s}^2\frac{k_T^2}{k_H^2}\delta r = -\frac{\rho'}{\rho}g. \tag{18.11}$$

We may pause at this moment and note that, since the fluid is assumed to behave as if it were incompressible ($\delta\rho = 0$) on a "short" time scale, we have $\rho'/\rho = -\delta r\, d\ln\rho/dr = -\delta r A$, since $A = d\ln\rho/dr$ is the appropriate value of the quantity A (see §17.2) for an incompressible fluid ($\Gamma_1 = \infty$). Thus, we have

$$s^2 = (k_H^2/k_T^2)Ag, \tag{18.12}$$

where $-Ag$ is the square of the Brunt-Väisälä frequency (see §17.2). This is the same result we obtained in §17.12 for high-order g modes.

We may note that, if we had omitted the pressure fluctuation P' from the equation of motion, we would have obtained $s^2 = Ag$, which is precisely in agreement with the discussion in §17.2, where P' was omitted. As has been stated by Gough (1977a), the pressure fluctuations tend to divert fluid motions away from the vertical, into the horizontal direction.

The above result, eq. (18.12), would also have been obtained if the element had moved adiabatically (that is, in accordance with eq. [5.36a]), rather than incompressibly. However, more work would have been required, and the assumption would have had to be made that the pressure scale height was large compared to the wavelength of the disturbance; but this assumption has already been made from the outset.

However, in order to come to any conclusions about stability, we must allow the density of the fluid to change ($\delta\rho \neq 0$) over long periods of time, as a result of nonadiabatic effects (heat gains and losses). In order to find $\delta\rho$ (or ρ'), we must use the energy equation. We use the form (see Chap. 5)

$$\frac{d}{dt}\left(\frac{\delta T}{T}\right) = \frac{\Gamma_2 - 1}{\Gamma_2}\frac{d}{dt}\left(\frac{\delta P}{P}\right) + \frac{1}{c_P T}\delta\left(\epsilon - \frac{1}{\rho}\nabla \cdot \mathbf{F}\right), \tag{18.13}$$

where c_P is the specific heat per unit mass at constant pressure. Using the relation (5.16) between Lagrangian and Eulerian variations, dropping P' (assumption [c]), noting that $d/dt = \partial/\partial t$ because the unperturbed fluid is static, we may write eq. (18.13) as

$$s\frac{T'}{T} = s\mathcal{S}\delta r + (c_P T)^{-1}\delta(\epsilon - \rho^{-1}\nabla \cdot \mathbf{F}), \tag{18.14}$$

where (see Ledoux 1974)

$$\mathcal{S} \equiv \frac{\Gamma_2 - 1}{\Gamma_2}\left(\frac{d \ln P}{dr}\right) - \frac{d \ln T}{dr} \tag{18.15}$$

(see also eq. [18.31] below).

We must now express $\delta(\epsilon - \rho^{-1}\nabla \cdot \mathbf{F})$ as a function of ρ'/ρ, T'/T, etc. We shall assume that the system in its unperturbed state is in thermal equilibrium, so that $(\epsilon - \rho^{-1}\nabla \cdot \mathbf{F})_0 = 0$. Then there is no difference between the Lagrangian and Eulerian variations, and

$$\delta\left(\epsilon - \frac{1}{\rho}\nabla \cdot \mathbf{F}\right) = \left(\epsilon - \frac{1}{\rho}\nabla \cdot \mathbf{F}\right)' = \epsilon' + \frac{\rho'}{\rho}\epsilon_0 - \frac{1}{\rho}\nabla \cdot \mathbf{F}', \qquad (18.16)$$

where ϵ_0 is the nuclear energy generation rate per unit mass in the unperturbed model, and we have once more (in the second term on the right side of eq. [18.16]) made use of the assumption that the unperturbed model is in thermal equilibrium.

To evaluate ϵ', we assume that it is a function of ρ', T', and X', the Eulerian variation of the mass fraction of hydrogen:

$$\epsilon' = \epsilon\left(\lambda\frac{\rho'}{\rho} + \nu\frac{T'}{T} + \alpha\frac{X'}{X}\right), \qquad (18.17)$$

where ϵ now denotes the nuclear energy generation rate per unit mass in the unperturbed model (we shall henceforth omit the zero subscript), λ and ν are, respectively, the density and temperature exponents in the energy generation law (see eq. [7.12]), and

$$\alpha \equiv \left(\frac{\partial \ln \epsilon}{\partial \ln X}\right)_{\rho,T}. \qquad (18.18)$$

For $\nabla \cdot \mathbf{F}'$, we assume a diffusion-type of equation (see eq. [17.112]), in which K is the thermal (perhaps radiative, see §18.1) conductivity, which will here be treated as a constant. (However, if we wanted to include the density and temperature dependence of K, we would only need to replace λ in the following by $\lambda - [\partial \ln K/\partial \ln \rho]_T$ and ν by $\nu - [\partial \ln K/\partial \ln T]_\rho$.) In evaluating $\nabla \cdot \mathbf{F}'$, we must write $\nabla \cdot \mathbf{F}' = -K\nabla \cdot \nabla[(T'/T)T] = -KT\nabla^2(T'/T) + 2\mathbf{F} \cdot \nabla(T'/T) + (T'/T)\nabla \cdot \mathbf{F}$. Now, in accordance with our assumption regarding all equilibrium quantities as being essentially spatially constant, we may neglect $\mathbf{F}$ in the second term in this last equality, since $\mathbf{F}$ is proportional to ∇T. However, we must retain $\nabla \cdot \mathbf{F}$ in the last term above, for otherwise we would be losing essential physics. (An analogous situation arose in Chapter 13 in connection with the Baker one-zone model, in which we treated all physical variables in the zone as constant except L_r in the energy equation; if we had kept L_r constant, we should have lost essential physics.) At any rate, we eventually obtain an expression for $\delta(\epsilon - \rho^{-1}\nabla \cdot \mathbf{F})$ in terms of ρ'/ρ, T'/T, and X'/X. Substituting this expression into the energy equation (18.14), we then have a relation among T'/T, ρ'/ρ, X'/X, and δr. This relation is

$$\left\{s - \frac{[\epsilon(\nu - 1) - (KT/\rho)k_T^2]}{c_P T}\right\}\frac{T'}{T}$$

$$= s\mathcal{S}\delta r + \frac{[\epsilon(\lambda + 1)]}{c_P T}\frac{\rho'}{\rho} + \frac{\epsilon\alpha}{c_P T}\frac{X'}{X}. \qquad (18.19)$$

To get another relation among T'/T, ρ'/ρ, and the like, we may use the equation of state. Taking its Eulerian variation, we have

$$\frac{P'}{P} = 0 = \chi_\rho \frac{\rho'}{\rho} + \chi_T \frac{T'}{T} + \chi_\mu \frac{\mu'}{\mu}, \qquad (18.20)$$

where we have set $P' = 0$ (assumption [c]); χ_ρ, χ_T, and χ_μ were defined in §17.2; and μ denotes mean molecular weight.

We may now express μ'/μ, and X'/X in terms of δr, by assuming that the time scale of the variations is short compared with the time (roughly, the nuclear time) required for nuclear processes to alter the composition of a mass element. Hence $\delta\mu/\mu$ and $\delta X/X = 0$, or

$$\frac{\mu'}{\mu} + \delta r \frac{d \ln \mu}{dr} = 0, \qquad (18.21a)$$

$$\frac{\mathrm{X}'}{\mathrm{X}} + \delta r \frac{d \ln X}{dr} = 0. \qquad (18.21b)$$

We now wish to combine the above equations in such a way as to obtain a dispersion relation for $\mathcal{S}$. We observe that we have obtained three basic equations: the combined mass equation and equation of motion, eq. (18.11); the energy equation, eq. (18.19); and the equation of state, eq. (18.20). If we regard X'/X and μ'/μ as expressed in terms of δr via eqs. (18.21a,b), then there are just three unknowns: ρ'/ρ, T'/T, and δr. We may regard these equations as three simultaneous equations in the above three unknowns. These equations are, moreover, linear and homogeneous. Therefore, the determinant of the coefficients of the three unknowns in these equations must vanish if there is to be a nontrivial solution. Evaluating this determinant leads to the desired dispersion relation, which is a cubic in $\mathcal{S}$. After some simplification, this equation may be written as

$$\mathcal{S}^3 - \left\{\frac{[\epsilon(\nu - 1) - (KT/\rho)k_T^2]}{c_P T} - \frac{\chi_T}{\chi_\rho}\frac{[\epsilon(\lambda + 1)]}{c_P T}\right\}\mathcal{S}^2$$
$$- \left[\frac{\chi_T}{\chi_\rho}\mathcal{S} - \frac{\chi_\mu}{\chi_\rho}\frac{d \ln \mu}{dr}\right]\frac{k_H^2}{k_T^2} g\mathcal{S}$$
$$+ \left\{- \frac{[\epsilon(\nu - 1) - (KT/\rho)k_T^2]}{c_P T}\frac{\chi_\mu}{\chi_\rho}\frac{d \ln \mu}{dr}\right.$$
$$\left. + \frac{\chi_T}{\chi_\rho}\frac{\epsilon\alpha}{c_P T}\frac{d \ln X}{dr}\right\}\frac{k_H^2}{k_T^2} g = 0. \qquad (18.22)$$

However, we note an identity. By using the relation (which follows from the equation of state of the stellar material)

$$\frac{d \ln P}{dr} = \chi_\rho \frac{d \ln \rho}{dr} + \chi_T \frac{d \ln T}{dr} + \chi_\mu \frac{d \ln \mu}{dr}, \tag{18.23}$$

together with the definition of $\mathscr{S}$ (eq. [18.15]) and some thermodynamic identities, we can show that

$$\frac{\chi_T}{\chi_\rho} \mathscr{S} - \frac{\chi_\mu}{\chi_\rho} \frac{d \ln \mu}{dr} \equiv \frac{d \ln \rho}{dr} - \frac{1}{\Gamma_1} \frac{d \ln P}{dr} \equiv A, \tag{18.24}$$

where A was defined in §17.2.

We consider first the very small "secular" root of eq. (18.22). This value of s is so small that s^2 and s^3 can be neglected. This root is

$$\mathscr{S}_{\text{secular}} = -\frac{1}{(-A)} \left\{ \frac{[\epsilon(\nu - 1) - (KT/\rho)k_T{}^2]}{c_P T} \frac{(-\chi_\mu)}{\chi_\rho} \frac{d \ln \mu}{dr} + \frac{\chi_T}{\chi_\rho} \frac{\epsilon\alpha}{c_P T} \frac{d \ln X}{dr} \right\}. \tag{18.25}$$

Remembering that A is negative for stability against convection and that χ_μ is normally negative for realistic equations of state, we see from eq. (18.25) that, for no nuclear energy production ($\epsilon = 0$), s is negative (stability) for $d \ln \mu/dr < 0$, which is normally the case in stellar evolution. However, for $d \ln \mu/dr > 0$, we see that s is positive, indicating instability (on a thermal time scale) against slow, convective-like motions. This instability is sometimes referred to as the Rayleigh-Taylor instability. It has also been referred to as a thermohalene instability (see, e.g., Ulrich 1972), and was referred to as "quasi-convection" in Cox and Giuli (1968, 13.3; see also §S.23.2 in this same reference). In §13.3 of this latter reference a physical explanation of this instability was also given. This explanation was based on the fact that, in a region of convective stability, an element of matter displaced slowly from its static position would experience no net forces, such as unbalanced buoyant forces, which would lead to convective instability in the ordinary sense. Nevertheless, in a region of spatially varying chemical composition, such an element will in general have a different temperature than its immediate surroundings. This temperature difference will, through gains or losses of heat, lead to further motions of the element, and hence to further density changes. The element will thus either slowly return to its original position (stability) or else move further away (instability).

If, on the other hand, $\epsilon \neq 0$, then the above conclusions might be altered. More specifically, let us assume, for example, that an element has slowly moved outward from its equilibrium position. If the element has moved slowly enough, then it will always be in pressure equilibrium with its surroundings (this is just assumption [c]). Moreover, if the motion is slow enough, there can be no unbalanced buoyant forces, and so the density of the element must be the same as that of its surroundings. Now, the mean molecular weight of the element will be the same as that at the level from which the element originated. Therefore, if, as is usually the case, the mean molecular weight of the unperturbed surroundings decreases outward ($d \ln \mu/dr < 0$), then the mean molecular weight of the element will be greater than that of its surroundings during its outward excursion. Hence, if the equation of state is anything like the perfect gas law, then the element will be *hotter than* its surroundings. Without nuclear energy sources, the element would simply lose heat to its surroundings, cool, contract, and so eventually return to its place of origin. However, if nuclear sources are present, and if they are sufficiently temperature-sensitive (that is, if ν is large enough) or of large enough scale (if $\epsilon[\nu - 1] > [KT/\rho]k_T^2$), then the enhanced temperature of the element (relative to its surroundings) will lead to an enhanced rate of energy production, perhaps to further heating, and so perhaps to further outward motion. In other words, the material might be *unstable* against slow mixing, as is in fact shown by eq. (18.25).

Now, what about the term in eq. (18.25) involving $d \ln X/dr$? This term gives a negative (stabilizing) contribution to $\mathscr{s}$ if $d \ln X/dr > 0$, which is the normal situation in an evolving star. Of course, $d \ln X/dr$ and $d \ln \mu/dr$ are related. The relation between them is, assuming that μ is a function only of the mass fractions of hydrogen, helium, and everything else, denoted, respectively, by, X, Y, and Z (see eq. [15.17], e.g., of Cox and Giuli 1968); and that Z remains constant,

$$\frac{d \ln \mu}{dr} = -\frac{5}{4}\mu X \frac{d \ln X}{dr}. \tag{18.26}$$

Hence, an outward increase in X will tend to cause μ to *decrease outward.* The origin of the sign of the term referred to at the beginning of this paragraph is easy to understand. The element has the chemical composition characteristic of its level of origin. Hence, if X in the unperturbed surroundings *increases outward,* and if the element moves outward, then clearly the value of X in the element will be smaller than its value in the immediate surroundings of the element. Since hydrogen is (by assumption) the main nuclear fuel, then a decreased value of X in the element (relative to its surroundings) will lead to a diminished rate of nuclear energy production in the element (again relative to its surroundings). Hence, the element will tend, from this effect, to cool, contract, and so return to its place of origin, that is, the motion will tend to be stable (negative contribution to $\mathscr{s}$). (For applications of these results to possible mixing in the sun, for example, see §19.6.)

We consider next the roots of eq. (18.22) corresponding to high-order g modes, for which $|\mathscr{s}|$, though perhaps small, is nevertheless much larger than $|\mathscr{s}_{\text{secular}}|$. We first consider the adiabatic limit, and we assume that nonadiabatic effects are small. This limit may be realized by formally taking $\epsilon = 0$ and $K = 0$ in eq. (18.22). Then the coefficients of even powers of $\mathscr{s}$ (including zero) vanish. We denote the roots of eq. (18.22) in this limit by $\mathscr{s}_0$. Assuming that $\mathscr{s}_0 \neq 0$, we have

$$\mathscr{s}_0 = \pm i \frac{k_H}{k_T} (-Ag)^{1/2}, \tag{18.27}$$

where $(-Ag)^{1/2}$ is the Brunt-Väisälä frequency (see §17.2).

We now write

$$\mathscr{s} = \mathscr{s}_0 + \Delta\mathscr{s}, \tag{18.28}$$

where we assume that $|\Delta\mathscr{s}| \ll |\mathscr{s}_0|$. Neglecting squares and higher powers of $|\Delta\mathscr{s}|$, we may solve for $\Delta\mathscr{s}$. However, noting that $\chi_\rho c_P = \Gamma_1 c_V$ (see §4.2) and using the identity (18.24), we obtain

$$\Delta\mathscr{s} = \frac{1}{2}\left\{\frac{\mathcal{S}}{A}\frac{\chi_T}{\chi_\rho}\frac{[\epsilon(\nu - 1) - (KT/\rho)k_T{}^2]}{c_P T} - \frac{1}{A}\frac{\epsilon\alpha}{c_P T}\frac{\chi_T}{\chi_\rho}\frac{d\ln X}{dr} - \frac{\chi_T}{\chi_\rho}\frac{[\epsilon(\lambda + 1)]}{c_P T}\right\}, \tag{18.29}$$

which essentially agrees with the result found by Osaki (1975). If there are no nuclear energy sources ($\epsilon = 0$), we have

$$\Delta\mathscr{s} = -\frac{1}{2}\frac{\mathcal{S}}{A}\frac{\chi_T}{\chi_\rho}\frac{(KT/\rho)k_T{}^2}{c_P T}. \tag{18.30}$$

We note that we may write (see eq. [18.15])

$$\mathcal{S} = \frac{1}{\lambda_P}(\nabla - \nabla_{\text{ad}}), \tag{18.31}$$

where λ_P is the pressure scale height, $\nabla \equiv d\ln T/d\ln P$, and $\nabla_{\text{ad}} \equiv (\Gamma_2 - 1)/\Gamma_2$. This expression should be compared with that for A (eq. [17.15]). In regions of uniform chemical composition, A and $\mathcal{S}$ are identical if $\chi_T/\chi_\rho = 1$ (as would be true for a simple perfect gas) and eq. (18.29) shows that any oscillations present will eventually die out. The same conclusion will obtain if the chemical composition is nonuniform but the material is stable against convection in the ordinary sense, that is, if $A < 0$, and $\mathcal{S} < 0$ (the material is subadiabatic); for then $\Delta\mathscr{s} < 0$. However, if $A < 0$, but if $\mathcal{S} > 0$ (the material is superadiabatic), then, according to eq. (18.29), $\Delta\mathscr{s} > 0$ and we would have the phenomenon of "overstable convection" (see below), first discovered by Kato (1966). This phenomenon would be likely to occur in a semi-convective zone. A physical explanation of this phenomenon has been given in Cox and Giuli (1968, §S.23.2). Briefly, an element of matter displaced inward (adiabatically), for example, from its eqiulibrium position will be, if $A < 0$, *less dense* than its surroundings. The resulting buoyant force will then push the element back up (outward), and the material will be stable against convection in the ordinary sense. However, if $\mathcal{S} > 0$, the inward-displaced element will be *cooler* than its surroundings. This relative coolness of the element will cause it to absorb heat from its surroundings. This absorption of heat will result in an expansion of the element and, accordingly, in a tendency for the density of the element to be lowered still further. The outward buoyancy force will thus be slightly enhanced by these heat gains. The ensuing outward velocity will thus be slightly larger than it would have been, were it not for the above-mentioned heat gains. Hence, an oscillating element will, under the assumed conditions, receive an extra "kick" on both its outward and inward excursions, and the oscillatory motion will clearly be overstable. This overstable oscillatory motion is what is termed "overstable convection."

The above treatment would also have yielded the results of Dilke and Gough (1972) regarding the resolution of the solar neutrino problem if the appropriate physics had been put in; that is, if it had been assumed that ϵ' in eq. (18.17) was a function also of the abundance of ^{3}He (see Ledoux 1974). Note that, now, the term containing $d \ln X/dr$ contributes to instability if $d \ln X/dr$ is positive in a region of stability against convection ($A < 0$). (If ^{3}He had been included, a similar term, with the same sign, in the abundance of ^{3}He would have appeared.) The physical interpretation of this result is as follows. We consider a region which is stable against convection in the ordinary sense ($A < 0$), and in which, further, $\mathcal{S} < 0$ (the material is subadiabatic), as will normally be the case. Then, an element which has moved slightly inward (adiabatically), say, will be *less dense* and *hotter* than its surroundings. The enhanced temperature in the element will lead to an enhanced rate of energy production therein, and this effect will tend to produce overstability, as shown by eq. (18.29) (term containing ν). Also, if $d \ln X/dr > 0$, then an *inward-moving* element will have more hydrogen than its surroundings, and this excess of hydrogen will lead to an enhancement of the rate of energy production in the element. This effect will, likewise, produce a tendency toward overstability, as shown by eq. (18.29). This effect of the gradients of hydrogen and ^{3}He abundances is the basic cause of the overstability found by Dilke and Gough (1972) and Christensen-Dalsgaard, Dilke, and Gough (1974).

Note that, if $A < 0$ but $\mathcal{S} > 0$, the nuclear term (that containing ν) in eq.

(18.29) works in the opposite sense from the term (containing K) that gives rise to overstable convection.

18.2b. ARBITRARY FREQUENCIES

We now do not assume that $|\mathscr{s}|$ is necessarily small, so that we are now no longer necessarily justified in adopting assumptions (b) and (c) stated in §18.2a. However, as we are still dealing with high-order modes, we are still justified in adopting assumption (a).

The radial component of the equation of motion (eq. [18.7]) is accordingly not altered. However, the mass equation (given previously by eq. [18.10]) is altered, and it now reads

$$\frac{k_H^2}{\mathscr{s}^2}\frac{P'}{\rho} = -\frac{\rho'}{\rho} - \delta r \frac{d \ln \rho}{dr} - ik_r \delta r. \tag{18.32}$$

Just as before, this equation may be used to eliminate P'/ρ from the radial component of the equation of motion. We obtain for the combined mass equation and equation of motion, in place of eq. (18.11),

$$\mathscr{s}^2 \frac{k_T^2}{k_H^2} \delta r = \left(ik_r \frac{\mathscr{s}^2}{k_H^2} - g\right)\frac{\rho'}{\rho} + \frac{\mathscr{s}^2}{k_H^2} ik_r \delta r \frac{d \ln \rho}{dr}. \tag{18.33}$$

The energy equation is the same as before (see eq. [18.14]), except that it now contains, in addition, P'/P. If we make the same assumptions as before about $\delta(\epsilon - [\nabla \cdot \mathbf{F}]/\rho)$, the resulting energy equation will now be a relation among P'/P, T'/T, ρ'/ρ, X'/X, and δr (see eq. [18.19]).

The equation of state (cf. eq. [18.20]) (with $P'/P \neq 0$) provides a relation among P'/P, T'/T, ρ'/ρ, and μ'/μ.

The mass equation (see eq. [18.23]) has already been seen to be a relation among P'/P, ρ'/ρ, and δr.

Finally, eqs. (18.21a,b) are two relations among, respectively, μ'/μ and δr, and X'/X and δr.

The preceding four equations may then be used to express P'/P, T'/T, μ'/μ, and X'/X in terms of ρ'/ρ and δr. Substituting this last expression (which is fairly long and complicated) into the combined mass equation and equation of motion, eq. (18.33), yields a fifth degree equation in $\mathscr{s}$. We now simplify the resulting equation by making use of some thermodynamic identities, among other things, and of the short-wavelength approximation (wavelength of the disturbance much smaller than any other length of interest, except the scale of variation of X or μ). We obtain, after considerable algebraic mainpulation,

$$\frac{1}{k_T^2}\frac{\rho}{\Gamma_1 P}\mathscr{s}^5 - \frac{1}{k_T^2}\frac{\rho}{\Gamma_1 P}\frac{[\epsilon(\nu-1)-(KT/\rho)k_T^2]}{c_V T}\mathscr{s}^4 + \mathscr{s}^3$$
$$+\left\{\frac{\chi_T}{\chi_\rho}\frac{[\epsilon(\lambda+1)]}{c_P T} - \frac{[\epsilon(\nu-1)-(KT/\rho)k_T^2]}{c_P T}\right\}\mathscr{s}^2$$
$$-\frac{k_H^2}{k_T^2}Ag\mathscr{s} - \frac{k_H^2}{k_T^2}g\left\{\frac{\chi_\mu}{\chi_\rho}\frac{[\epsilon(\nu-1)-(KT/\rho)k_T^2]}{c_P T}\frac{d\ln\mu}{dr}\right.$$
$$\left.-\frac{\chi_T}{\chi_\rho}\frac{\epsilon\alpha}{c_P T}\frac{d\ln X}{dr}\right\} = 0. \qquad (18.34)$$

The adiabatic limit may be realized, formally, by setting ϵ and K both equal to zero in eq. (8.34). Then the coefficients of all even powers of $\mathscr{s}$ (including zero) vanish. If we assume that $\mathscr{s} \neq 0$, we are then left with a quadratic in $\mathscr{s}^2$ which is, in fact, the same quadratic that was discussed in §17.12 (except that there it was assumed that the magnitude of the square of the Brunt-Väisälä frequency, $|Ag|$, was much smaller than either $k_H^2\Gamma_1 P/\rho$ or $k_r^2\Gamma_1 P/\rho$). In particular, for $|\mathscr{s}|$ large, we have (as was obtained in §17.12),

$$\mathscr{s}^2 \approx -k_T^2\frac{\Gamma_1 P}{\rho} = -(k_r^2 + k_H^2)\frac{\Gamma_1 P}{\rho}. \qquad (18.35)$$

We may note that, as is physically plausible and as may also be shown, taking $|\mathscr{s}|$ large is the same as taking the adiabatic limit.

For $|\mathscr{s}|$ small, eq. (18.34) leads to results which are identical to those obtained in §18.2.a.

For $|\mathscr{s}|$ large, we may proceed as in §18.2.a, and set $\mathscr{s} = \mathscr{s}_0 + \Delta\mathscr{s}$, where $|\Delta\mathscr{s}| \ll \mathscr{s}_0$, and where $\mathscr{s}_0$ is given by eq. (18.35). Neglecting squares and higher powers of $|\Delta\mathscr{s}|$, we obtain

$$\Delta\mathscr{s} = \frac{1}{2}\left\{\frac{[\epsilon(\nu-1)-(KT/\rho)k_T^2]}{c_V T}\left(1-\frac{\chi_\rho}{\Gamma_1}\right) + \frac{\chi_T}{\Gamma_1}\frac{[\epsilon(\lambda+1)]}{c_V T}\right\}. \qquad (18.36)$$

This result shows, plausibly enough, that, in the absence of thermonuclear energy generation ($\epsilon = 0$), the amplitude of a high-frequency disturbance will steadily diminish with time (stability), as a result of heat conduction. On the other hand, any thermonuclear energy generation which may be present ($\epsilon \neq 0$) will, as may be expected, act in the opposite sense, and will tend to cause the amplitude of the disturbance to increase with time (overstability).

18.3. RECENT WORK

In this section we shall briefly review some recent work on the calculation of linear, nonradial, nonadiabatic oscillations of spherical stellar models.

Perhaps the first investigation of this kind was carried out by Christensen-Dalsgaard and Gough (1975), as applied to an evolved (present) solar model. The main purpose of this investigation was to check the results of an earlier, quasi-adiabatic calculation of the vibrational stability of a solar model against low-l (in particular, $l = 1$), low-order g^+ modes of nonradial oscillation (Christensen-Dalagaard, Dilke, and Gough 1974). This model had been found in this earlier investigation to be overstable in the g_1^+, $l = 1$ mode, as a result of the gradient of the ^{3}He abundance in the model (see §18.2 for a physical explanation of this effect). Instability in this mode has important implications with regard to the solar neutrino problem (see Christensen-Dalsgaard, Dilke, and Gough 1974 and references therein; Bahcall and Sears 1972; Bahcall 1977; and §18.2).

This recent investigation showed that, throughout almost the entire volume of the model the nonadiabatic eigenfunctions were almost indistinguishable from the corresponding adiabatic eigenfunctions. However, nonadiabatic effects were found to be appreciable in the very outermost layers, that is, in a thin layer only about 300 km thick lying just beneath the photosphere. It is in this region that our knowledge of the structure of the sun is probably most incomplete. Convective transfer also probably takes place here, and convective flux variations were (understandably) neglected by the authors. They found severe damping in this thin shell; this damping was found to be comparable in magnitude to the driving in the interior. The authors concluded that, because of the many uncertainties in our knowledge of these outermost layers, the vibrational stability of the sun toward low-order g^+ modes could not at the present time be definitively determined.

Perhaps the next investigation along these lines was summarized in the very important paper of Ando and Osaki (1975). They solved the equations of linear, nonradial, nonadiabatic pulsation theory in the Cowling approximation (this approximation was found by the authors to be well justified in this application) as applied to a model of the solar envelope. This paper contains many interesting and useful results, but we can review only a few of them here. One interesting feature is that the authors treated the radiation transfer by a kind of Eddington approximation which is a considerable improvement over the diffusion approximation. This Eddington approximation is described in an appendix in Unno (1965). Unno and Spiegel (1966) have shown that this approximation agrees with the results of an accurate radiative transfer treatment both in the optically thin and optically thick limits. However, this Eddington approximation is of course still an approximation to an exact radiative transfer treatment.

The authors found that the lower p modes were trapped (see §17.10) in the outer solar convection zone, but were evanescent in and above the photosphere. The essential reason for the trapping is that the Brunt-Väisälä frequency N (see §17.2) becomes very small (zero if the convective zone is assumed to be adiabatic) in a convective region, and rises to rather large values just exterior to the convective region (see Fig. 1 of Ando and Osaki 1975). On the other hand, the critical acoustic frequency S_l (see eq. [17.89]) is also smaller than the (rather large) angular oscillation frequencies of interest to the authors, even for the rather large spherical harmonic orders considered ($10 \leq l \leq 1500$). Thus, part of the convective region is an A (acoustic) region. In and above the photosphere, however, the angular frequencies of interest are generally smaller than N, yet larger than S_l. These photospheric (and somewhat higher) regions are therefore evanescent, according to the discussion in §17.10.

It is interesting that most of these trapped p modes were found to be overstable by the authors. The main destabilizing agent was the κ-mechanism of Baker and Kippenhahn (1962) (also see Chapter 9 of this book), although a mechanism discussed also by Moore and Spiegel (1966) was found to play a small role. Use of the diffusion approximation was found to produce more instability than was produced by the Eddington approximation. Reasons for this difference are also given by the authors. Most of the pulsation energy in these lower p modes was concentrated into a thin layer (roughly 1000 km thick) lying just beneath the photosphere. Because of the rapid decrease of density in the equilibrium model with height, however, this pulsation energy decreased to nearly zero just above the photosphere.

Perhaps the most interesting of their results was the finding that the most unstable of these trapped p modes lay in a long, mountain-range-like region in the ($\sigma - k_H$) (angular frequency-horizontal wave number) plane, centered about a frequency corresponding to a period of 300 s, and representing a wide range of horizontal wavelengths, say from ~1000 km to slightly more than ~300,000 km. These numbers are in agreement with those observed in the five-minute oscillation originally discovered by Leighton, Noyes, and Simon (1962), and described more recently by, for example, Musman and Rust (1970); Deubner (1974); and Fossat, Ricort, Aime, and Roddier (1974). Recent observational evidence (Rhodes, Ulrich, and Simon 1977 and references therein) has apparently definitively confirmed the nonradial p-mode character of the five-minute oscillation. Other models of the five-minute solar oscillation can be found in the review article by Stein and Leibacher (1974). See also some of the other references given in Ando and Osaki (1975).

In this investigation the interaction between convection and pulsation was understandably ignored, because of the absence of a reliable time-dependent theory of convection. The authors admitted that this interaction could conceivably be important and could modify their results. They also emphasized the sensitivity of their results to the equilibrium model of the solar envelope employed. They also pointed out that possible energy losses in the form of running waves escaping into the chromosphere and corona, not considered in their model, might be important.

In their next paper (Ando and Osaki 1977) they took into account this possible leakage of energy by running waves propagating into the chromosphere and corona. The conclusions were essentially the same as in the earlier paper. In other words, this leakage of energy is evidently not very severe, at least as far as their stability results as found earlier are concerned. In particular, this leakage is not serious for the waves of long horizontal wavelength, and is nonexistent for the waves of short horizontal wavelength, because these short waves are nonpropagating in the corona.

The methods of Ando and Osaki (1975) were then used by Ando (1976) to examine the possibility of overstability of acoustic nonradial oscillation modes with large l (between 10 and 1500) in late-type stars over a wide region of the Hertzsprung-Russell diagram. He found that the wave-trapping as occurs in the sun (see earlier in this section) in the hydrogen convection zone was a fairly common phenomenon, particularly in those stars that were cool enough to have a well-developed hydrogen convection zone. He also found that these trapped acoustic modes were generally overstable, due primarily to the κ-mechanism of Baker and Kippenhahn (1962) (see also Chapter 9). For stars somewhat on the cool side of the Cepheid instability strip (see Fig. 3.1), the growth rate η (fractional amplitude increase per period) for the maximally unstable modes with l values in the above range, obeyed the approximate interpolation formula $\eta \propto g^{-0.5}\, T_e^{6}$, where g denotes surface gravity and T_e denotes effective temperature. He suggested that the resulting chromospheric activity might explain the Wilson-Bappu effect (Wilson and Bappu 1957). (However, these general ideas regarding the Wilson-Bappu effect do not seem to be consistent with those of Ayres, Linsky, and Shine 1975.)

The calculations of Shibahashi and Osaki (1976a,b) were not nonadiabatic (actually, stability analyses were carried out in the quasi-adiabatic approximation), but the results of these calculations, in the writer's opinion, were sufficiently important that they should be at least mentioned here. In Shibahashi and Osaki (1976a), nonradial oscillations of large l were studied in stars of masses 15 $M_\odot$ and 30 $M_\odot$ ($M_\odot$ = solar mass) having semi-convective zones. One of the very interesting results of this study is that for large l ($\gtrsim 8$) the nonradial oscillations for this type of star were found to fall into one of two fairly distinct classes: either predominantly g modes trapped in the μ-gradient region in the stellar core, or predominantly p modes trapped in the stellar envelope. Moreover, this dichotomy increases as l increases. (See Shibahashi 1979 and Unno, Osaki, Ando, and Shibahashi 1979 for further discussion of this effect.)

In their stability analyses, they found that the g modes of large l (≈ 15) which were trapped in the semi-convection zones of varying μ were in fact overstable. This overstability corresponds physically to the overstable convection found earlier by Kato (1966) by a local analysis and discussed in §18.2. As is correctly pointed out by the authors, the question of overstable convection is actually a question of the overstability of high-order g modes, and a correct analysis must be a global one. This result may be considered an answer to earlier criticisms of Kato's (1966) result by Gabriel (1969) and Auré (1971), who pointed out that Kato's result was based on a local analysis. The trapping in the region of the μ-gradient means that the amplitudes of the nonradial g modes are large just in those regions where the amplitudes must be relatively large in order for Kato's mechanism to work. However, the authors pointed out that this overstability probably has nothing (directly, at least) to do with the instability of the β Cephei stars (see Chapter 3), primarily because this latter instability seems to involve small l ($\approx$two or perhaps even zero). In this paper also is presented a new mode classification scheme based on the aforementioned dichotomy of trapped core g modes versus trapped envelope p modes. This scheme is intended to reflect more clearly the physical nature of the nonradial oscillations present in the star (see also Wolff 1979).

In their second paper, Shikahashi and Osaki (1976b) investigated the overstability of massive stars ($M = 5\ M_\odot$, $11\ M_\odot$, $20\ M_\odot$, and $40\ M_\odot$) toward nonradial oscillations of g modes trapped in the region of the μ-gradient (see earlier), where hydrogen shell burning was also occurring, this nuclear burning being the cause of any possible overstability. Only some of their models were found to be overstable, the $20\ M_\odot$ model for one value of l ($=10$), and the $40\ M_\odot$ model for a few values of l (11 to 40). Moreover, the growth times of the unstable modes were found to be relatively long, of the same order as the evolution times, and all unstable modes had become stabilized by the time the models became red giants. The overall conclusion of the authors was that the few cases of overstability found due to shell hydrogen burning probably had nothing *directly* to do with the observed instability of the β Cephei stars (see Chapter 3).

The reasons that the authors often found stability in their models are interesting and are explained in some detail in their paper (Shibahashi and Osaki 1976b), but space limitations prevent us from discussing the results further.

There have of late been a number of nonradial, nonadiabatic oscillation calculations of different types of stars, mostly by Osaki and Dziembowski. In particular, Osaki (1977) investigated the vibrational stability against nonradial oscillations of Cepheid-type stars. One of Osaki's (1977) most interesting findings derived ultimately from the g-mode character of nonradial modes of normal frequencies in the interior regions of highly centrally condensed stellar models, such as Cepheids are thought to be. (The reasons for this dual character of nonradial oscillations in centrally concentrated models were given in §17.10.) The *g*-mode character of such modes was also realized and pointed out by Dziembowski (1971), who showed that in these highly condensed central regions the nonradial (essentially *g* mode) oscillations had very small vertical wavelengths, with more than a thousand nodes in these central regions. Largely as a result of these short wavelengths, nonadiabatic effects are very important for these waves, even in the deep stellar interior. These nonadiabatic effects were estimated by Dziembowski (1971) by use of the quasi-adiabatic approximation, and were found to be so large that the damping times for these interior waves were only a few times larger than typical giant-star pulsation periods. This finding led Dziembowski (1971) to the conclusion that nonradial oscillations were either not excited, or only weakly excited, in Cepheid-like stars. In a subsequent investigation Dziembowski (1975) also concluded on similar grounds that nonradial modes were either not excited at all or only weakly excited in models of horizontal branch stars which might represent RR Lyrae stars.

These short damping times were known to Osaki (1977), who also realized that such short-wave *g* modes travel very slowly (propagation velocity $\propto$ [wavelength]^2, see §17.12). He therefore concluded that these short-wavelength waves would be damped out via nonadiabatic effects long before they could reach the stellar center and be reflected back to form standing waves. Thus, nonradial oscillations of giant-star models could not, Osaki (1977) reasoned, be of the usual standing-wave type. The waves running toward the stellar center were essentially absorbed: the central stellar regions formed a sink for nonradial oscillations. He then reformulated the inner boundary condition to accommodate this loss of pulsation energy by means of waves running toward the center.

With the running-wave effects of the central regions taken into account by means of an inner boundary condition, Osaki (1977) was thus able to treat the stellar envelope as essentially an isolated pulsating unit for p modes, effectively trapped within the envelope (called $\bar{p}$ modes by Osaki 1977), just as can be done for purely radial pulsations in giant-star models (see §9.2b).

Osaki (1977) found that, because of the above leakage of pulsation energy into the stellar interior, nonradial oscillations of small l were stable. However, these oscillations became unstable for larger values of l ($\geq$6 for what Osaki called the $\bar{f}_1$ mode and $l \geq 4$ for the $\bar{p}_1$ mode), as a result of driving by the hydrogen and helium ionization zones (see Chapter 10). Moreover, the growth rates of these large-l nonradial unstable envelope p_1

modes were of the same order as the growth rates for purely radial oscillations of similar star models.

These nonradial oscillations, even though found by Osaki (1977) to be overstable in a linear analysis, are not normally observed (at least directly) in Cepheids. Osaki (1977) suggested that the reason that such nonradial oscillations are not observed in Cepheids is that nonlinear effects tend (for some reason) to favor the radial modes. He also suggested that these overstable high-order nonradial modes might explain the "microturbulence" or "macroturbulence" inferred in the spectra of giant and supergiant stars, as had earlier been suggested by Lucy (1976) for α Cygni and still earlier by Serkowski (1970) for U Mon (an RV Tauri variable) and by Shawl (1974) for o Ceti (Mira) (see also Schwarzschild 1975). Small oscillations on short time scales (~ months) have been observed in almost all supergiants (Abt 1957; Rosendahl and Snowden 1971; Maeder and Rufener 1972; Fernie 1976).

In an important paper the entire problem of radial and nonradial, nonadiabatic oscillations in linear theory in various kinds of stars, including giants and supergiants, was discussed in a very thorough and very fundamental way by Dziembowski (1977b).

To treat the quasi-adiabatic case more accurately than in the past, he obtained asymptotic solutions of the (linearized) radiative flux and energy equations, valid for radial or nonradial modes in the deep stellar interior. These asymptotic solutions were based on methods developed by several Russian mathematicians (see, e.g., Feschenko, Shkil', and Nikolenko 1967). By means of these asymptotic solutions, he found that in many cases the usual quasi-adiabatic analysis could give completely wrong answers for nonradial oscillations in the deep stellar interior, even though nonadiabatic effects may be very small there. In particular, the amplitudes of high-order g modes trapped in the deep interior of a highly concentrated giant star model may be considerably smaller than is predicted by the quasi-adiabatic theory. As Dziembowski (1977b) pointed out, these very small interior amplitudes of high-order trapped g modes could equivalently be interpreted as Osaki (1977) had done, as running waves lost to the interior. In fact, both authors independently arrived at the same kind of interior boundary condition which was intended to take these effects into account.

Dziembowski (1977b) also discussed high-order g modes trapped in the μ-gradient region in the interior of evolved models and arrived at essentially the same conclusions that Shibahashi and Osaki (1976a) had reached (see earlier in this section), namely that such waves would have fairly large amplitudes in this region (the Kato overstable convection phenomenon).

Dziembowski (1977b) concluded, overall, that nonradial oscillations could be excited in the envelopes of several kinds of stars, by the same envelope ionization mechanisms that are supposed to be responsible for the pulsations of Cepheids, RR Lyrae variables, and the like (see Chap. 10), and with about the same growth rates as for radial pulsations. In particular, he found that p_1 modes with small l were stable in the envelopes of horizontal branch models which might represent RR Lyrae variables. However, those p_1 modes with larger values of l ($\approx$6–7) were overstable, with about the same growth rates as for radial modes, a result that sounds reminiscent of the results found by Osaki (1977) for a giant star model that might represent a Cepheid. Dziembowski (1977b) also found that nonradial f modes with large values of l ($\sim$100 to 200) were overstable at effective temperatures much hotter (up to roughly 10,000°K) than for stars in the RR Lyrae instability strip on the Hertzsprung-Russell diagram.

Dziembowski (1977b) also investigated a supergiant model which was supposed to represent α Cygni, and found (in agreement with earlier results by Lucy 1976) that this model was overstable only to high-l nonradial oscillations. However, he found that lower-l modes were excited in a cooler supergiant, lying close to the Cepheid instability strip on the Hertzsprung-Russell diagram.

In the same paper Dziembowski also investigated models of δ Scuti stars for radial and nonradial vibrational instability (see Fitch 1976b and Breger 1979 for reviews of these stars). He found (as he had earlier, see Dziembowski 1975) that these models were overstable to many modes, both radial and nonradial, and all with comparable growth rates. He suggested that mode coupling may well occur among these many excited modes. Such coupling might account (in part, at least) for the complicated observed behavior of many of these objects (see Chapter 3 and Breger 1979).

Dziembowski (1977c) calculated the light variations which a star undergoing nonradial oscillations would exhibit.

Dziembowski (1977a), using a fully nonadiabatic, nonradial method of analysis, investigated the stability against both radial and nonradial oscillations of models of white dwarf stars. The purpose of this work was to study theoretically the variable DA white dwarfs (or ZZ Ceti stars) whose properties have recently been summarized by McGraw and Robinson (1976), McGraw (1977), Nather (1978), and Van Horn (1978) (see Chapter 3). McGraw and Robinson (1976) had suggested, on the basis of an earlier result of Vauclair (1971), that the basic cause of the variability was the same as for the ordinary Cepheid-like variable stars—namely, He^+ ionization in the envelope (see Chapter 10) (however, see J. P. Cox and Hansen 1979).

The models used by Dziembowski (1977a) to test this suggestion were characterized by envelopes rich in hydrogen and helium. These envelopes were, however, too cool for nuclear "burning" of hydrogen or helium to occur. The main result was that, although there was always a considerable destabilizing effect of the He^+ ionization zone, nevertheless the radial modes, as well as all nonradial modes of low l and relatively low order, were found to be stable. It is the g modes of low $l(= 1$ or $2)$ and relatively low order (roughly, $\lesssim$20–25) which span the range of observed periods of the DA white dwarfs; the f and p modes have periods that are far too short. Nonradial g modes with large l (hundreds to thousands) and orders around 10 to 20 were, on the other hand, found to be violently unstable, with growth times of the order of days. The large-l f modes were overstable, as were (in some cases) the large-l p modes. Thus, Dziembowski (1977a) concluded that the DA variable white dwarfs may form an extension of the Cepheid instability strip to stars with the highest effective temperatures and surface gravities known (with the possible exception of neutron stars, which we do not consider here).

However, those modes which are self-excited do not have the periods observed in these stars, the periods of the self-excited modes being too short. (In §17.12 it was stated that, for g modes with large l, the period was almost independent of l [except for the homogeneous model]. However, the periods of g modes will generally increase as l decreases; see, for instance, Brickhill 1975 or Van Horn 1976.) Therefore, Dziembowski (1977a) suggested that those low-l stable modes with long periods in about the observed range were excited by a nonlinear coupling to the high-l unstable modes with short periods. He presented some plausibility arguments for this view and urged the development of a theory of mode coupling for nonradial oscillations (some discussion of mode coupling for radial oscillations may be found in Ledoux and Walraven 1958, §§86–88). The theoretical situation with regard to the variable white dwarfs is discussed in the very interesting, very thorough paper by Van Horn (1978).

In a subsequent calculation Dziembowski (1979) found that low-l g modes with about the observed periods were indeed excited by the Stellingwerf bump mechanism (see §§10.1 and 13.1).

Sienkowicz and Dziembowski (1978) investigated the vibrational stability of accreting white dwarfs in close binary systems. The authors found that all modes, both radial and nonradial, were overstable at high accretion rates. The cause of the overstability in all cases was found to be nuclear driving in the hydrogen- or helium-burning shell source, located near the stellar surface. The g modes were more strongly excited by some three orders of magnitude than the radial, f, or p modes. (The reason that the g modes were so strongly excited is that they have appreciable amplitudes in white dwarfs only in the outer layers [see Brickhill 1975]. Throughout most of the volume of a white dwarf the Brunt-Väisälä frequency $N = (-Ag)^{1/2}$ [see §17.2] is nearly zero, whereas the critical acoustic frequency S_l [see §17.10] is fairly large. For the *g*-mode angular frequencies σ of interest, we have that $N < \sigma < S_l$ throughout most of the volume of a white dwarf. Therefore, according to the discussion in §17.10, almost the entire interior of a white dwarf is an evanescent region for *g* modes.) These overstable *g* modes have growth rates of the order of months and periods in the 10–50 s range. Nevertheless, the authors do not propose this *g*-mode overstability of accreting white dwarfs as an explanation of the oscillations observed during the outbursts of dwarf novae (see Chapter 3).

Saio and Cox (1979b) have obtained results for the linear, nonradial, nonadiabatic oscillations of models of massive main sequence stars and of massive stars only slightly evolved off the main sequence. The nonadiabatic periods were almost exactly the same as those computed adiabatically. Also, the nonadiabatic eigenfunctions were practically the same as the adiabatic eigenfunctions throughout almost the entire model. The main purpose of this work was to examine the effectiveness of the Stellingwerf bump mechanism (see §§10.1 and 13.1) on the excitation of nonradial oscillations.

19

Miscellaneous Topics

In the present chapter we shall discuss briefly some of the effects of *not* making some of the many simplifying assumptions we have made throughout nearly all of the earlier chapters. Many of these complicating effects have not been fully worked out yet, and some are at present being actively pursued by researchers in the field. Our treatment of the various topics in this chapter will accordingly be for the most part rather brief and perhaps superficial; the present chapter should be considered, more than anything else, as a guide to the literature.

For example, the effects of viscosity (molecular or any other kind, such as radiative) on stellar oscillations have not yet been worked out, in the most general case, and have been discussed in some detail only for purely radial oscillations (see, e.g., Ledoux and Walraven 1958, §§47, 49, 71; Cox and Giuli 1968, Chap. 27 and references therein). Some exploratory steps in connection with the oscillations of partially crystalline white dwarfs have been taken by Van Horn and Savedoff (1976). A recent and fairly thorough discussion of the oscillatory properties of such structures has been carried out by Hansen and Van Horn (1979).

Perhaps the most extensive work has been done on the effects of stellar rotation on oscillations, and this subject will be considered in §19.1. The effects of large-scale magnetic fields on stellar oscillations are qualitatively different and have been less thoroughly studied; these are considered in §19.2. The very difficult, important, and still unsolved problem of the interaction between stellar pulsation and convection is discussed briefly in §19.3. The effects of thermal imbalance on stellar pulsation have been investigated fairly exhaustively in recent years, and these effects are discussed in §19.4.

The effects of general relativity have been investigated fairly thoroughly in the linear, adiabatic approximation, and some discussion, as well as references to the literature, are given in §19.5. Those investigations, where the effects of rotation and general relativity have been considered simultaneously, will be discussed and referred to in §19.5. Unless we say so explicitly, we shall assume throughout this chapter that nonrelativistic physics applies.

The subject of secular stability is a special case of pulsation theory, and is considered briefly in §19.6.

Finally, we shall attempt in §19.7 to assess the overall relevance and significance of pulsation theory to the study of stars and stellar evolution.

19.1. EFFECTS OF ROTATION

19.1a. GENERAL CONSIDERATIONS

The inclusion of rotation into a consideration of stellar pulsation complicates matters greatly. Rotation alters and greatly complicates even the underlying, unperturbed (nonpulsating) star. These latter complications are legion and are discussed in, for example, the review articles by Mestel (1965); Lebovitz (1967); Strittmatter (1969); Roxburgh (1970); Fricke and Kippenhahn (1972); Van Horn (1973); Ostriker (1978); (see also Wrubel 1958, p. 38; Schwarzschild 1958, §21; numerous articles in Slettebak 1969; von Zeipel 1924; and Baker and Kippenhahn 1959). An excellent monograph, dealing with all aspects of rotating stars, has recently been published (Tassoul 1979). Much earlier, the uniform (rigid-body) rotation of an incompressible spheroid was studied by Bryan (1889); see also Love (1889). One example of these complications is that, as was first shown by von Zeipel (1924) (see also Baker and Kippenhahn 1959), a rotationally distorted star cannot in general be in thermal equilibrium. These departures from thermal equilibrium normally give rise to slow "meridional currents" whose velocity is usually of order {[stellar radius/Kelvin-Helmoltz time (see Chapter 2)] × [ratio of centrifugal acceleration to gravitational acceleration]} (e.g., Sweet 1950; Öpik 1951). These meridional currents generate other problems which have been discussed by, for example, Osaki (1972, 1966). In addition, rotation gives rise to restrictions on the way in which the angular velocity of rotation can vary in a star. For example, according to Goldreich and Schubert (1967) and Fricke (1968) (see also James and Kahn 1971), the angular momentum per unit mass cannot decrease outward in a star.

In fact, rotating fluids in general are complicated and have themselves given rise to an extensive literature (see, e.g., Chandrasekhar 1961; Greenspan 1968; Batchelor 1967, Chap. 7).

It may be noted that most of the above complicating effects, in the stellar case, derive ultimately from the nonsphericity of the star, which is in turn a result of the centrifugal force per unit mass $\varpi\Omega^2$, where ϖ is the perpendicular distance from the rotation axis and Ω is the angular rotation velocity.[1] Since this centrifugal force $\propto \Omega^2$, then this force, together with

[1]The use of the symbol Ω to denote angular velocity throughout this section and the next should not be confused with the use of the same symbol to denote dimensionless pulsation frequency in earlier chapters (especially Chapters 8 and 17).

the consequent nonsphericity, may be negligible for a sufficiently slowly rotating star.

Besides the above effects, the mass elements in a star which is *pulsating* as well as *rotating* are subject, during their oscillations, to two additional forces besides the usual forces: *centrifugal* forces and *Coriolis* forces. As stated above, the former are proportional to Ω^2, whereas the latter are proportional only to the *first* power of Ω. Thus, the Coriolis forces cannot be neglected even in a star that is rotating so slowly that centrifugal forces can be neglected. Nevertheless, in view of these two new types of forces, it is perhaps not surprising that entirely new phenomena can occur in a star that is both rotating and pulsating. For example, the toroidal modes (see §17.3) become important and give rise to Rossby-like waves with non-vanishing frequencies, as has been shown by Papaloizou and Pringle (1978). In fact, these authors have shown that, if the star is rotating so slowly that departures from sphericity are negligible, the angular frequency of such modes is given by $2m\Omega/l(l + 1)$, where l and m are the usual indices of a spherical harmonic. These new modes have been called *r* modes by Papaloizou and Pringle.

It is obvious that there can be no purely radial oscillations in an arbitrarily rapidly rotating star. Nevertheless, if the star is not rotating too rapidly, the Lagrangian displacement of a mass element during its oscillations can be mostly radial, with tangential components which are small compared to the radial component (this statement is obvious from eq. [82.12] of Ledoux and Walraven 1958 or from eq. [52] of Lynden-Bell and Ostriker 1967). Such oscillations are called quasi-radial, and they would become purely radial oscillations in the limit of vanishingly small stellar rotational velocity.

Almost all investigations of the effects of rotation on stellar pulsations have assumed that the oscillations were *adiabatic,* with the Lagrangian pressure and temperature variations given by eqs. (5.36). This assumption will also be made, unless stated otherwise, in the remainder of this subsection. It turns out that rotation (at least in only slightly rotationally distorted stars) always has a *stabilizing* influence on quasi-radial oscillations, in the sense that the critical value of the appropriate adiabatic exponent Γ_1 below which dynamical instability sets in, is lowered below the value 4/3 (see Chapter 8) for such oscillations (see, e.g., Ledoux 1945; Cowling and Newing 1949; Ledoux and Walraven 1958, §82; Chandrasekhar 1972; J. P. Cox 1974a, 1976a and references therein). This stabilizing effect has, in fact, been shown by Lebovitz (1970) to be valid for any kind of differential rotation, not just rigid-body rotation, at least for Γ_1 constant. For further discussion of the dynamical stability of rotating stars, see, for example, Bisnovatyi-Kogan and Blinnikov (1974).

19.1b. THE VARIATIONAL PRINCIPLE FOR A ROTATING STAR

It was stated earlier (see Chapter 15) that the eigenfunctions and eigenvalues of the linear adiabatic wave equation (LAWE) still obeyed a variational principle for a rotating star, but that this variational principle had, in this latter case, to be modified somewhat. We shall consider this variational principle here, following in large part the excellent discussion of Lynden-Bell and Ostriker (1967). The most important part of this modification is, as we shall see, that the eigenfrequency ω must be purely real in order that the variational principle apply. Clement (1964) was the first to show that a variational principle exists for the eigenfunctions and eigenvalues of the LAWE for a uniformly rotating star, but he did not show that this variational principle applies only to purely real ω.

We also showed earlier (see §15.2) that the eigenfrequencies ω were either purely real or purely imaginary for a nonrotating star. For a rotating star, however, it can happen that ω is complex (see below). Thus, for example, in a rotating star we have the possibility of the rather strange phenomenon of overstable convection, to be discussed below (see also Ledoux and Walraven 1958, §82).

To deal with these matters, we consider the linearized equation of motion in the general form of eq. (15.18). We now write

$$\frac{d}{dt} = \frac{\partial}{\partial t} + \mathbf{v}_0 \cdot \nabla \equiv \frac{\partial}{\partial t} + M, \tag{19.1}$$

where $\mathbf{v}_0$ denotes the fluid velocity in the unperturbed (nonoscillating) configuration, and where this equation defines the operator M. If the flows are steady, then $\partial \mathbf{v}_0/\partial t = 0$ and the operators $\partial/\partial t$ and M commute. If we also assume that $\boldsymbol{\zeta}(\mathbf{r},t) = \boldsymbol{\xi}(\mathbf{r})e^{i\omega t}$ where $\boldsymbol{\zeta}(\mathbf{r},t) \equiv \delta\mathbf{r}$ is the Lagrangian displacement of a mass element, the LAWE becomes

$$-\omega^2\boldsymbol{\xi} + 2i\omega M\boldsymbol{\xi} + M^2\boldsymbol{\xi} = -\mathcal{P}(\boldsymbol{\xi}) - \mathcal{V}(\boldsymbol{\xi}), \tag{19.2}$$

where the operators $\mathcal{P}$ and $\mathcal{V}$ were defined in §15.3.

We may write this equation, somewhat analogously to Lynden-Bell and Ostriker (1967), as

$$-\omega^2\mathbf{A}(\boldsymbol{\xi}) + \omega\mathbf{B}(\boldsymbol{\xi}) + \mathbf{C}(\boldsymbol{\xi}) = 0, \tag{19.3}$$

where the operators $\mathbf{A}$, $\mathbf{B}$, and $\mathbf{C}$ are defined by the relations

$$\mathbf{A}(\boldsymbol{\xi}) \equiv \boldsymbol{\xi}, \tag{19.4a}$$

$$\mathbf{B}(\boldsymbol{\xi}) \equiv 2iM\boldsymbol{\xi}, \tag{19.4b}$$

$$\mathbf{C}(\boldsymbol{\xi}) \equiv M^2\boldsymbol{\xi} + \mathcal{P}(\boldsymbol{\xi}) + \mathcal{V}(\boldsymbol{\xi}). \tag{19.4c}$$

Forming the scalar product of eq. (19.3) and $\boldsymbol{\xi}^* \rho \, d\tau$ ($d\tau$ being a volume element and an asterisk denoting the complex conjugate) and integrating over the entire volume V of the configuration, we have

$$-a\omega^2 + b\omega + c = 0, \tag{19.5}$$

where

$$a \equiv \int_V \boldsymbol{\xi}^* \cdot \mathbf{A}(\boldsymbol{\xi}) \rho d\tau = \int_V \boldsymbol{\xi}^* \cdot \boldsymbol{\xi} \rho d\tau, \tag{19.6}$$

and so forth. (Note that a is just the oscillatory moment of inertia, denoted by J in Chapters 8 and 16; see also eq. [15.16].)

Now, it has been shown by Lynden-Bell and Ostriker (1967) that the operators **A**, **B**, and **C** are all Hermitian in the sense of eq. (15.11). It then follows that the numbers a, b, and c are all real.

We now regard $\boldsymbol{\xi}(\mathbf{r})$ as any sufficiently regular vector function of $\mathbf{r}$ (that is, not necessarily as a solution of the LAWE, eq. [19.3]), and ω simply as a solution of the quadratic equation (19.5) (that is, not necessarily an eigenfrequency of the LAWE) (the corresponding quantities were denoted, respectively, by $\mathbf{u}[\mathbf{r}]$ and Σ in §15.2). We subject $\boldsymbol{\xi}(\mathbf{r})$ to a small variation $\Delta\boldsymbol{\xi}$ at each point $\mathbf{r}$, keeping $\mathbf{r}$ and all unperturbed quantities fixed during the variation. The variation of eq. (19.5) then yields an expression involving $\Delta\boldsymbol{\xi}$, $\Delta\boldsymbol{\xi}^*$, $\Delta\mathbf{A}(\boldsymbol{\xi})$, $\Delta\mathbf{B}(\boldsymbol{\xi})$, $\Delta\mathbf{C}(\boldsymbol{\xi})$, and $\Delta\omega$, where $\Delta\omega$ is the corresponding variation in ω. However, it follows from the linearity of **A**, **B**, and **C** that $\Delta\mathbf{A}(\boldsymbol{\xi}) = \mathbf{A}(\Delta\boldsymbol{\xi})$, etc. Also, because of the Hermiticity of **A**, **B**, and **C**, $\boldsymbol{\xi}^*$ and $\Delta\boldsymbol{\xi}$ can be interchanged in this expression. Finally, we note that, whereas **A** and **C** are purely real, **B** is purely imaginary. The final result may be put in the form

$$(-2a\omega + b)\Delta\omega + \int_V (\Delta\boldsymbol{\xi}^*) \cdot \{-\omega^2\mathbf{A}(\boldsymbol{\xi}) + \omega\mathbf{B}(\boldsymbol{\xi}) + \mathbf{C}(\boldsymbol{\xi})\} \rho d\tau$$
$$+ \int_V (\Delta\boldsymbol{\xi}) \cdot \{-\omega^2\mathbf{A}(\boldsymbol{\xi}^*) - \omega\mathbf{B}(\boldsymbol{\xi}^*) + \mathbf{C}(\boldsymbol{\xi}^*)\} \rho d\tau = 0. \tag{19.7}$$

Now, if $\boldsymbol{\xi}$ is a solution of the LAWE (eq. [19.3]), then the first integral in eq. (19.7) vanishes. However, in general $\boldsymbol{\xi}^*$ is *not* a solution of the equation

$$-\omega^2\mathbf{A}(\boldsymbol{\xi}^*) - \omega\mathbf{B}(\boldsymbol{\xi}^*) + \mathbf{C}(\boldsymbol{\xi}^*) = 0; \tag{19.8}$$

rather, $\boldsymbol{\xi}^*$ is a solution of the complex conjugate of eq. (19.3). It follows, then, that ω does *not* obey a variational principle in general.

However, if ω is real, then eqs. (19.8) and the complex conjugate of eq. (19.3) are the same, so that the quantity in curly brackets in the second integral in eq. (19.7) also vanishes, and so we have $\Delta\omega = 0$; that is, ω in this case obeys a variational principle.

We have therefore proved the following: if ω is real, those solutions of the LAWE (eq. [19.3]) for a star having internal (steady) motions for which ω is an extremum ($\Delta\omega = 0$) with respect to arbitrary small variations of $\boldsymbol{\xi}(\mathbf{r})$, are the *eigenfunctions* $\boldsymbol{\xi}_k(\mathbf{r})$, and the associated eigenvalues are the corresponding *adiabatic eigenfrequencies* ω_k, both for the kth mode.

Note that some of the conclusions of §15.2 follow as a special case of the present considerations.

It has recently been shown by Schutz (1979) that, by a small modification of the arguments given by Lynden-Bell and Ostriker (1967), on which the present discussion is based, a variational principle can be constructed which applies to both real and complex eigenvalues.

The solution of the quadratic equation (19.5) is

$$\omega = \frac{b}{2a} \pm \left(\frac{b^2}{4a^2} + \frac{c}{a}\right)^{1/2}. \tag{19.9}$$

It is obvious from the definition that a is positive definite, while b and c, though always real, are not necessarily always positive. A negative value of c corresponds to dynamical instability in a non-rotating star ($b = 0$) (see §15.2). On the other hand, in the case of a rotating star a negative value of c can imply a complex value of ω. A complex value of ω would be manifested as overstable convection, which will be discussed from a more physical standpoint in §19.1c. Note that eq. (19.9) reveals, to some extent, the stabilizing influence of rotation: c must be *sufficiently* negative before ω will be complex; whereas, without rotation *any* negative value of c will render the star dynamically unstable.

19.1c. SPLITTING OF FREQUENCIES OF NONRADIAL OSCILLATIONS BY ROTATION

In Chapters 17 and 18 we pointed out that the azimuthal spherical harmonic index m never appeared in any of the equations of linear, nonradial, adiabatic or nonadiabatic oscillations of static spherical stars. Therefore, the eigenfrequencies of nonradial oscillations of such stars are $(2l + 1)$-fold degenerate for each mode (see §17.3). In a rotating star, however, this degeneracy is completely lifted; what would be the eigenfrequency for a given value of l in a nonrotating star is split into $2l + 1$ sublevels, corresponding to $m = -l, -(l - 1), \ldots, (l - 1), l$. This splitting is symmetrical (for slow rotation) about the frequency corresponding either to $m = 0$, or to no rotation, and the frequency difference between successive sublevels increases as the rotation speed of the star increases. (The splitting is quite analogous to the Zeeman effect in atomic physics.)

In the following discussion we shall revert back to our old notation of Chapter 8 and write J in place of a (see eq. [19.6]): $J \equiv a$. We shall also only retain terms proportional to the *first* power of the angular rotation velocity Ω of the star. If the Ω^2 terms were to be kept, the appropriate integrations would have to be carried out over the *rotationally distorted* unperturbed (nonoscillating) configuration (see, e.g., Simon 1969; Smeyers and Denis 1971). Writing

$$\omega = \sigma_0 + \sigma', \tag{19.10}$$

where σ_0 is the angular oscillation frequency of the non-rotating object and σ' is the correction for the effects of rotation, we have from eqs. (19.9) and (19.6)

$$\sigma' = \frac{b}{2a} + O(\Omega^2) = \frac{i}{J} \int_V \boldsymbol{\xi}^* \cdot (M\boldsymbol{\xi})\rho d\tau + O(\Omega^2), \tag{19.11}$$

where the operator M was defined in eq. (19.1), and O means "of the order of."

We shall write

$$\mathbf{v}_0 = \mathbf{v}_M + v_\phi \mathbf{e}_\phi, \tag{19.12}$$

where in a system of cylindrical coordinates (ϖ, ϕ, z)

$$\mathbf{v}_M = v_\varpi \mathbf{e}_\varpi + v_z \mathbf{e}_z \tag{19.13}$$

as the velocity solely in the meridional plane; v_ϖ, etc., are the three appropriate components of $\mathbf{v}_0$; $v_\phi \mathbf{e}_\phi$ is the velocity solely in the ϕ direction; and the **e**'s denote unit vectors along the respective axes. If this ϕ-motion is caused by rotation, we have

$$v_\phi = \varpi\Omega, \tag{19.14}$$

where the axis of rotation is parallel to the z axis. If we take into account the fact that the base unit vectors may change direction upon differentiation, and if we assume a ϕ-dependence of the components $\xi_{\varpi,\phi,z} \propto e^{im\phi}$ of $\boldsymbol{\xi}$, we then obtain, after only a small amount of manipulation,

$$M\boldsymbol{\xi} = (\mathbf{v}_M \cdot \nabla)\boldsymbol{\xi} + im\Omega\boldsymbol{\xi} + \boldsymbol{\Omega} \times \boldsymbol{\xi}. \tag{19.15}$$

If the motions in the unperturbed system represent pure rotation, that is, if we neglect meridional currents, then all we need do is set $\mathbf{v}_M = 0$ in eq. (19.15). For completeness, we also note that, for pure rotation ($\mathbf{v}_M = 0$),

$$M^2\boldsymbol{\xi} = M(M\boldsymbol{\xi}) = -m^2\Omega^2\boldsymbol{\xi} + 2im\Omega\boldsymbol{\Omega} \times \boldsymbol{\xi} + \boldsymbol{\Omega} \times (\boldsymbol{\Omega} \times \boldsymbol{\xi}), \tag{19.16}$$

which is clearly $O(\Omega^2)$.

Inserting eq. (19.15) into eq. (19.11), we have for the (first order in Ω)

correction to the adiabatic angular frequency due to rotation

$$\sigma' = -m\frac{1}{J}\int_V \Omega\boldsymbol{\xi}^* \cdot \boldsymbol{\xi}\rho d\tau + \frac{i}{J}\int_V \boldsymbol{\xi}^* \cdot (\boldsymbol{\Omega} \times \boldsymbol{\xi})\rho d\tau + O(\Omega^2). \qquad (19.17)$$

This expression agrees, for Ω = constant, with the results of Cowling and Newing (1949), Ledoux (1951), and Ledoux and Walraven (1958, §82) when account is taken of the fact that eq. (19.17) was derived in an inertial coordinate system, rather than in a rotating coordinate system. The second integral in eq. (19.17) will also be found to be proportional to m when account is taken of the fact that each component of $\boldsymbol{\xi}$ is assumed to be proportional to exp $(im\phi)$; note that ξ_ϕ is proportional also to m (see remark just before eq. [19.15]; also the discussion leading to eq. [19.18] below). We note that, to order of magnitude, $|\sigma'| \sim |\Omega|$.

For a uniformly rotating star (Ω = constant), it is sometimes advantageous (and, in fact, has been done in the above references) to derive the expression for the frequency splitting by working in a coordinate system that is rotating with constant angular speed at the angular velocity Ω. In this case only the second term on the right side of eq. (19.17) is obtained as the correction, due to uniform rotation, to the angular oscillation frequency as seen in the rotating coordinate system. This result is understandable when account is taken of the fact that nonradial oscillations with $m \neq 0$ represent azimuthal running waves, propagating in the sense of increasing ϕ for $m < 0$, and decreasing ϕ for $m > 0$.

We note from eq. (19.17) that, for $m = 0$, *all first order (in Ω) terms vanish*. The second term on the right side vanishes for $m = 0$, for then $\boldsymbol{\xi}$ is purely real. Thus, the oscillation frequency of any axially symmetric perturbation (this includes quasi-radial oscillations as a special case) is not affected at all to first order in Ω by rotation; only second (and higher) order terms in Ω have any effect on the oscillation frequencies of such perturbations.

Next, we note that all first order (in Ω) terms on the right side of eq. (19.17) are purely real. The second term is purely real, for one can easily show that $\boldsymbol{\xi}^* \cdot (\boldsymbol{\Omega} \times \boldsymbol{\xi})$ is purely imaginary. It can also be shown that all second order (in Ω) terms on the right side of eq. (19.17) are purely real. Thus, it is true that corrections to the eigenfrequency due to rotation are oscillatory at least through terms $O(\Omega^2)$.

This reality (at least through terms $O[\Omega^2]$) of σ' gives us an interesting view of overstable convection. If σ_0 in $\omega = \sigma_0 + \sigma'$ is purely imaginary (dynamical instability in the nonrotating state of the star), then ω is accordingly complex, representing damped or growing oscillations. As was pointed out in Chapter 17, dynamically unstable g modes have a close connection with convection; in fact, such a dynamical instability is usually interpreted as convective instability. Thus, σ_0 determines the growth or damping time of what would be ordinary convection in a nonrotating star, but becomes overstable convection in a rotating star. This growth or damping time is determined primarily by the properties of the convectively unstable region, and essentially does not depend (at least directly) on the rotation period of the star. On the other hand, the oscillation period of this overstable convection is determined solely by σ', which in turn depends only on the rotation rate and the structure of the star.

Physically, the oscillatory nature of the motion is strictly a manifestation of the Coriolis force $\propto -\mathbf{\Omega} \times \mathbf{v}_0$. Thus, for example, a dynamically unstable rising or sinking element of matter experiences a Coriolis force due to the rotation; this Coriolis force causes the element to swerve sideward, so that it moves in a circular path, with a certain periodicity, rather than straight "up" or "down."

The first-order (in Ω) splitting of the frequencies of nonradial oscillations of slowly, *differentially* rotating stars has been computed by Hansen, Cox, and Van Horn (1977). The spherical coordinate components ξ_r, ξ_θ, ξ_ϕ of $\boldsymbol{\xi}$ are usually expressed in terms of two functions $a(r)$ and $b(r)$ of r alone, where $a(r)$ and $b(r)$ are defined in Ledoux and Walraven (1958, §82). (The present a and b should not be confused with the a and b of §19.1b.) Physically, a and b are proportional, respectively, to the radial and tangential components of $\boldsymbol{\xi}$ (see §17.3). Also, $a(r)$ and $b(r)$ are the solutions of the ordinary adiabatic, nonradial oscillation differential equations of a static spherical star as given in §17.5. If Ω is constant throughout the star, then one can show that, as seen in an inertial frame,

$$\sigma' = -m\Omega(1 - C), \tag{19.18}$$

where

$$C = \frac{\int_0^R \rho r^2 \, dr[2ab + b^2]}{\int_0^R \rho r^2 \, dr[a^2 + l(l+1)b^2]}. \tag{19.19}$$

The quantity C clearly depends on the structure of the star and on its adiabatic oscillation properties, but usually turns out to be less than unity in value. It therefore follows from eq. (19.18) that, normally, prograde modes of nonradial oscillation ($m < 0$), for example, have somewhat higher frequencies (shorter periods) in rotating than in nonrotating stars. (Part of this result comes about because σ' is the perturbation of the angular frequency as seen in the inertial frame. As seen in a frame rotating with the angular velocity Ω, however, the perturbation of the angular frequency would be $\sigma' + m\Omega = m\Omega C$; consequently, prograde modes [$m <$

0] would have smaller frequencies than retrograde ones, as seen in this frame.)

It is interesting to note that, for g modes in white dwarfs, $|a(r)| \ll |b(r)|$ (Brickhill 1975), so that, as may be seen from eq. (19.19), $C \approx 1/[l(l + 1)]$ for such modes of such stars. It can be shown that the correction accounting for rotation to the phase velocity of an azimuthal running wave of nonradial oscillation, as seen in the rotating coordinate system, of a white dwarf is then approximately $-\Omega/(l[l + 1])$ (Wolff 1977).

19.1d. NONADIABATIC EFFECTS

Very little work has been done on *nonadiabatic* oscillations in rotating stars. Perhaps of some interest in this connection is the paper of Aizenman and Cox (1975b), who obtained a formal solution of the linearized, nonradial, nonadiabatic oscillation equations for the stability coefficient κ (see Chapters 9 and 16) of a steadily, otherwise arbitrarily rotating star. It is interesting that the resulting expression for κ included several thermal imbalance terms (see §19.4). But, after all, this result should not be surprising, since, as we have stated above, a rotating star cannot be in thermal equilibrium; consequently, such terms should be present.

Perhaps the most important result of this investigation is that the usual C integral (see §16.2) still enters as, normally, the major contributor to κ. (This C should not be confused with the C of §19.1a.) More specifically, Aizenman and Cox obtained the following expression for the stability coefficient κ:

$$-\kappa = \frac{C_r}{2K} + \text{thermal imbalance terms,} \qquad (19.20)$$

where C_r is the real part of the C integral (see eq. [16.21]), and K is defined in their paper. They showed that K was proportional to the total kinetic energy of the rotating, pulsating star. In fact, they showed that the relatively simple physical interpretation embodied in eq. (8.49) also applied to a pulsating system having arbitrary but steady motions in its unperturbed (nonoscillating) configuration.

In order to be able to say something about the stability of the rotationally split m sublevels of nonradial oscillations, Hansen, Cox, and Carroll (1978) computed the vibrational stability, in the quasi-adiabatic approximation, of these sublevels, assuming that the rotation was so slow that the distortion of the star from sphericity was negligible. They also made the Cowling approximation (see §17.9), merely for computational convenience, and used some of the results of Aizenman and Cox (1975b). The main result, as applied to upper main sequence B stars and to white dwarfs, was that prograde modes ($m < 0$) were always slightly less stable than retrograde modes ($m > 0$).

Some attempts to determine some of the effects of (slow) rotation on *nonadiabatic,* nonradial oscillations were described by Saio, Cox, Hansen, and Carroll (1979).

19.2. EFFECTS OF MAGNETIC FIELDS

Mass elements in stars that are pervaded by large-scale magnetic fields are subject not only to the usual forces, such as gravity, but also to those of purely electromagnetic origin.

Just as in the case of rotating stars, the basic "static" model about which the oscillations are supposed to be taking place is itself, in the case of stars possessing large-scale magnetic fields, subject to considerable uncertainty, and has been discussed in a number of papers (see, e.g., Ledoux and Walraven 1958, §83; Deutsch 1958; Davies 1968; Sargent 1964; Ledoux 1965; Cowling 1965; Ledoux and Renson 1966; Monaghan 1965, 1966a,b, 1968a,b,c; Roxburgh 1963, 1966; Cameron 1970; Wright 1969; Monaghan and Robson 1971; Wolff 1976; Mestel 1978). In fact, many magnetic field configurations assumed in the past for magnetic stars may be subject to local instabilities, according to the recent work of Tayler and collaborators (Tayler 1973a,b; Markey and Tayler 1973, 1974; Wright 1973).

Nevertheless, a number of papers dealing with the oscillations of magnetic stars have been written. To date, all such studies have been limited to the linear, adiabatic approximation, but even so the difficulties are formidable. A good discussion of some of these problems is given in Ledoux and Walraven (1958, §83). The boundary conditions are also complicated and involve certain subtleties (see, e.g., Smeyers 1976). Many references to the literature in this field may be found in the review papers by J. P. Cox (1974a, 1976a). See in particular Goossens (1972, 1976a,b) and Goossens, Smeyers, and Denis (1976). A study of some of the combined effects of rotation and of magnetic fields on stellar pulsation has been made by Chanmugam (1979).

We shall not go into the details of the matter here, but note only that special cases suggest that the presence of a magnetic field will lift the degeneracy of nonradial oscillations only partially, as contrasted with rotation, which lifts this degeneracy completely. That is, each frequency of given l is split up into only $l + 1$ sublevels, corresponding to the absolute value of m: $|m| = 0, 1, \ldots, l$.

This partial lifting of the degeneracy of the m sublevels by a magnetic field (as contrasted to complete lifting in the case of rotation) may be understood as follows. Nonradial oscillations with $m \neq 0$ correspond to azimuthal running waves (cf. eq. [17.36′]), where the direction of motion of the wave obviously depends on the sign of m. In the case of rotation, the direction of propagation of the wave is either prograde or retrograde, and so the sign of m has a physical effect. In the case of a magnetic field, if the nonpulsating system is axially symmetric, there is clearly no physical distinction between $+m$ and $-m$; that is, only the magnitude $|m|$ matters. Therefore only $|m|$ enters into the equations. We note also that special cases suggest that the order of magnitude of the frequency perturbation due to the presence of a magnetic field is $|\sigma'_{\text{mag}}| \sim |\sigma_0| E_{\text{mag}}/E_{\text{grav}}$, where σ_0 is the unperturbed angular frequency (that is, the angular frequency in the absence of a magnetic field); E_{mag} is the total magnetic energy in the star, of order $E_{\text{mag}} \sim H^2R^3$, where R is the stellar radius and H is the magnetic field strength; and E_{grav} is the gravitational energy of the star, of order $E_{\text{grav}} \sim GM^2/R$.

Taking $|\sigma'_{\text{rot}}| \sim |\Omega|$, where σ'_{rot} is the perturbation in the angular oscillation frequency due to rotation, it is shown in Hansen, Cox, and Van Horn (1977) that the condition that $|\sigma'_{\text{mag}}| \ll |\sigma'_{\text{rot}}|$ may be expressed in the order-of-magnitude form

$$E_{\text{mag}} \ll (E_{\text{rot}}E_{\text{grav}})^{1/2}, \tag{19.21}$$

where E_{rot} is the rotational kinetic energy of the star, of order $E_{\text{rot}} \sim MR^2\Omega^2$. There it is stated that frequency splitting due to magnetic fields is likely to be much smaller than splitting due to rotation unless the magnetic fields are very strong indeed, say some 10^5 gauss for upper main sequence stars or some 10^{10} gauss for cooling white dwarfs, where representative parameter values have been used in both cases.

19.3. EFFECTS OF CONVECTION

In dealing with the effects of convection on pulsating stars, one usually considers two main aspects of these effects. First, they may affect its unperturbed, or static, structure. Second, convection may have a more-or-less direct effect on the pulsations themselves. This division of effects of convection into two separate parts is appropriate if the pulsations are small, that is, if they are regarded as only a perturbation on an otherwise static star. However, in the case of finite-amplitude pulsations, it is not so clear that such a conceptual separation of effects is warranted (see below).

The first of these effects—effect on the static model—is usually treated by use of a mixing-length theory of some sort, such as that of Vitense (1953) or Böhm-Vitense (1958) (see also Cox and Giuli 1968, Chap. 14; Gough 1977a) or some variation thereof (e.g., Spiegel 1963; Parsons 1969; Ulrich 1970a,b,c, 1976; see also Nordlund 1974, 1976). Such static models have been constructed as models of pulsating stars by, for example, Baker and Kippenhahn (1965); Iben (1971b); Tuggle and Iben (1973); Cogan (1970, 1977); Henden and Cox (1976); A. N. Cox, Deupree, King, and Hodson (1977); Saio, Kobayashi, and Takeuti (1977). When considering the pulsations of such a model, the convective flux *variations* are usually either ignored (as in essentially all of the above references) or treated in a greatly simplified fashion (as suggested by, e.g., A. N. Cox, Brownlee, and Eilers 1966). Such modifications of the static structure of the star by convection, in fact, have been called upon to account for certain properties of some kinds of pulsating stars (in particular, the masses of the "beat Cepheids," as summarized in the review paper of J. P. Cox 1978).

A few attempts to incorporate time dependence in models of convection have been made by Gough (1967, 1977b), Unno (1967), and Castor (1968a); see also Fraley (1968), Ulrich (1970a,b,c, 1976), Spiegel (1971, 1972) and Wood (1973). However, most of these are based on some sort of mixing-length model, along with the attendant necessary assumptions.

The problem of time-dependent convection has, on the other hand, been approached in a very fundamental way by Deupree in a series of papers (Deupree 1975a,b, 1976a,b,c, 1977a,b,c,d). He has solved numerically the nonlinear equations of mass, momentum, and energy conservation, in two spatial dimensions and time, in a convectively unstable region. Because of computer time limitations, he could follow the motion only of the largest convection cells. He treated the break-up of these large eddies into smaller ones and the subsequent turbulent cascade which eventually shows up as heat, by the introduction of an eddy viscosity coefficient. Perhaps the most limiting of Deupree's assumptions is that convection can be adequately described in two (as opposed to three in nature) spatial dimensions. Nevertheless, the calculations are still formidable. Yet they may be, essentially, physically correct, and they may, in fact, have provided us with the first physical explanation as to how convection may cause a return to stability on the red side of the Cepheid instability strip (see below).

It is widely believed that the existence of convection in the stellar envelope is what terminates instability of the red side of the instability strip, as has been suggested by Baker and Kippenhahn (1965), among others. The relation between envelope convection and the red edge of the Cepheid instability strip has been discussed by Böhm-Vitense and Nelson (1976) on the basis of the mixing-length theory of convection. A red edge of the instability strip was obtained by Baker and Gough (1967) by use of a variant of the mixing-length theory. However, the calculations of Deupree yielded a red edge of the instability strip, at least for RR Lyrae variables, and presumably also for Cepheids, without any appeal to a phenomenological convection theory such as a mixing-length theory. Although a number of features of Deupree's work may (perhaps legitimately) be criticized, still we have a calculation of the red edge that is in reasonable agreement with observations. The dependence of this computed red edge on various factors such as luminosity, composition, and overtone pulsations has been examined by Deupree (1977a,b,c). A recent calculation of the red edge by Baker and Gough (1979), based on Gough's (1977b) theory, is essentially in agreement with Deupree's work.

From this work a physical picture emerges as to how convection may terminate instability at the red edge of the instability strip. Deupree has found that the convective heat flux is greatest at about the instant of maximum compression (near minimum stellar radius) of the convective regions. Thus, convection (which is most efficient in the hydrogen ionization zone) causes the energy that has been dammed up by the operation of the kappa and gamma mechanisms in the He^+ ionization zone (see Chapter 10) to leak out at this phase. Thus, the driving effects of the ionization zone(s) are essentially undone by the convection. According to Deupree, only a small amount of convection is needed to effectively throttle pulsations on the red side of the instability strip.

This picture certainly seems plausible. It remains to be seen whether further calculations will show that this picture is basically incorrect or incomplete.

The conceptual separation of effects of convection discussed at the beginning of this section—namely, the effect of convection on the static model and the direct effect on the pulsations themselves—of course never enters directly in Deupree's approach. In fact, he has found that there is considerable coupling between pulsations and convection, in the sense that they appear to be mutually destructive phenomena: the existence of pulsations tends to prevent convection from becoming very extensive or well developed, while convection tends to quench pulsations.

Deupree (1977a) has, in fact, used the above property of convection and pulsation together to argue that convection should not be important except very near the red edge of the instability strip. One would then conclude that the details of the light curves and so on, as computed with purely radiative models, should be essentially unaffected by convection. Such a conclusion follows from Castor's (1968b) work on the phase lag, described in Chapter 11, as was there pointed out.

19.4. EFFECTS OF THERMAL IMBALANCE

In nature, the "static" structure of a star will in general be a function of time. The reason is that all stars must evolve (see Chapter 4), and the only question then is the *rate* at which the static state of the star is changing. For most purposes in stellar pulsation theory this rate is so small compared with rates associated with the pulsations that the static state of the star may be considered truly static as an excellent approximation. This approximation is usually made in stellar pulsation theory, and has in fact been made in all the other parts of this book.

The static state of the star will usually be changing relatively slowly, such as on a Kelvin or nuclear time scale. However, in some cases the static model may be changing fairly rapidly. Examples may be found in relatively late evolutionary phases, for example in stars undergoing thermal runaways of various kinds (e.g., the helium flash, which occurs when helium burning begins near or at the center in a degenerate region of a relatively low-mass [~one solar mass] star, see, e.g., Härm and Schwarzschild 1964; or the fairly abrupt ignition of helium in a thin shell source, see, e.g., Gingold and Faulkner 1974).

For the above "practical" reasons it would be useful to know whether or not these effects of slow changes in the static model are important for the vibrational characteristics of the model; and, if these effects are indeed important, how to calculate them. Such slow changes in the static model are usually referred to as thermal imbalance effects, because they will be present whenever the model is not in thermal equilibrium, that is, whenever $\epsilon - (\nabla \cdot \mathbf{F})/\rho \neq 0$, where ϵ is the net rate per unit mass of thermonuclear energy generation, $\mathbf{F}$ is the net flux of energy, and ρ is the mass density (since a star must evolve, as was shown in Chapter 4, a star can never be precisely in thermal equilibrium). Although considerable work (see below) has now shown that these thermal imbalance effects are usually small compared with other effects, still a theory of stellar pulsation would not be complete if they could not (in principle, at least) be taken into account.

Perhaps the earliest work on these effects was that of Thomas (1931), which is described by Ledoux (1958, §13, p. 648). This approach was based on a consideration of the total energy of a star, defined as the sum of the kinetic, gravitational, and thermal energies, and called the dynamical energy of the star by Ledoux (1958).

The problem of the radial pulsations of stars in thermal imbalance, based on such an energy equation, was then examined by Kato and Unno (1967). Some of Kato and Unno's results were discussed by Okamoto (1967) and applied to pre-main sequence gravitationally contracting stars. He found that these effects were sometimes important.

The problem was later treated, again on the basis of such an energy equation, by Simon (1970, 1971). Simon and Sastri (1972) applied some of their own results to certain high-order modes of some highly evolved stellar models. Axel and Perkins (1971) used an energy equation as a basis for an examination of the problem; Unno (1967, 1968) approached it from the standpoint of a general variational principle; and Davey (1970) discussed it, but not on the basis of an energy equation, for a special case.

However, the use of such "energy" approaches for this problem was criticized by Cox, Hansen, and Davey (1973) on the grounds that these approaches were potential sources of ambiguities. For example, as was shown in Chapter 4, there are at least two possible forms of an energy equation for a star: one based on the total energy, as defined above, of the star; the other based only on the sum of the kinetic and gravitational energies. It was pointed out by these authors that use of the first of the above energy equations led to the appearance of intrinsically second-order (in smallness) quantities, which seemed to complicate matters even more. On the other hand, no such intrinsically second-order quantities appeared with use of the second of the above energy equations; only products of first-order quantities were involved. Also, the widespread (essentially universal) use of the quasi-adiabatic approximation in this problem was objected to in that paper.

Cox, Hansen, and Davey (1973) then derived the general, linearized, third-order (in time) nonadiabatic equation of motion for the radial pulsations of a star in thermal imbalance. In this equation the "static," or unperturbed, model was assumed to be an explicit function of time. The time dependence adopted for the oscillating quantities was, for example,

$$\frac{\delta r}{r_0} \equiv \zeta(m,t) = \xi(m,t) \exp\left[\int_0^t \omega(t')\, dt'\right], \tag{19.22}$$

where subscript zero denotes the "static" value of the relevant quantity, and m denotes interior mass. The amplitude $\xi(m,t)$ and the eigenfrequency $\omega(t)$ were both regarded as functions of time because the static model was assumed to be changing (slowly) in time. The authors also assumed that $t_s/t_{ff} \gg 1$, where t_s is a long time, of the order of the Kelvin or nuclear time (see Chapter 2), characteristic of the slow, evolutionary changes of the static model; and t_{ff} is the free-fall time (see Chapter 2), generally of the order of the pulsation period Π. (The above assumption will be discussed briefly below.) Using the method given by Ledoux (1963) (see also Chapter 9) and Cox and Giuli (1968, Chap. 27), the authors then obtained integral expressions, accurate to $O(t_{ff}/t_s)$, for ω, written as

$$\omega = i\sigma - \kappa_{\rm I}. \tag{19.23}$$

Here σ is the oscillatory angular frequency, not affected to the above order by thermal imbalance effects; and κ_I is the stability coefficient (cf. Chap. 9), the subscript "I" meaning that this stability coefficient is appropriate for the *relative* pulsation amplitude $\delta r/r_0$. These integral expressions are rather complicated integrals, over the entire stellar mass, of quantities characteristic of the static model and their time derivatives, multiplied by quadratic functions of the (strictly speaking, unknown) eigenfunctions of the pulsating model, where these eigenfunctions are, in principle, solutions of the linearized, nonadiabatic equation of motion (analogous integral expressions were given in Chapter 9).

One of the simplest of all possible cases concerns the adiabatic pulsations ($\delta P/P_0 = \Gamma_{1,0}\delta\rho/\rho_0$) of the homogeneous (spatially constant density) model, slowly contracting homologously, that is, with $r_0^{-1}\,\partial r_0/\partial t$ a function only of time, where r_0 is the radius interior to which is a certain amount of mass; and also pulsating homologously, that is, with ξ constant in both space and time. For this case we have for the thermal imbalance part of κ_I

$$\kappa_I(\text{thermal imbalance}) = -\frac{6\Gamma_{1,0} - 5}{\tau}, \tag{19.24}$$

where $\tau \equiv -R_0^{-1}dR_0/dt$, R_0 being the "equilibrium" radius of the static model, and $\Gamma_{1,0}$ has been assumed constant both in space and time. Thus, for this simple case, the thermal imbalance effects are seen to be *destabilizing* for *contraction*.

The above energy approaches were discussed further by Davey and Cox (1974), who presented an explicit expression (accurate to the second order in smallness) for the rate of change of the sum of the kinetic and gravitational energies of a star in thermal imbalance. As was stated earlier by J. P. Cox, Hansen, and Davey (1973), this equation contained no intrinsically second-order quantities, only products of first-order quantities. Also, the quasi-adiabatic approximation was not explicitly invoked.

There it was shown that the stability coefficient, for the above energy, say κ_E, based on this expression, agreed precisely with results obtained previously by use of an energy approach by Kato and Unno (1967), Simon (1971), and Simon and Sastri (1972), in those cases where a comparison was possible, but not with the results of the small-amplitude approach of Cox, Hansen, and Davey (1973) nor with those of Unno (1968).

On these grounds Davey and Cox (1974) proposed a "modified" energy approach, based on the average rate of change of the kinetic energy alone, and showed that this modified energy approach led to exact agreement with the small-amplitude approach of Cox, Hansen, and Davey (1973), at least for quasi-adiabatic pulsations. However, it was pointed out by Demaret (1976) that the demonstration by Davey and Cox (1974) of the equivalence of the results of their modified energy approach and those of the small-amplitude approach was based on an insufficiently general assumed time dependence of the pulsating quantities. Thus, it was correctly concluded by Demaret (1976) that there are, in principle at least, *two* stability coefficients for a star in thermal imbalance, one for the pulsation amplitude and one for the total pulsation energy. He showed, moreover, that these two stability coefficients were connected by a simple relation (see below).

It was also pointed out by Davey and Cox (1974) that the stability coefficient for *relative* variations, $\delta r/r_0$, denoted above by κ_{I}; and the stability coefficient for *absolute* variations, δr, denoted by κ_{II}; were different. For example, in a contracting star a given *absolute* variation will appear to grow more rapidly when considered as a *relative* variation. These authors also argued that κ_{II} was more fundamental and more physical than κ_{I}.

The value of the thermal imbalance contribution to κ_{II} for the quasi-adiabatic, homologous oscillations of the homogeneous model contracting homologously with constant $\Gamma_{1,0}$ is

$$\kappa_{\mathrm{II}}(\text{thermal imbalance}) = -\frac{6\Gamma_{1,0} - 9}{4\tau}, \tag{19.25}$$

where τ was defined just after eq. (19.24).

In J. P. Cox, Davey, and Aizenman (1974) the small-amplitude approach was discussed further. In particular, it was shown that those integrals involving temporal changes in the pulsation amplitude (the ξ of eq. [19.22]) were negligible to $O(t_{ff}/t_s)$, for the stability coefficient for stars in thermal imbalance. There it was also shown that the assumption usually made in studies of stars in thermal imbalance, that is, that $t_{ff}/t_s \ll 1$ (see earlier), was equivalent to the assumption of quasi-adiabatic oscillations. Moreover, it was shown that these assumptions were essentially always justified for all stars except large, low-mass, red supergiants.

The connection referred to above between κ_{E} and κ_{II} was in fact already contained in the paper by Cox, Davey, and Aizenman (1974), as has been pointed out by Simon (1977). The relation between these two stability coefficients is $\kappa_{\mathrm{E}} = \kappa_{\mathrm{II}} - (d[J\Sigma^2]/dt)/(2J\Sigma^2)$, where the quantity $J\Sigma^2$ is defined in Cox, Hansen, and Davey (1973) (see also eq. [9.44]). As pointed out by Simon (1977), this becomes, to $O(t_{ff}/t_s)$,

$$\kappa_{\mathrm{E}} = \kappa_{\mathrm{II}} - \frac{1}{\sigma}\frac{d\sigma}{dt}. \tag{19.26}$$

It was shown by Simon (1977) that the stability coefficient for the *velocity* associated with the pulsations is identical to κ_{E}. Thus, the foregoing energy approaches really reveal information about the *velocity* associated with the pulsations, and velocity has the virtue of being a more-or-less directly observable quantity.

Simon (1977) also pointed out that the stability coefficient for velocity could in some cases indicate instability, while the stability coefficient for amplitude could at the same time indicate stability; or vice versa. He discussed stability in general and presented a broadened definition of pulsational stability for stars in thermal imbalance.

The above partial differential equation for the small oscillations of a star in thermal imbalance was solved directly for some simple cases by Aizenman and Cox (1974), using the asymptotic methods originated and developed by several Russian mathematicians (Krylov and Bogoluibov 1937; Bogoluibov and Mitropolsky 1961; Mitropolsky 1964; Feschenko, Shkil', and Nikolenko 1967). These results led to exact agreement with those of the earlier integral expressions for all cases examined. Solutions by use of these asymptotic techniques were independently obtained by Demaret (1974, 1975).

The problem of the quasi-adiabatic, radial oscillations of stars in thermal imbalance was also discussed from the standpoint of the differential equation for the pulsation amplitude by Davey (1974). He showed that, as long as $t_{ff}/t_s \ll 1$, the functions entering into the integral expressions derived earlier by Cox, Hansen, and Davey (1973) and by Cox, Davey, and Aizenman (1974) were the solutions of the linear, adiabatic wave equation (see Chapter 8), *without thermal imbalance effects.* This result had in fact for expediency been assumed to be valid by the above authors, and used by them to obtain some of their numerical results. Davey (1974) also directly obtained solutions of this differential equation by use of a power-series approach, as in Rosseland (1949, Chap. 3), for some of the cases examined by Aizenman and Cox (1974) by means of the more powerful, and much more complicated, asymptotic methods. Exact agreement was obtained in every case examined, even in one case involving nonhomologous "slow" motion ($r_0^{-1}\partial r_0/\partial t$ a function of interior mass m *as well as* of time).

Sastri and Simon (1973) studied the effects of thermal imbalance on several radial modes of a prenova model. They used an approach based on the dynamical energy, defined above. Cox (1974b) investigated the effects of thermal imbalance on the pulsations of a thermally unstable, shell helium-burning stellar model constructed by Gingold and Faulkner (1974), using a small-perturbation approach based on integral expressions (see above). In both cases the models were greatly out of thermal balance and the slow motion was far from homologous. These thermal imbalance effects were found not to be negligible, but they were smaller than other effects, at least as found by Cox (1974b). In the work of Sastri and Simon (1973), these thermal imbalance effects rendered some of the modes stable against small oscillations.

In Aizenman and Cox (1975a) a general, rather physical discussion of the problem of the pulsations of stars in thermal imbalance was presented. These thermal imbalance effects were conceptually separated into two parts: "dynamical" and "nonadiabatic" contributions. The dynamical contribution contained no nonadiabatic effects; that is, it was present even if the oscillations were assumed to be perfectly adiabatic. This term is given by the relation

$$\kappa_{\mathrm{II,dyn}} = \frac{1}{2}\frac{d\omega/dt}{\omega} = \frac{1}{2}\frac{d\sigma/dt}{\sigma}, \tag{19.27}$$

where ω was given in eq. (19.23) and σ is the instantaneous oscillation angular pulsation frequency. The last equality in eq. (19.27) is accurate to $O(t_{ff}/t_s)$. This contribution arises solely from the scale change in the inertial term in the momentum equation, and was first isolated by Unno (1968). The dynamical term contains the entire difference in stability coefficients for relative versus absolute variations (see above).

The other contribution, the nonadiabatic part, arises solely from non-adiabatic effects (exchanges of heat), independently of any effects brought about by the slow motion of the static model. This contribution is the same for relative as for absolute variations. As was explained in Aizenman and Cox (1975a), this term is generally *destabilizing* for *contraction*, and its effects may be easily understood on physical grounds in terms of small phase shifts between the local pressure and density variations δP and $\delta\rho$, respectively, during the oscillations (see Chapter 9 of this book). Thus, if a region is contracting, it is generally losing heat, and $(\epsilon - \partial L_r/\partial m)_0$ will be negative. This local loss of heat resulting from thermal imbalance in the static model appears to the pulsating star as a source of heat, and causes δP to lag slightly behind $\delta\rho$, thus leading to a tendency for the pulsations to be amplified (see Chapter 9).

Such a conceptual separation of the thermal imbalance effects into two separate contributions was also effected independently by Demaret (1974, 1975) who called them "isentropic" and "nonisentropic," respectively.

It was also shown by Aizenman and Cox (1975a) that, to $O(t_{ff}/t_s)$, exactly the same equations apply to the *nonradial* oscillations of a star in thermal imbalance, as to the *radial* oscillations of such a star.

In an important paper by Buchler (1978), the problem of the oscillations of a star in thermal imbalance was discussed from the standpoint of the two-time method described in Cole (1968) (see also §13.4). As Buchler pointed out, this method has, among other things, the advantage that one need not limit oneself to a *linearized* theory, as has been the case with all previous discussions of this problem.

Buchler also concluded, in agreement with Demaret (1976) and Simon (1977) (see also the above discussion in connection with eq. [19.26]), that two separate and sometimes apparently contradictory stability coefficients exist for a star in thermal imbalance. As stated previously, one of these stability coefficients is based on the pulsation *amplitude,* the other on the pulsation *energy* (or velocity).

Buchler (1978) discussed the sometimes contradictory nature of the above two stability coefficients, and concluded that this situation exists because the problem of the vibrational stability of stars in thermal imbalance had not previously been properly posed. He attributed the situation mainly to the difficulty of defining time averages of quantities that were not strictly periodic. He proposed that the stability coefficient be based on the *action* (per unit mass), say $I = \eta^2\omega$, where η is the amplitude of δr. As partial justification for this proposal, he showed that I was an adiabatic invariant, in the sense that $dI/dt = 0$ in adiabatic motion. This adiabatic invariance of I is completely consistent with the above discussion of the dynamical or isentropic contribution to the stability coefficient. This conclusion may easily be understood, for, if I does not change in adiabatic motion, then $\eta \propto \omega^{-1/2}$, which would also be concluded on the basis of the dynamical or isentropic contribution to the stability coefficient alone (see eq. [19.27]).

As Buchler (1978) pointed out, his proposal is tantamount to retaining only the nonadiabatic or nonisentropic part of the stability coefficient. In fact, it can easily be shown that a stability coefficient based on the action (per unit mass) I is identical to $\kappa_{\mathrm{II,na}}$, the nonadiabatic part of the stability coefficient for the radius amplitude. On this basis we may reproduce a result derived by Aizenman and Cox (1975a, eq. [65]), for nonradial oscillations (radial oscillations may be considered a special case of this, by simply replacing $\rho^{-1}\nabla \cdot \mathbf{F}$ by $\partial L_r/\partial m$ [m = interior mass], throughout in these equations). We have

$$\kappa_{\mathrm{II,na}} = -\frac{C_r}{2\sigma^2 J} + \frac{D}{2\sigma^2 J} + \frac{G_{1,r}}{2\sigma^2 J} + \frac{G_{3,r}}{2\sigma^2 J}, \tag{19.28}$$

where C_r is the real part of the C integral (see eq. [16.21]), J is the oscillatory moment of inertia (see eq. [16.19]), and D, G_1, and G_3 are rather complicated thermal imbalance integrals defined in the above reference. The integrals G_1 and G_3 would vanish for constant gammas, and D would vanish for a star in thermal equilibrium. For the simple case, considered earlier, of the quasi-adiabatic, homologous oscillations of the homogeneous model slowly contracting homologously with constant gammas, it can be shown that

$$\kappa_{\text{II,na}} = -3(\Gamma_{1,0} - 1)/(2\tau), \tag{19.29}$$

where τ was defined just after eq. (19.24).

Buchler (1978) also pointed out that, as had been suggested earlier by Simon (1974), stellar pulsations of a large enough amplitude could result in a feedback on the static structure of the star and so affect its evolution.

Finally, the results of several earlier works on the vibrational stability of stars in thermal imbalance were compared numerically by Vemury (1978). In general, there was excellent agreement among the results of the various prescriptions when similar assumptions were adopted. Thus, it is interesting to note that the energy method used by Sastri and Simon (1973), based on the dynamical energy of a star, and involving intrinsically second-order (in smallness) terms, led to results in substantial agreement with those of the energy method of Davey and Cox (1974) and of Demaret (1976). In the latter two methods only first-order (in smallness) quantities are involved (see above). Vemury (1978) also agreed with Demaret (1976) and Simon (1977) that there were at least two stability coefficients for a star in thermal imbalance (see above), and that they were simply related (see eq. [19.26]).

19.5. EFFECTS OF GENERAL RELATIVITY

Fortunately, effects of general relativity are nearly always negligibly small for most purposes in stellar pulsation theory. However, in some cases, such as pulsations of neutron stars or of white dwarfs (or of such extreme models as those examined by Thorne and Żytkow 1977), we really ought to use general relativity rather than Newtonian physics. However, as this latter kind of physics has been used everywhere else in this book, our remarks on effects of general relativity will be kept fairly brief.

The general-relativistic theory of stellar structure has been thoroughly discussed by Thorne (1967a); Zeldovich and Novikov (1971); Weinberg (1972); Misner, Thorne, and Wheeler (1973); Thorne (1977, 1978); and Thorne and Żytkow (1977).

Because purely radial oscillations cannot give rise to gravitational radiation and its associated complications, the general relativistic theory of such oscillations, in the linear, adiabatic approximation, has been rather thoroughly worked out. (We are using the term "adiabatic" above in the classical thermodynamic sense of implying no heat exchanges with the surroundings; see Chapter 4.) As was first shown by Chandrasekhar (1964a,b,c), the important variational property, as well as most of the other mathematical properties of such oscillations for the nonrelativistic case (see Chap. 8), have close analogues in general relativistic theory (see, e.g., Thorne 1967a and Bardeen, Thorne, and Meltzer 1966).

One very interesting result of studies such as the above concerns the *dynamical* stability of a spherical star in general relativity. Thus, such a star can be dynamically unstable, due to general relativistic effects, under conditions such that these effects are far too small to affect the static structure of the star. If Γ_1 is the relativistic generalization of the appropriate adiabatic exponent (e.g., Thorne 1967a, p. 282), then the criterion for dynamical stability, according to general relativity, is (Chandrasekhar 1964c)

$$\Gamma_1 - \frac{4}{3} > K\frac{2GM}{c^2R} = 4.24 \times 10^{-6}K\left(\frac{M}{M_\odot}\right)\left(\frac{R}{R_\odot}\right)^{-1}, \qquad (19.30)$$

where K is a dimensionless number whose value depends on the mass concentration of the star but which generally lies between about 0.5 and 1.5 (see also Bludman 1973a,b). Dynamical stability, including general relativistic effects, with numerous references to the literature, is also discussed by Bisnovatyi-Kogan and Blinnikov (1974). See also Heintzmann and Hillebrandt (1975), Hillebrandt and Steinmetz (1976), and Chanmugam (1977).

Radiation of gravitational waves is associated with nonradial oscillations having $l \geq 2$ (l is the index in the spherical harmonic $Y_l^m[\theta,\phi]$). The subject of linear, adiabatic, nonradial oscillations in general relativity has been discussed in a series of papers by Thorne and collaborators (Thorne 1967b, 1969a,b; Price and Thorne 1969; Campolattaro and Thorne 1970; Ipser and Thorne 1973; Ni 1973); see also Schutz (1972); Dedic and Tassoul (1974); and Will (1974). Radiation of gravitational wave energy from models of white dwarfs has been calculated by Osaki and Hansen (1973a). *Special* relativistic effects in radiative transfer in stars have been discussed by Davis (1974).

A variational principle for the linear, adiabatic, nonradial oscillations of fluid spheres in general relativity has been found by Detweiler and Ipser (1973). This variational principle has been applied to general relativistic stellar models by Detweiler (1975a,b), to infer some things about their nonradial oscillatory properties.

19.6. SECULAR STABILITY

The problem of the *secular stability* of a star is concerned with the question of whether or not a star is stable against small perturbations carried out so slowly that the star always remains in *hydrostatic equilib-*

rium (see Chapter 2) throughout the entire process. Thus, secular stability is concerned with the question of whether or not a star is stable against small departures from *thermal equilibrium*. Consequently, secular stability is seen to be a special case of pulsational stability. The equations describing secular stability can be obtained from those describing pulsational stability merely by dropping all dynamical (acceleration) terms. The relevant time scale for the secular stability problem is therefore a "slow" time scale, and significant changes in the physical system may be expected to occur, roughly, on something like a Kelvin time scale (see Chapter 2).

Numerous theoretical studies were carried out, and many papers written, on the subject of the secular stability of stars, primarily in the late 1960's and early 1970's. Fortunately, an excellent review paper on this subject has recently been published by Hansen (1978), which contains many references to the literature. Consequently, our discussion of the subject here will be very brief.

Practically all investigations of this subject have assumed spherically symmetric perturbations, as shall we, throughout this section, unless we specify otherwise. Only a few studies have dealt with nonradial perturbations; see, for example, Kippenhahn (1967) and Rosenbluth and Bahcall (1973). (See also Kippenhahn 1974 for a very clear discussion).[2]

Perhaps the mathematically simplest and most illuminating approach to this subject is the use of integral expressions, such as those in §9.3, for the eigenvalues. Assuming a time dependence $\exp(i\omega t)$ for all perturbed quantities, the appropriate cubic equation is eq. (9.43), where all quantities are defined in §9.3. Denoting by ω_1 the secular ("small") root, and assuming that $|\omega_1|^2 \ll |\Sigma^2|$, we obtain approximately

$$\omega_1 \approx \frac{iC}{\Sigma^2 J}. \tag{19.31}$$

This result shows that the motion will be aperiodic if C is purely real, and that the star will be secularly stable if C is positive, since we have shown in §8.11 that $\Sigma^2 J$ is always positive for a dynamically stable star ($\Gamma_1 > 4/3$), which is the only kind we consider here. (This present C should not be confused with the C of §19.1).

It has been shown by Hansen (1978) (see also Cox and Giuli 1968, Chap. 27) that $|1/\omega_1| \sim t_k$, the Kelvin time (see Chap. 2).

Moreover, if the motion is *homologous* ($\delta r/r$ independent of position in

[2]The question of slow mixing inside a star falls under this category. The question of slow mixing in the sun has been investigated by Rosenbluth and Bahcall (1973), who found that the sun is stable against mixing.

the star), then the usual Jeans criterion (Jeans 1928) can easily be obtained from eq. (19.31). Assuming that $\delta P/P$, $\delta\kappa/\kappa$, and $\delta\epsilon/\epsilon$ can be expressed in terms of $\delta\rho/\rho$ and $\delta T/T$ by simple expressions such as in eqs. (13.2), (13.3), and just before eq. (13.23) (all symbols have their usual meaning), we may easily evaluate the integral C in terms of n, s, λ, ν, χ_ρ, and χ_T, where $C > 0$ is the condition for secular stability. We may accordingly write the (generalized) Jeans condition as

$$3\lambda + \nu\left(\frac{4 - 3\chi_\rho}{\chi_T}\right) > -4 - 3n + (s + 4)\left(\frac{4 - 3\chi_\rho}{\chi_T}\right), \qquad (19.32)$$

which becomes $3\lambda + \nu > s - 3n$ for a perfect gas equation of state ($\chi_\rho = \chi_T = 1$). For example, for the typical values $\lambda = 1$, $s = 3$, and $n = 1$, the Jeans condition requires that $\nu > -3$, a very mild requirement indeed.

The physical meaning of the (generalized) Jeans condition is as follows. Consider a star initially in thermal equilibrium, $L = \int_M \epsilon \, dm$. Suppose that the star is compressed homologously by the small fractional amount $(-\xi)(=$ const.). The fractional rate of increase of nuclear energy generation $(\delta\epsilon/\epsilon)$ is then given by the left side of the inequality (19.32), after multiplication by $(-\xi)$; while the right side, again after multiplication by $(-\xi)$, gives the fractional rate of increase in the luminosity of the star $(\delta L/L)$. If the Jeans condition, eq. (19.32), is satisfied, then $\delta\epsilon/\epsilon > \delta L/L$, which means that the star is gaining heat. Hence, the total energy E (thermal and gravitational) of the star must be increasing. Since by the virial theorem (see Chapter 2). $E =$ const. $\times$ (gravitational energy) < 0, then increasing E implies an increase in the stellar radius R. But an increase in R is exactly opposite to the initial perturbation, which was a *compression* of the star (a *decrease* in R).

We may also interpret the above sequence of events in terms of the internal temperature of the star. Since this temperature generally varies inversely as some power of R in a homologous change (the first power if the equation of state is the perfect gas law), then *increasing* R implies a *decreasing* internal temperature. Hence, in the present case of a star which is secularly stable, we may say that the gain of heat accompanying the initial compression causes the star to expand and cool, and thus to tend to approach its original condition again.

In actuality, the situation is considerably more complicated than that described above, and the temporal behavior of a star often has an oscillatory component, as described more fully in Hansen (1978). Also, in the case of evolving stars, which are already changing their condition on something like a Kelvin time, the exact interpretation and significance of a secular stability analysis are not clear.

19.7. GENERAL SIGNIFICANCE OF STELLAR OSCILLATION THEORY

Perhaps the main importance of stellar oscillation theory is that it provides us, in principle at least, with an additional "probe" of stellar internal structure. The manner in which a star oscillates obviously depends on its internal structure and possibly also on other things, such as the presence or absence of a magnetic field, its rotation, and so on. Therefore, from observations about the way a star oscillates, one can hope to learn something more about its internal structure (and possibly about other things as well) than the information provided by the static characteristics of the star alone.

A good example of this last use of pulsation theory is embodied in the suggestion by Christensen-Dalsgaard and Gough (1976) that the recently observed small solar oscillations (see references in Christensen-Dalsgaard and Gough 1976, also H. A. Hill 1978 and H. A. Hill and Dziembowski 1979) be used to infer something about the solar interior, much as information from the vibrations caused by terrestrial earthquakes is used to infer something about the earth's interior.

Actually, this last application of *radial* pulsation theory has been used by astronomers for many years in connection with Cepheid variables and related kinds of variable stars (see Chapter 3). Thus, as was pointed out in Chapter 8, for the fundamental radial mode of a star with the adiabatic exponent $\Gamma_1 = 5/3$, the constant Q in the period-mean density relation must obey the relation $Q \leq 0\overset{d}{.}116$, the upper limit applying to the homogeneous model (ρ = const. throughout). For (somewhat realistic) centrally concentrated models, a representative value is $Q \approx 0\overset{d}{.}04$, the exact value depending on the details of the stellar model. The fact that empirical results are consistent with these expectations gives us some confidence that we have essentially correct values of at least some of the basic parameters of Cepheids. (However, it is interesting to note that, prior to the 1952 [Baade 1952] revision of the zero point of the Cepheid period-luminosity relation [see Fernie 1969 for a good account of the history of this subject; see also Baade 1963], Eddington had already noticed that something was wrong [Eddington 1932]. The observational information available at that time about Cepheids yielded values of Q close to the theoretical upper limit. On this basis Eddington concluded that Cepheids must be more nearly homogeneous than other stars. The new zero point corresponded to an increase in Cepheid luminosities by roughly a factor of four, and hence, for given color or effective temperature, to radii about twice as large as previously believed. These larger radii imply lower mean densities, and hence smaller values of Q, more in accord with theoretical expectations.)

At the present time there seems to be a "Cepheid mass discrepancy," which is especially severe for the beat (or "double-mode") Cepheids. This mass discrepancy is discussed in two recent review papers by A. N. Cox (1979) and by J. P. Cox (1978); see also J. P. Cox (1979) and A. N. Cox, Hodson, and King (1979), where numerous references to the literature are given. In short, the application of radial pulsation theory to Cepheids suggests that their masses may be some 20–40% smaller than expected from evolutionary considerations. These masses may be too small by a factor of two or three in the case of the beat Cepheids. The cause of this mass discrepancy is presently unknown, but it may eventually lead to new knowledge about pulsation theory, stellar properties, or both (e.g., see A. N. Cox 1978a,b; A. N. Cox, Hodson, and King 1979). At any rate, this difficulty provides us with an example of how pulsation theory may eventually lead to new knowledge about stars.

In the case of nonradial oscillations, recent observations and theories indicate that such oscillations may be a fairly common feature in at least many kinds of stars. For example, recent observations by McGraw (1977) (see also Robinson and McGraw 1976a,b; Robinson, Nather, and McGraw 1976; Nather 1978) suggest the existence of a new class of variable stars, the "ZZ Ceti" stars. These stars are thought to be ordinary white dwarfs with effective temperatures around 10^4K. The light variations may be due, as McGraw (1977) has suggested, to high-order g-mode oscillations.

As an example, M. Smith (1977) and Smith and McCall (1978) have recently observed more-or-less periodic variations in line profiles in the spectra of *B* stars (see also Buta and Smith 1979; M. A. Smith and Buta 1979; M. A. Smith 1979a). Smith concludes that these line profile variations can only be understood as resulting from nonradial oscillations, and his work even suggests tentative mode identifications in most cases. He has called these stars "line profile variable *B* stars," or "53 Persei stars," and they may also represent a new class of variable stars.

Obviously, some familiarity with nonradial oscillation theory is necessary in order to interpret observations of stellar variability in terms of nonradial oscillations.

M. Smith (1977) has also suggested that the β Cephei variables (see Chap. 3) may be related to the line profile variable *B* stars (see §3.2c). As yet, neither the cause(s) nor the nature of the variations of the β Cephei stars is understood (see Cox 1976b; Kato 1976 for reviews; for more recent developments, see Smith 1979a; Stellingwerf 1978; J. P. Cox and Stellingwerf 1979; Saio and Cox 1979a).

It has also been suggested by Lucy (1976), Osaki (1977), and Dziembowski (1977b) that very high-order nonradial oscillations might be rather commonplace in very luminous, hot stars. These authors have shown that such oscillations are often unstable in such stars. These high-order nonradial oscillations may show up observationally as macro- or micro-turbulence, a phenomenon which has long not been understood. Such oscillations may also play a role in the heating of stellar chromospheres and coronae.

Finally, the theory of nonradial oscillations may be expected to be of fundamental importance in the interpretation of oscillations of objects which are obviously not spherically symmetric. Examples of such objects would be those which have a clearly defined direction, as in the case of a rotating star or disk. Another example of such an object would be a star pervaded by a large-scale magnetic field. Perhaps the oscillations of dwarf novae after outburst (Robinson 1976) might be attributable to a rapidly rotating accretion disk, as has in fact been suggested by Patterson, Robinson, and Nather (1977). (However, see Stiening, Hildebrand, and Spillar 1979.) It has recently been shown by Van Horn, Wesemael, and Winget (1980) that vertical, nonaxisymmetric oscillations of accretion disks about white dwarfs indeed have periods in the observed range.

References

Also serves as author index. Numbers in parentheses indicate chapter and section of this book in which referenced work was primarily cited.

Abt, H. A. 1957, *Ap. J.,* **126,** 138 (18.3).

Abt, H. A. 1959, *Ap. J.,* **130,** 769 (3.1).

Adams, T. F., and Castor, J. I. 1979, *Ap. J.,* **230,** 826 (11.3).

Adams, T. F., Davis, C. G., and Keller, C. F. 1978, preprint (12.2).

Africano, S., *see* Smith, M. A., Africano, S., and Worden, S. P., 1979.

Aime, C., *see* Fossat, E., Ricort, G., Aime, C., and Roddier, F., 1974.

Aizenman, M. L., and Cox, J. P. 1974, *Ap. J.,* **194,** 663 (9.3, 19.4).

Aizenman, M. L., and Cox, J. P. 1975a, *Ap. J.,* **195,** 175 (19.4).

Aizenman, M. L., and Cox, J. P. 1975b, *Ap. J.,* **202,** 137 (19.1d).

Aizenman, M. L., Cox, J. P., and Lesh, J. R. 1975, *Ap. J.,* **197,** 399 (17.10).

Aizenman, M. L., Hansen, C. J., and Ross, R. R. 1975, *Ap. J.,* **201,** 387 (17.8).

Aizenman, M. L., and Perdang, J. 1971, *Astron. Astrophys.,* **15,** 200 (5.1).

Aizenman, M. L., and Perdang, J. 1972, *Astron. Astrophys.,* **17,** 190 (5.1).

Aizenman, M. L., and Smeyers, P. 1977, *Astrophys. and Sp. Sci.,* **48,** 123 (17.3, 18).

Aizenman, M. L., Smeyers, P., and Weigert, A. 1977, *Astron. Astrophys.,* **58,** 41 (17.8, 17.10).

Aizenman, M. L., *see* Cox, J. P., Davey, W. R., and Aizenman, M. L., 1974a; Hansen, C. J., Aizenman, M. L., and Ross, R. R., 1976.

Aleshin, V. I. 1964, *Sov. Astron.,* **8,** 154 (12.2).

Alfvén, H., and Fälthammar, C. G. 1963, *Cosmical Electrodynamics* (Oxford: Clarendon Press), 2nd ed. (4.6).

Aller, L. H. 1963, *Astrophysics: The Atmospheres of the Sun and Stars,* 2nd ed. (N. Y.: Ronald Press) (3).

Aller, L. H., and McLaughlin, D. B. 1965, eds., *Stellar Interiors* (Chicago: Univ. of Chicago Press).

Aller, L. H., *see* Goldberg, L., and Aller, L. H., 1943.

Ando, H. 1976, *Publ. Astron. Soc. Japan,* **28,** 517 (17.10, 18.3).

Ando, H., and Osaki, Y. 1975, *Publ. Astron. Soc. Japan,* **27,** 581 (17.6b, 17.10, 17.14, 18.1, 18.3).

Ando, H., and Osaki, Y. 1977, *Publ. Astron. Soc. Japan,* **29,** 221 (18.3).

Ando, H., *see* Unno, W., Osaki, Y., Ando, H., and Shibahashi, H., 1979.

Andrew, A. L. 1967, *Australian J. Phys.,* **20,** 363 (17.12).

Auré, J. L. 1971, *Astron. Astrophys.,* **11,** 345 (18.3).

Axel, L., and Perkins, F. 1971, *Ap. J.,* **163,** 29 (8.5, 19.4).

Ayres, T. R., Linsky, J. L., and Shine, R. A. 1975, *Ap. J.,* **195,** L121 (18.3).

Baade, W. 1926, *Astr. Nachr.,* **228,** 359 (3.3).

Baade, W. 1952, *Trans. IAU.,* **8,** 397 (19.7).

Baade, W. 1956, *Publ. A.S.P.,* **68,** 5 (3.1).

Baade, W. 1963, *Evolution of Stars and Galaxies,* ed. C. Payne-Gaposchkin (Cambridge, Mass.: Harvard University Press) (3.1, 19.7).

Bahcall, J. N. 1977, *Ap. J.,* **216,** L115 (18.3).

Bahcall, J. N., and Sears, R. L. 1972, *Ann. Rev. Astron. Astrophys.,* **10,** 25 (18.3).

Bahcall, J. N., *see* Rosenbluth, M. N., and Bahcall, J. N., 1973.

Baker, N. 1966, in *Stellar Evolution,* eds. R. F. Stein and A. G. W. Cameron (New York: Plenum Press), p. 333 (13, 13.1, 13.3).

Baker, N. 1973, in *Stellar Evolution,* eds. H.-Y. Chiu and A. Murial (Cambridge, Mass.: MIT Press) (12.2).

Baker, N., and Gough, D. O. 1967, *Ap. J.,* **72,** 784 (19.3).

Baker, N., and Gough, D. O. 1979, *Ap. J.,* **234,** 232 (10.1, 19.3).

Baker, N., and Kippenhahn, R. 1959, *Z. Astrophys.,* **48,** 140 (19.1a).

Baker, N., and Kippenhahn, R. 1962, *Z. Astrophys.,* **54,** 114 (9.2a, 9.2b, 10.1, 10.2, 11.1, 18.3).

Baker, N., and Kippenhahn, R. 1965, *Ap. J.,* **142,** 868 (8.4, 9.2b, 10.1, 10.2, 11.1, 11.4, 19.3).

Baker, N., and von Sengbusch, K. 1969, "Mitteilungen der Astronomischen Gesellschaft," No. 27, p. 162 (12.2).

Baker, N., and von Sengbusch, K. 1970, *Bull. Am. Astron. Soc.,* **2,** 181 (12.2).

Balona, J. M., and Stobie, R. S. 1979, *M.N.R.A.S.,* **187,** 217 (3.3).

Bappu, M. K. V., *see* Wilson, O. C., and Bappu, M. K. V., 1957.

Bardeen, J. M., Thorne, K. S., and Meltzer, D. W. 1966, *Ap. J.,* **145,** 505 (19.5).

Batchelor, G. K. 1967, *An Introduction to Fluid Dynamics* (Cambridge: At the University Press) (4, 4.2, 19.1a).

Bath, G. T., Evans, W. D., and Pringle, J. E. 1974, *M.N.R.A.S.,* **166,** 113 (3.2a).

Becker, R. H., *see* Lewin, W. H. G., Hoffman, J. H., Doty, J., Clark, G. W., Swank, J. H., Becker, R. H., Pravdo, S. H., and Serlemitsos, P. J., 1977.

Bellman, R. 1970, *Introduction to Matrix Analysis,* 2nd ed. (N. Y.: McGraw-Hill) (8.12b).

Bickley, W. G., *see* Temple, G., and Bickley, W. G., 1956.

Bierlaire, J., *see* Ledoux, P., Simon, R., and Bierlaire, J., 1955.

Bisnovatyi-Kogan, G. S., and Blinnikov, S. I. 1974, *Astron. Astrophys.,* **31,** 391 (19.1a, 19.5).

Blinnikov, S. I., *see* Bisnovatyi-Kogan, G. S., and Blinnikov, S. I., 1974.

Bludman, S. A. 1973a, *Ap. J.,* **183,** 637 (19.5).

Bludman, S. A. 1973b, *Ap. J.,* **183,** 649 (19.5).

Bogoluibov, N. N., and Mitropolsky, Yu. A. 1961, *Asymptotic Methods in the Theory of Non-Linear Oscillations* (transl. from Russian: Delhi, India: Hindustan Publ. Co.) (19.4).

Bogoluibov, N. N., *see* Krylov, M. M., and Bogoluibov, N. N., 1937.

Böhm-Vitense, E. 1958, *Zs. f. Ap.,* **46,** 108 (19.3).

Böhm-Vitense, E., and Nelson, G. D. 1976, *Ap. J.,* **210,** 741 (19.3).

Born, M., and Wolf, E. 1959, *Principles of Optics* (N. Y.: Pergamon Press) (17.12).

Boury, A., *see* Noels, A., Boury, A., Scuflaire, R., and Gabriel, M., 1974.

Bouw, G. D., *see* Parsons, S. B., and Bouw, G. D., 1971.

Breger, M. 1979, *Publ. A.S.P.,* **91,** 5 (3.1, 18.3).

Brickhill, A. J. 1975, *M.N.R.A.S.,* **170,** 405 (3.2a, 18.3, 19.1c).

Brickhill, A. J., *see* Warner, B., and Brickhill, A. J., 1974.

Brown, R. H., *see* Hanbury Brown, R., 1974.

Brown, T. M., Stebbins, R. T., and Hill, H. A. 1976, in *Proc. of the Los Alamos Solar and Stellar Pulsation Conf.,* eds., A. N. Cox and R. G. Deupree, p. 1 (1).

Brown, T. M., Stebbins, R. T., and Hill, H. A. 1978, *Ap. J.,* **223,** 324 (1).

Brown, T. M., *see* Hill, H. A., Stebbins, R. T., and Brown, T. M., 1975.

Brownlee, R. R., *see* Cox, A. N., Brownlee, R. R., and Eilers, D. D., 1966.

Bryan, G. H. 1889, *Phil. Mag.* (5), **27,** 254 (19.1a).

Buchler, J. R. 1978, *Ap. J.,* **220,** 629 (12.2, 19.4).

Burbidge, G., and Burbidge, M. 1967, *Quasi-Stellar Objects* (San Francisco: W. H. Freeman) (1).

Burbidge, G., *see* Perry, J. J., Burbidge, E. M., and Burbidge, G. R., 1978.

Burbidge, M., *see* Burbidge, G., and Burbidge, M., 1967; Perry, J. J., Burbidge, E. M., and Burbidge, G. R., 1978.

Buta, R., and Smith, M. A. 1979, *Ap. J.,* **232,** 213 (3.2c, 3.4, 17.6b, 19.7).

Buta, R., *see* Smith, M. A., and Buta, R., 1979.

Cameron, A. G. W. 1970, *Ann. Rev. Astron. and Astrophys.,* **8,** 179 (3, 19.2).

Cameron, A. G. W., *see* Stein, R. F., and Cameron, A. G. W., 1966.

Campbell, L., and Jacchia, L. 1941, *The Story of Variable Stars* (Philadelphia: The Blakiston Company) (1).
Campolattaro, A., and Thorne, K. S. 1970, *Ap. J.*, **159,** 847 (19.5).
Canuto, V. 1977, *Annals N. Y. Acad. Sci.*, **302,** 514 (3).
Carroll, B. W., *see* Hansen, C. J., Cox, J. P., and Carroll, B. W., 1978; Saio, H., Cox, J. P., Hansen, C. J., and Carroll, B. W., 1979.
Castor, J. I. 1966, Ph.D. Thesis, California Institute of Technology (11.1).
Castor, J. I. 1968a, unpublished manuscript (19.3).
Castor, J. I. 1968b, *Ap. J.*, **154,** 793 (11.1, 11.2, 11.3, 11.4, 19.3).
Castor, J. I. 1970, unpublished lecture notes (13, 13.4).
Castor, J. I. 1971, *Ap. J.*, **166,** 109 (8.3, 8.12b, 9.2a, 9.2b, 11.1, 11.4).
Castor, J. I. 1972, *Ap. J.*, **178,** 779 (4.3).
Castor, J. I., Davis, C. G., and Davison, D. K. 1977, Los Alamos Rept. LA-6664 (12.2).
Castor, J. I., *see* Adams, T. F., and Castor, J. I., 1979; Cox, J. P., Castor, J. I., and King, D. S., 1972.
Caudell, T. P., *see* Hill, H. A. Caudell, T. P., and Rosenwald, R. D., 1977; Hill, H. A., Rosenwald, R. D., and Caudell, T. P., 1977.
Chandrasekhar, S. 1939, *An Introduction to the Study of Stellar Structure* (Chicago: University of Chicago Press) (also available as a Dover paperback, 1957) (8.3, 8.12c, 17.5b).
Chandrasekhar, S. 1961, *Hydrodynamic and Hydromagnetic Stability* (Oxford: Clarendon Press) (5.1, 17.2, 17.3, 19.1a).
Chandrasekhar, S. 1963, *Ap. J.*, **138,** 896 (15.2).
Chandrasekhar, S. 1964a, *Phys. Rev. Lett.*, **12,** 114 (8.10, 19.5).
Chandrasekhar, S. 1964b, *Phys. Rev. Lett.*, **12,** 437 (8.10, 19.5).
Chandrasekhar, S. 1964c, *Ap. J.*, **140,** 417 (8.10, 19.5).
Chandrasekhar, S. 1964d, *Ap. J.*, **139,** 644 (15.2, 17.7).
Chandrasekhar, S. 1972, *Observatory*, **92,** 116 (19.1a).
Chandrasekhar, S., and Fermi, E. 1952, *Ap. J.*, **118,** 116 (4.6).
Chandrasekhar, S., and Lebovitz, N. R. 1964, *Ap. J.*, **140,** 1517 (15.3).
Chanmugam, G. 1977, *Ap. J.*, **217,** 799 (19.5).
Chanmugam, G. 1979, *M.N.R.A.S.*, **187,** 769 (19.2).
Chiu, H.-Y. 1967, *Stellar Physics* (Waltham, Mass.: Blaisdell) (4.3).
Chiu, H.-Y., and Murial, A. 1973, eds., *Stellar Evolution* (Cambridge, Mass.: MIT Press).
Christensen-Dalsgaard, J. 1976, *M.N.R.A.S.*, **174,** 87 (17.8).
Christensen-Dalsgaard, J. 1979, *M.N.R.A.S.*, in press (Preface, III, 17, 17.8, 17.9, 17.10, 17.12, 18).
Christensen-Dalsgaard, J., Dilke, F. W. W., and Gough, D. O. 1974, *M.N.R.A.S.*, **169,** 429 (7.6, 18.2a, 18.3).

Christensen-Dalsgaard, J., and Gough, D. O. 1975, *Mem. Soc. R. Sci. Liège,* **8,** 309 (18.3).

Christensen-Dalsgaard, J., and Gough, D. O. 1976, *Nature,* **259,** 89 (19.7).

Christy, R. F. 1962, *Ap. J.,* **136,** 887 (10.2, 12.2).

Christy, R. F. 1964, *Rev. Mod. Phys.,* **36,** 555 (11.1, 12.2).

Christy, R. F. 1966a, *Ann. Rev. Astron. Astrophys.,* **4,** 353 (1).

Christy, R. F. 1966b, *Ap. J.,* **144,** 108 (5.1).

Christy, R. F. 1967, in *Aerodynamic Phenomena in Stellar Atmospheres* (IAU Symposium No. 28), ed. R. N. Thomas (London: Academic Press), p. 105 (1).

Christy, R. F. 1968, *Q. J. R. Astron. Soc.,* **9,** 13 (1, 3.3).

Christy, R. F. 1969a, *J. R. Astron. Soc. Can.,* **63,** 299 (1).

Christy, R.F. 1969b, *J. R. Astron. Soc. Can.,* **64,** 8 (1).

Christy, R. F. 1970, *J. R. Astron. Soc. Can.,* **64,** 8 (1).

Clark, G. W., *see* Lewin, W. H. G., Hoffman, J. H., Doty, J., Clark, G. W., Swank, J. H., Becker, R. H., Pravdo, S. H., and Serlemitsos, P. J., 1977.

Clayton, D. D. 1968, *Principles of Stellar Evolution and Nucleosynthesis* (N. Y.: McGraw-Hill) (4.3, 5.2).

Clement, M. 1964, *Ap. J.,* **140,** 1045 (15.3, 19.1b).

Cline, T. L., and Desai, U. D. 1976, *Astr. Space Sci.,* **42,** 17 (1).

Cocke, W. J., Disney, M. J., and Taylor, D. J. 1969, *Nature,* **221,** 529 (1).

Cogan, B. C. 1970, *Ap. J.,* **162,** 139 (3.2b, 8.6, 19.3).

Cogan, B. C. 1977, *Ap. J.,* **211,** 890 (3.2b, 19.3).

Cogan, B. C. 1978a, *Ap. J.,* **221,** 635 (3.2b, 3.3).

Cogan, B. C. 1978b, *Ap. J.,* **225,** L39 (3.2b).

Cole, J. D. 1968, *Perturbation Methods in Applied Mathematics* (Waltham, Mass.: Blaisdell) (12.3, 13.4, 19.4).

Condon, E. U., and Shortley, G. H. 1935, *The Theory of Atomic Spectra* (Cambridge: Cambridge University Press) (17.3).

Cowling, T. G. 1941, *M.N.R.A.S.,* **101,** 367 (17.7, 17.8, 17.9).

Cowling, T. G. 1951, *Ap. J.,* **114,** 272 (18.2).

Cowling, T. G. 1957, *Magnetohydrodynamics* (N. Y.: Interscience) (4.6).

Cowling, T. G. 1965, in *Stellar Interiors,* eds., L. H. Aller and D. B. McLaughlin (Chicago: University of Chicago Press), p. 425 (19.2).

Cowling, T. G., and Newing, R. A. 1949, *Ap. J.,* **109,** 149 (19.1a, 19.1c).

Cox, A. N. 1978a, *Sky and Telescope,* **55,** 115 (February 1978) (3.2b, 19.7).

Cox, A. N. 1978b, in *Current Problems in Stellar Pulsation Instabilities,* eds., D. Fischel, J. R. Lesh, and W. M. Sparks (3.2b, 19.7).

Cox, A. N. 1979, *Ap. J.,* 229 (3.3, 3.4).

Cox, A. N., Brownlee, R. R., and Eilers, D. D. 1966, *Ap. J.*, **144,** 1024 (8.12b, 12.2, 19.3).

Cox, A. N., and Cox, J. P. 1967, *Sky and Telescope,* **33,** 278 (1).

Cox, A. N., and Cox, J. P. 1976, in *Multiple Periodic Variable Stars* (IAU Colloq. No. 29), ed. W. S. Fitch (Dordrecht: Reidel) (3.2b).

Cox, A. N., and Davis, C. G. 1975, *Dudley Obs. Rep.*, **9,** 297 (3.3).

Cox, A. N., and Deupree, R. G. 1976, eds., *Proceedings of the Solar and Stellar Pulsation Conference* (LASL Report UC-34b) (1, 12.2).

Cox, A. N., Deupree, R. G., King, D. S., and Hodson, S. W. 1977, *Ap. J.*, **214,** L127 (3.2b, 19.3).

Cox, A. N., Hodson, S. W., and Davey, W. R. 1976, in *Proc. Los Alamos Solar and Stellar Pulsation Conf.*, eds., A. N. Cox and R. G. Deupree, p. 188 (12.2).

Cox, A. N., Hodson, S. W., and King, D. S. 1979, *Ap. J.*, **230,** L109 (3.2b, 19.7).

Cox, A. N., King, D. S., and Tabor, J. E. 1973, *Ap. J.*, **184,** 201 (10, 10.3).

Cox, A. N., and Stewart, J. N. 1970, *Ap. J. Suppl.*, **19,** 243 (4.3).

Cox, A. N., and Tabor, J. E. 1976, *Ap. J. Suppl.*, **31,** 271 (4.3).

Cox, A. N., see Cox, J. P., Cox, A. N., Eilers, D. D., and King, D. S., 1967; Cox, J. P., Cox, A. N., and Olsen, K. H., 1963; Cox, J. P., Cox, A. N., Olsen, K., King, D. S., and Eilers, D. D., 1966; Henden, A. A., and Cox, A. N., 1976; King, D. S., Wheeler, J. C., Cox, J. P., Cox, A. N., and Hodson, S. W., 1979.

Cox, J. P. 1955, *Ap. J.*, **122,** 286 (7.6, 10.1, 10.2).

Cox, J. P. 1958, *Ap. J.*, **127,** 194 (10.2).

Cox, J. P. 1959, *Ap. J.*, **130,** 296 (10.3).

Cox, J. P. 1963, *Ap. J.*, **138,** 487 (9.2a, 9.2b, 10.2, 10.3, 11.1, 11.4).

Cox, J. P. 1967, in *Aerodynamic Phenomena in Stellar Atmospheres* (IAU Symposium No. 28), ed. R. N. Thomas (London: Academic Press), p. 3 (1, 2.1, 8.8, 8.12c, 9.2b).

Cox, J. P. 1974a, *Rep. Prog. Phys.*, **37,** 563 (1, 3, 3.1, 4.3, 5.1, 8.12c, 8.15, 9, 9.1, 10, 10.2, 10.3, 11, 11.1, 11.2, 11.4, 12, 12.1, 12.2, 13.4, 19.1a, 19.2).

Cox, J. P. 1974b, *Ap. J.*, **192,** L85 (19.4).

Cox, J. P. 1975, *Mem. Soc. Roy. Sci. Liège,* Coll. 8°, 6e Ser., 8, **129** (1, 12.2).

Cox, J. P. 1976a, *Ann. Rev. Astron. Astrophys.*, **14,** 247 (1, 7.6, 17, 17.2, 17.8, 19.1a, 19.2).

Cox, J. P. 1976b, in *Proc. Los Alamos Solar and Stellar Pulsation Conf.*, eds., A. N. Cox and R. G. Deupree, p. 127 (19.7).

Cox, J. P. 1978, in *Current Problems in Stellar Pulsation Instabilities*, eds., D. Fischel, J. R. Lesh, and W. M. Sparks (3.2b, 19.3, 19.7).

Cox, J. P. 1979, *Bull. Astron. Soc. India*, **7**, 4 (1, 3.2b).

Cox, J. P., Castor, J. I., and King, D. S. 1972, *Ap. J.*, **172**, 423 (10.3).

Cox, J. P., Cox, A. N., Eilers, D. D., and King, D. S. 1967, paper presented at the Thirteenth General Assembly of the IAU, Prague, Czechoslovakia (11.1).

Cox, J. P., Cox, A. N., and Olsen, K. H. 1963, *Astron. J.*, **68**, 276 (12.2).

Cox, J. P., Cox, A. N., Olsen, K. H., King, D. S., and Eilers, D. D. 1966, *Ap. J.*, **144**, 1038 (10.1, 12.2).

Cox, J. P., Davey, W. R., and Aizenman, M. L. 1974, *Ap. J.*, **191**, 439 (9.3, 19.4).

Cox, J. P., Eilers, D. D., and King, D. S. 1967, *Astron. J.*, **72**, 294 (11.1).

Cox, J. P., and Giuli, R. T. 1968, *Principles of Stellar Structure* (N. Y.: Gordon and Breach) (1, 2.1, 2.2, 4.1, 4.2b, 4.2c, 4.3, 5.2, 7.5, 8.3, 8.4, 8.6, 8.7, 8.12c, 9, 9.1, 9.3, 9.4, 10, 10.1, 10.3, 17.2, 17.9, 17.10, 17.12, 18.2a, 19, 19.3, 19.4, 19.6).

Cox, J. P., and Hansen, C. J. 1979, in *White Dwarfs and Degenerate Variable Stars* (IAU Colloq. No. 53), eds., M. P. Savedoff and H. M. Van Horn (3, 3.2b, 10, 10.3, 18.3).

Cox, J. P., Hansen, C. J., and Davey, W. R. 1973, *Ap. J.*, **182**, 885 (9.3, 19.4).

Cox, J. P., and King, D. S. 1970, unpublished calculations (11.4).

Cox, J. P., and King, D. S. 1972, in *Evolution of Population II Stars* (Dudley Obs. Rep. No. 4), ed. A. G. D. Philip (10.3).

Cox, J. P., King, D. S., and Stellingwerf, R. F. 1972, *Ap. J.*, **171**, 93 (2.1, 3.2b, 3.3, 8.6).

Cox, J. P., and Stellingwerf, R. F. 1979, *Publ. A.S.P.*, **91**, 319 (10, 10.3, 19.7).

Cox, J. P., and Whitney, C. A. 1958, *Ap. J.*, **127**, 561 (10.2).

Cox, J. P., *see* Aizenman, M. L., and Cox, J. P., 1974; 1975a,b; Aizenman, M. L., Cox, J. P., and Lesh, J. R., 1975; Cox, A. N., and Cox, J. P., 1967, 1976; Davey, W. R., and Cox, J. P., 1974; Hansen, C. J., Cox, J. P., and Carroll, B. W., 1978; Hansen, C. J., Cox, J. P., and Herz, M. A., 1972; Hansen, C. J., Cox, J. P., and van Horn, M. A., 1977; King, D. S., and Cox, J. P., 1968; King, D. S., Cox, J. P., Eilers, D. D., and Davey, W. R., 1973; King, D. S., Hansen, C. J., Ross, R. R., and Cox, J. P., 1975; King, D. S., Wheeler, J. C., Cox, J. P., and Hodson, S. W., 1978; King, D. S., Wheeler, J. C., Cox, J. P., Cox, A. N., and Hodson, S. W., 1979; Saio, H., and Cox, J. P., 1979a,b; Saio, H., Cox, J. P., Hansen, C. J., and Carroll, B. W., 1979.

Davey, W. R. 1970, Ph.D. Thesis, University of Colorado (8.12b, 9.2b, 19.4).
Davey, W. R. 1973, *Ap. J.*, **179,** 235 (9.2b).
Davey, W. R. 1974, *Ap. J.*, **194,** 687 (9.3, 19.4).
Davey, W. R., and Cox, J. P. 1974, *Ap. J.*, **189,** 113 (19.4).
Davey, W. R., *see* Cox, A. N., Hodson, S. W., and Davey, W. R., 1976; Cox, J. P., Davey, W. R., and Aizenman, M. L., 1974; Cox, J. P., Hansen, C. J., and Davey, W. R., 1973; King, D. S., Cox, J. P., Eilers, D. D., and Davey, W. R., 1973.
Davidson, A., Henry, J. P., Middleditch, J., and Smith, H. E. 1972, *Ap. J.*, **177,** L97 (1).
Davies, G. F. 1968, *Aust. J. Phys.*, **21,** 294 (19.2).
Davis, C. G. 1974, *Ap. J.*, **187,** 175 (19.5).
Davis, C. G., and Davison, D. K. 1978, *Ap. J.*, **122,** 929 (12.2).
Davis, C. G., *see* Castor, J. I., Davis, C. G., and Davison, D. K., 1978; Cox, A. N., and Davis, C. G., 1975; Adams, T. F., Davis, C. G., and Keller, C. F., 1978.
Davison, D. K., *see* Castor, J. I., Davis, C. G., and Davison, D. K., 1978.
Dedic, H., and Tassoul, J.-L. 1974, *Ap. J.*, **188,** 173 (19.5).
Defouw, R. J. 1973, *Ap. J.*, **182,** 215 (5.1, 18.2).
Demaret, J. 1974, *Ap. Space Sci.*, **31,** 305 (5.4c, 19.4).
Demaret, J. 1975, *Ap. Space Sci.*, **33,** 189 (19.4).
Demaret, J. 1976, *Ap. Space Sci.*, **45,** 31 (19.4).
Demarque, P. 1973, in *Variable Stars in Globular Clusters and in Related Systems* (IAU Colloq. No. 21), ed. J. D. Fernie (Dordrecht: Reidel) (1).
Denis, J., see Goossens, M., Smeyers, P., and Denis, J., 1976; Smeyers, P., and Denis, J., 1971.
Derrickson, J. H., *see* Fishman, G. J., Watts, J. W., Jr., and Derrickson, J. H., 1978.
Desai, U. D., *see* Cline, T. L., and Desai, U. D., 1976.
Detre, L., ed. 1968, *Nonperiodic Phenomena in Variable Stars* (Dordrecht: Reidel) (1).
Detweiler, S. L. 1975a, *Ap. J.*, **197,** 203 (19.5).
Detweiler, S. L. 1975b, *Ap. J.*, **201,** 440 (19.5).
Detweiler, S. L., and Ipser, J. R. 1973, *Ap. J.*, **185,** 685 (19.5).
Deubner, F. L. 1974, *Solar Phys.*, **39,** 31 (18.3).
Deupree, R. G. 1974, Ph.D. Dissertation, University of Toronto (10.1, 17.7).
Deupree, R. G. 1975a, *Ap. J.*, **198,** 419 (10.1, 19.3).
Deupree, R. G. 1975b, *Ap. J.*, **201,** 183 (10.1, 19.3).

Deupree, R. G. 1976a, *Ap. J.*, **205,** 286 (10.1, 19.3).

Deupree, R. G. 1976b, in *Proc. Los Alamos Solar and Stellar Pulsation Conf.*, eds., A. N. Cox and R. G. Deupree, p. 222 (10.1, 19.3).

Deupree, R. G. 1976c, in *Proc. Los Alamos Solar and Stellar Pulsation Conf.*, eds., A. N. Cox and R. G. Deupree, p. 229 (10.1, 19.3).

Deupree, R. G. 1977a, *Ap. J.*, **211,** 509 (10.1, 11.4, 19.3).

Deupree, R. G. 1977b, *Ap. J.*, **214,** 502 (10.1, 19.3).

Deupree, R. G. 1977c, *Ap. J.*, **215,** 232 (10.1, 19.3).

Deupree, R. G. 1977d, *Ap. J.*, **215,** 620 (10.1, 19.3).

Deupree, R. G., *see* Cox, A. N., and Deupree, R. G., 1976; Cox, A. N., Deupree, R. G., King, D. S., and Hodson, S. W., 1977.

Deutsch, A. J. 1958, *Handbuch der Physik,* ed. S. Flügge (Berlin: Springer-Verlag), **51,** 689 (1, 19.2).

DeWitt, C., *see* Thorne, K., 1967a.

Dilke, F. W. W., and Gough, D. O. 1972, *Nature,* **240,** 262 (18.2, 18.2a).

Dilke, F. W. W., *see* Christensen-Dalsgaard, J., Dilke, F. W. W., and Gough, D. O., 1974.

Disney, M. J., *see* Cocke, W. J., Disney, M. J., and Taylor, D. J., 1969.

Doty, J., *see* Lewin, W. H. G., Hoffman, S. H., Doty, J., Clark, G. W., Swank, J. H., Becker, R. H., Pravdo, S. H., and Serlemitsos, P. J., 1977.

Duffey, G. H. 1973, *Theoretical Physics: Classical and Modern Views* (Boston: Houghton Mifflin Co.) (4.2b).

Duval, P., and Karp, A. H. 1978, *Ap. J.*, **222,** 220 (3.4).

Dziembowski, W. 1971, *Acta Astron.*, **21,** 289 (17.5, 17.5b, 17.6b, 17.10, 18.3).

Dziembowski, W. 1975, *Mem. Soc. Roy. Sci. Liège,* **8,** 287 (17.10, 18.3).

Dziembowski, W. 1977a, *Acta Astron.*, **27,** 1 (18.3).

Dziembowski, W. 1977b, *Acta Astron.*, **27,** 95 (17.10, 18.3, 19.7).

Dziembowski, W. 1977c, *Acta Astron.*, **27,** 203 (3.4, 18.3).

Dziembowski, W. 1979, in *White Dwarfs and Degenerate Variable Stars* (IAU Colloq. No. 53), eds., M. P. Savedoff and H. M. Van Horn (18.3).

Dziembowski, W., *see* Sienkowicz, R., and Dziembowski, W., 1978; Hill, H. A., and Dziembowski, W., 1979.

Eardley, D. M., and Press, W. H. 1975, *Ann. Rev. Astron. Astrophys.*, **13,** 381 (1).

Eckart, C. 1960, *Hydrodynamics of Oceans and Atmospheres* (London: Pergamon Press) (17.9, 17.10).

Eddington, A. S. 1918a, *M.N.R.A.S.*, **79,** 2 (1, 8).

Eddington, A. S. 1918b, *M.N.R.A.S.*, **79,** 177 (1, 8).

Eddington, A. S. 1926, *The Internal Constitution of the Stars* (Cambridge: Cambridge University Press; also available as a Dover paperback, 1959) (1, 2.1, 5.1, 9.4, 10.1, 10.2).

Eddington, A. S. 1932, *M.N.R.A.S.*, **92,** 471 (19.7).

Eddington, A. S. 1941, *M.N.R.A.S.*, **101,** 182 (10.2, 11.4).

Eddington, A. S. 1942, *M.N.R.A.S.*, **102,** 154 (10.2, 11.4).

Eddy, J. A. 1978, ed., *The New Solar Physics* (Boulder: Westview) (1).

Efremov, Yu. N., *see* Kukarkin, B. V., Kholopov, P. N., Efremov, Yu. N., Kukarkina, N. P., Kurochkin, N. E., Medvedeva, G. I., Perova, N. B., Fedorovitch, V. P., and Frolov, M. S., 1969; Kukarkin, B. V., Kholopov, P. N., Efremov, Yu. N., Kukarkina, N. P., Kurochkin, N. E., Medvedeva, G. I., Perova, N. B., Pskovsky, Yu. P., Fedorovitch, V. P., and Frolov, M. S., 1974.

Eggleton, P., Mitton, S., and Whelan, J. 1976, eds., *Structure and Evolution of Close Binary Systems* (IAU Symp. No. 73) (Dordrecht, Boston: Reidel).

Eilers, D. D., *see* Cox, A. N., Brownlee, R. R., and Eilers, D. D., 1966; Cox, J. P., Cox, A. N., Eilers, D. D., and King, D. S., 1967; Cox, J. P., Cox, A. N., Olsen, K. H., King, D. S., and Eilers, D. D., 1966; Cox, J. P., Eilers, D. D., and King, D. S., 1967; King, D. S., Cox, J. P., Eilers, D. D., and Davey, W. R., 1973.

Eisenfeld, J. 1969, *J. Math. Annal. Appl.*, **26,** 357 (17.3).

Epstein, I. 1950, *Ap. J.*, **112,** 6 (8.13, 10.2, 17.13).

Evans, N. R. 1976, *Ap. J.*, **209,** 135 (3.3).

Evans, W. D., *see* Bath, G. T., Evans, W. D., and Pringle, J. E., 1974; Strong, I., Klebesadel, R. W., and Evans, W. D., 1975.

Fälthammar, C. G., *see* Alfvén, H., and Fälthammar, C. G., 1963.

Faulkner, D. J. 1977a, *Ap. J.*, **216,** 49 (3.2b).

Faulkner, D. J. 1977b, *Ap. J.*, **218,** 209 (2.1, 3.2b, 8.6).

Faulkner, D. J., *see* Gingold, R. A., and Faulkner, D. J., 1974.

Fedorovitch, V. P., *see* Kukarkin, B. V., Kholopov, P. N., Efremov, Yu. N., Kukarkina, N. P., Kurochkin, N. E., Medvedeva, G. I., Perova, N. B., Fedorovitch, V. P., and Frolov, M. S., 1969; Kukarkin, B. V., Kholopov, P. N., Efremov, Yu. N., Kukarkina, N. P., Kurochkin, N. E., Medvedeva, G. I., Perova, N. B., Pskovsky, Yu. P., Fedorovitch, V. P., and Frolov, M. S., 1974; Kukarkin, B. V., Kholopov, P. N., Fedorovitch, V. P., Frolov, M. S., Kukarkina, N. P., Kurochkin, N. E., Medvedeva, G. I., Perova, N. B., and Pskovsky, Yu. D., 1976.

Feltz, K. A., Jr., *see* McNamara, D. H., and Feltz, K. A., Jr., 1978.

Fermi, E., *see* Chandrasekhar, S., and Fermi, E., 1952.

Fernie, J. D. 1967, *A. J.*, **72,** 1327 (3.1).

Fernie, J. D. 1968, *Ap. J.*, **151,** 197 (3.3).

Fernie, J. D. 1969, *Publ. A.S.P.*, **81,** 707 (3.1, 19.7).

Fernie, J. D. 1973, ed., *Variable Stars in Globular Clusters and in Related Systems* (IAU Colloq. No. 21) (Dordrecht: Reidel).

Fernie, J. D. 1976, *Publ. A.S.P.*, **88,** 116 (18.3).

Feschenko, S. F., Shkil', N. I., and Nikolenko, L. D. 1967, *Asymptotic Methods in the Theory of Linear Differential Equations* (transl. from Russian; N. Y.: American Elsevier) (18.2, 19.4).

Feshbach, H., see Morse, P. M., and Feshbach, H., 1953.

Fischel, D., and Sparks, W. M. 1975, eds., *Cepheid Modeling* (NASA Report SP-383) (1, 12.2).

Fischel, D., Lesh, J. R., and Sparks, W. M. 1978, eds., *Current Problems in Stellar Pulsation Instabilities* (Washington, D.C.: NASA) (1).

Fishman, G. J., Watts, J. W., Jr., and Derrickson, J. H. 1978, *Ap. J.*, **223,** L13 (1).

Fitch, W. S. 1970, *Ap. J.*, **161,** 669 (3.2b).

Fitch, W. S. 1976a, ed., *Multiple Periodic Variable Stars* (29th IAU Colloquium) (Dordrecht: Reidel) (1).

Fitch, W. S. 1976b, in *Multiple Periodic Variable Stars* (IAU Colloq. No. 29), ed. W. S. Fitch (Dordrecht: Reidel) (18.3).

Fitch, W. S., and Szeidl, B. 1976, *Ap. J.*, **203,** 616 (3.2b).

Forman, W., Jones, C. A., and Liller, W. 1972, *Ap. J.*, **177,** L103 (1).

Forsyth, A. R. 1929, *A Treatise on Differential Equations* (N. Y.: MacMillan) (17.10).

Fossat, E., Ricort, G., Aime, C., and Roddier, F. 1974, *Ap. J.*, **193,** L97 (18.3).

Fraley, G. S. 1968, *Astrophys. Sp. Sci.*, **2,** 96 (12.2, 19.3).

Franklin, P. 1940, *A Treatise on Advanced Calculus* (N. Y.: John Wiley & Sons) (9.2b).

Fricke, K. 1968, *Z. Astrophys.*, **68,** 317 (19.1a).

Fricke, K. J., and Kippenhahn, R. 1972, *Ann. Rev. Astron. Astrophys.*, **10,** 45 (19.1a).

Frogel, J. A., *see* Stothers, R., and Frogel, J. A., 1967.

Frolov, M. S., *see* Kukarkin, B. V., Kholopov, P. N., Efremov, Yu. N., Kukarkina, N. P., Kurochkin, N. E., Medvedeva, G. I., Perova, N. B., Fedorovitch, V. P., and Frolov, M. S., 1969; Kukarkin, B. V., Kholopov, P. N., Efremov, Yu. N., Kukarkina, N. P., Kurochkin, N. E., Medvedeva, G. I., Perova, N. B., Pskorsky, Yu. P., Fedorovitch, V. P., and Frolov, M. S., 1974; Kukarkin, B. V., Kholopov, P. N., Fedorovitch, V. P., Frolov, M. S., Kukarkina, N. P., Kurochkin, N. E., Medvedeva, G. I., Perova, N. B., and Pskovsky, Yu. P., 1976.

Gabriel, M. 1969, *Astron. Astrophys.*, **1,** 321 (18.3).

Gabriel, M., *see* Noels, A., Boury, A., Scuflaire, R., and Gabriel, M., 1974.

Gaposchkin, S. 1972, *Smithsonian Astrophys. Obs. Spec. Rep.*, No. 310 (3.1).

Gaposchkin, S., *see* Payne-Gaposchkin, C., and Gaposchkin, S., 1963; 1965; 1966.

Gascoigne, S. C. B. 1969, *M.N.R.A.S.*, **146,** 1 (3.1).

Geyer, V. 1970, *Astron. Astrophys.*, **5,** 116 (3.1).

Gingold, A. R., and Faulkner, D. J. 1974, *Ap. J.*, **188,** 145 (19.4).

Gingold, A. R., *see* Rodgers, A. W., and Gingold, A. R., 1973.

Ginzburg, V. L., and Zheleznyakov, V. V. 1975, *Ann. Rev. Astron. Astrophys.*, **13,** 511 (1).

Giuli, R. T., *see* Cox, J. P., and Giuli, R. T., 1968.

Glasby, J. S. 1975, *The Nebular Variables* (N. Y. and Oxford: Pergamon Press) (1).

Goertzel, G., and Tralli, N. 1960, *Some Mathematical Methods of Physics* (N. Y.: McGraw-Hill) (8.10, 17.3).

Goldberg, L., and Aller, L. H. 1943, *Atoms, Stars, and Nebulae* (Philadelphia: Blakiston) (3.1).

Goldreich, P., and Schubert, G. 1967, *Ap. J.*, **150,** 571 (18.2, 19.1a).

Goldstein, H. 1950, *Classical Mechanics* (Cambridge: Addison-Wesley Press) (5.4).

Goossens, M. 1972, *Astrophys. Sp. Sci.*, **16,** 386 (19.2).

Goossens, M. 1976a, *Astrophys. Sp. Sci.*, **43,** 9 (19.2).

Goossens, M. 1976b, *Astrophys. Sp. Sci.*, **44,** 397 (19.2).

Goossens, M., and Smeyers, P. 1974, *Astrophys. Sp. Sci.*, **26,** 137 (17.10, 17.13).

Goossens, M., Smeyers, P., and Denis, J. 1976, *Astrophys. Sp. Sci.*, **39,** 257 (19.2).

Gough, D. O. 1967, *Astron. J.*, **72,** 799 (13, 19.3).

Gough, D. O. 1977a, in *Problems of Stellar Convection* (IAU Colloq. No. 38), ed. J. P. Zahn (Dordrecht: Reidel) (18.2a, 19.3).

Gough, D. O. 1977b, *Ap. J.*, **214,** 196 (19.3).

Gough, D. O. 1977c, in *The Solar Output and Its Variation,* ed. O. R. White (Boulder: Colorado Associated University Press), p. 451 (1).

Gough, D. O., *see* Baker, N., and Gough, D. O., 1967; 1979; Christensen-Dalsgaard, J., Dilke, F. W. W., and Gough, D. O., 1974; Christensen-Dalsgaard, J., and Gough, D. O., 1975; 1976.

Gursky, H. 1977, *Ann. N. Y. Acad. Sci.*, **302,** 197 (1).

Greenspan, H. P. 1968, *The Theory of Rotating Fluids* (Cambridge: Cambridge University Press) (1, 19.1a).

Hanbury Brown, R. 1974, *The Intensity Interferometer* (London: Halsted Press) (3.3).

Hansen, C. J. 1973, ed., *Physics of Dense Matter* (Dordrecht, N. Y.: Reidel).

Hansen, C. J. 1978, *Ann. Rev. Astron. Astrophys.*, **16,** 15 (19.6).

Hansen, C. J. 1979, in *Proceedings of the Workshop on Nonradial and Nonlinear Stellar Instabilities,* eds., H. Hill and W. Dziembowski (N. Y.: Springer-Verlag) (3, 3.2a).

Hansen, C. J., Aizenman, M. L., and Ross, R. R. 1976, *Ap. J.,* **207,** 736 (17.10).

Hansen, C. J., Cox, J. P., and Carroll, B. W. 1978, *Ap. J.,* **226,** 210 (19.1d).

Hansen, C. J., Cox, J. P., and Herz, M. A. 1972, *Astron. Astrophys.,* **19,** 144 (5.1).

Hansen, C. J., Cox, J. P., and Van Horn, H. M. 1977, *Ap. J.,* **217,** 151 (19.1c, 19.2).

Hansen, C. J., and Van Horn, H. M. 1979, *Ap. J.,* **233,** 253 (17.4, 19).

Hansen, C. J., *see* Aizenman, M. L., Hansen, C. J., and Ross, R. R., 1975; King, D. S., Hansen, C. J., Ross, R. R., and Cox, J. P., 1975; Osaki, Y., and Hansen, C. J., 1973a,b; Pollack, J., and Hansen, C. J., 1970; Cox, J. P., and Hansen, C. J., 1979; Saio, H., Cox, J. P., Hansen, C. J., and Carroll, B. W., 1979.

Haramundanis, K., *see* Payne-Gaposchkin, C., and Haramundanis, K., 1970.

Härm, R., and Schwarzschild, M. 1964, *Ap. J.,* **139,** 594 (19.4).

Härm, R., and Schwarzschild, M. 1972, *Ap. J.,* **172,** 403 (5.1).

Härm, R., *see* Kruskal, M., Schwarzschild, M., and Härm, R., 1977.

Harwit, M. 1973, *Astrophysical Concepts* (N. Y.: John Wiley) (3).

Hawardin, T., *see* Stobie, R. S., and Hawardin, T., 1972.

Heintzmann, H., and Hillebrandt, W. 1975, *Astron. Astrophys.,* **28,** 51 (19.5).

Henden, A. A., and Cox, A. N. 1976, in *Proc. Los Alamos Solar and Stellar Pulsation Conf.,* eds., A. N. Cox and R. G. Deupree, p. 167 (19.3).

Henry, J. P., *see* Davidson, A., Henry, J. P., Middleditch, J., and Smith, H. E., 1972.

Herbig, G. H. 1962, *Adv. Astron. Astrophys.,* **1,** 47 (1, 3).

Herbig, G. H. 1978, *Ap. J.,* **217,** 693 (3).

Herbst, W., Hesser, J. E., and Ostriker, J. P. 1974, *Ap. J.,* **193,** 679 (3.2a).

Herz, M. A., *see* Hansen, C. J., Cox, J. P., and Herz, M. A., 1972.

Hesser, J. E., *see* Herbst, W., Hesser, J. E., and Ostriker, J. P., 1974.

Hewish, A. 1970, *Ann. Rev. Astron. Astrophys.,* **8,** 265 (1, 3).

Hildebrand, R. H., *see* Stiening, R. F., Hildebrand, R. H., and Spillar, E. J., 1979.

Hill, H. A. 1978, in *The New Solar Physics,* ed. J. A. Eddy (Boulder, Colorado: Westview) (1, 8.4, 17.7, 19.7).

Hill, H. A., Caudell, T. P., and Rosenwald, R. D. 1977, *Ap. J.,* **213,** L81 (17.6).

Hill, H. A., and Dziembowski, W. 1979, eds., *Proceedings of the Workshop on Nonradial and Nonlinear Stellar Instabilities* (N. Y.: Springer-Verlag) (1, 8.4, 19.7).

Hill, H. A., Rosenwald, R. D., and Caudell, T. P. 1977, *Ap. J.,* **225,** 304 (17.6b).

Hill, H. A., Stebbins, R. T., and Brown, T. M. 1975, in *Proc. Fifth International Conf. on Atomic Masses and Fundamental Constants,* eds., J. H. Sanders and A. H. Wapstra (N. Y.: Plenum Press) (1).

Hill, H. A., *see* Brown, T. M., Stebbins, R. T., and Hill, H. A., 1976; 1978.

Hill, S. J. 1970, Ph.D. Thesis, University of Colorado (3.3).

Hill, S. J., and Willson, L. A. 1979, *Ap. J.,* **229,** 1029 (10.2).

Hillebrandt, W., and Steinmetz, K. O. 1976, *Astron. Astrophys.,* **53,** 283 (19.5).

Hillebrandt, W., *see* Heintzmann, H., and Hillebrandt, W., 1975.

Hiltner, W. A., and Mook, D. E. 1970, *Ann. Rev. Astron. Astrophys.,* **8,** 139 (1).

Hodson, S. W., *see* Cox, A. N., Deupree, R. G., King, D. S., and Hodson, S. W., 1977; Cox, A. N., Hodson, S. W., and Davey, W. R., 1976; King, D. S., Wheeler, J. C., Cox, J. P., and Hodson, S. W., 1978; King, D. S., Wheeler, J. C., Cox, J. C., Cox, A. N., and Hodson, S. W., 1979; Cox, A. N., Hodson, S. W., and King, D. S., 1979.

Hoffman, J. H., *see* Lewin, W. H. G., Hoffman, J. H., Doty, J., Clark, G. W., Swank, J. H., Becker, R. H., Pravdo, S. H., and Serlemitsos, P. J. 1977.

Hoffmeister, C. 1971, *Veränderliche Sterne,* under collaboration by G. A. Richter and W. Wenzel (Leipzig: Johann Ambrosius Barth) (1, 3).

Hofmeister, E. 1967, *Z. Ap.,* **65,** 194 (3.1).

Huguenin, G. R., *see* Taylor, J. H., and Huguenin, G. R., 1971.

Hummer, D. G., *see* Mihalas, D., Kunasz, P. B., and Hummer, D. G., 1975.

Hurley, M., Roberts, P. H., and Wright, K. 1966, *Ap. J.,* **143,** 535 (8.12c, 8.15, 17.7, 17.8).

Iben, I., Jr. 1971a, *Publ. A.S.P.,* **83,** 697 (1).

Iben, I., Jr. 1971b, *Ap. J.,* **166,** 131 (9.2a, 9.2b, 9.3, 11.4, 19.3).

Iben, I., Jr., *see* Tuggle, R. S., and Iben, I., Jr., 1973.

Ince, E. L. 1944, *Ordinary Differential Equations* (N. Y.: Dover Publications) (8.5, 8.7, 8.8, 8.9, 8.10, 17.9, 17.10).
Ipser, J. R., and Thorne, K. S. 1973, *Ap. J.,* **181,** 181 (19.5).
Ipser, J. R., *see* Detweiler, S. L., and Ipser, J. R., 1973.
Ishizuka, T. 1967, *Publ. Astron. Soc. Japan,* **19,** 495 (13).

Jacchia, L., *see* Campbell, L., and Jacchia, L., 1941.
Jacobs, K. C., *see* Smith, E. V. P., and Jacobs, K. C., 1973.
Jackson, J. D. 1962, *Classical Electrodynamics* (N. Y.: John Wiley & Sons, Inc.) (17.3).
James, R. A., and Kahn, F. D. 1971, *Astron. Astrophys.,* **12,** 332 (19.1a).
Jeans, J. H. 1928, *Astronomy and Cosmogony* (Cambridge: Cambridge University Press; also available as a Dover paperback, 1961) (13.1, 19.6).
Jenkins, F. A., and White, H. E. 1957, *Fundamentals of Optics,* 3rd ed. (N. Y.: McGraw-Hill) (17.12).
Jones, C. A., *see* Forman, W., Jones, C. A., and Liller, W., 1972.
Joos, G. 1932, *Theoretical Physics* (N. Y.: Hafner Publishing Co.) (4.5).

Kahn, F. D., *see* James, R. A., and Kahn, F. D., 1971.
Kamijo, F., *see* Unno, W., and Kamijo, F., 1966.
Kaniel, S., and Kovetz, A. 1967, *Phys. Fluids,* **10,** 1186 (17.3).
Karp, A. H. 1975a, *Ap. J.,* **201,** 641 (3.3, 3.4).
Karp, A. H. 1975b, in *Cepheid Modeling,* eds., D. Fischel and W. M. Sparks (Washington, D.C.: NASA), p. 99 (11.4).
Karp, A. H., *see* Duval, P., and Karp, A. H., 1978.
Kato, S. 1966, *Publ. Astron. Soc. Japan,* **18,** 374 (18.2, 18.2a, 18.3).
Kato, S. 1976, in *Multiple Periodic Variable Stars* (IAU Colloq. no. 29), ed. W. S. Fitch (Dordrecht: Reidel) (19.7).
Kato, S., and Unno, W. 1967, *Publ. Astron. Soc. Japan,* **19,** 1 (19.4).
Keeley, D. A. 1970a, *Ap. J.,* **161,** 643 (10.2).
Keeley, D. A. 1970b, *Ap. J.,* **161,** 657 (10.2).
Keller, C. F., and Mutschlecner, J. P. 1971, *Ap. J.,* **167,** 127 (11.4).
Keller, C. F., *see* Adams, T. F., Davis, C. G., and Keller, C. F., 1978.
Kholopov, P. N., *see* Kukarkin, B. V., Kholopov, P. N., Efremov, Yu. N., Kukarkina, N. P., Kurochkin, N. E., Medvedeva, G. I., Perova, N. B., Fedorovitch, V. P., and Frolov, M. S., 1969; Kukarkin, B. V., Kholopov, P. N., Efremov, Yu. N., Kukarkina, N. P., Kurochkin, N. E., Medvedeva, G. I., Perova, N. B., Pskovsky, Yu. P., Fedorovitch, V. P., and Frolov, M. S., 1974; Kukarkin, V. B., Kholopov, P. N., Fedorovitch, V. P., Frolov, M. S., Kukarkina, N. P., Kurochkin, N. E., Medvedeva, G. I., Perova, N. B., and Pskovsky, Yu. P., 1976.

King, D. S., and Cox, J. P. 1968, *Publ. A.S.P.*, **80,** 365 (1, 10.2, 11.1).

King, D. S., Cox, J. P., Eilers, D. D., and Davey, W. R. 1973, *Ap. J.*, **182,** 859 (8.13, 10.1).

King, D. S., Hansen, C. J., Ross, R. R., and Cox, J. P. 1975, *Ap. J.*, **195,** 467 (3.2b, 8.6).

King, D. S., Wheeler, J. C., Cox, J. P., and Hodson, S. W. 1978, *BAAS*, **10,** 633 (9.2c).

King, D. S., Wheeler, J. C., Cox, J. P., Cox, A. N., and Hodson, S. W. 1979, in *Proceedings of the Workshop on Nonradial and Nonlinear Stellar Instabilities*, eds., H. Hill and W. Dziembowski (N. Y.: Springer-Verlag) (9.2c).

King, D. S., *see* Cox, A. N., King, D. S., and Tabor, J. E., 1973; Cox, J. P., Castor, J. I., and King, D. S., 1972; Cox, J. P., Cox, A. N., Eilers, D. D., and King, D. S., 1967; Cox, J. P., Cox, A. N., Olsen, K. H., King, D. S., and Eilers, D. D., 1966; Cox, A. N., Deupree, R. G., King, D. S., and Hodson, S. W., 1977; Cox, J. P., Eilers, D. D., and King, D. S., 1967; Cox, J. P., and King, D. S., 1970; 1972; Cox, J. P., King, D. S., and Stellingwerf, R. F., 1972; Cox, A. N., Hodson, S. W., and King, D. S., 1979.

Kippenhahn, R. 1967, *Z. Astrophys.*, **67,** 271 (19.6).

Kippenhahn, R. 1974, in *Late Stages of Stellar Evolution* (IAU Symp. No. 66), ed. R. J. Tayler (Dordrecht: Reidel), p. 20 (19.6).

Kippenhahn, R., Rahe, J., and Strohmeier, W. 1977, *The Interaction of Variable Stars with Their Environment* (IAU Colloq. No. 42) (Nürnberg; Astronomisches Inst. d. Universität Erlagen) (1).

Kippenhahn, R., *see* Baker, N., and Kippenhahn, R., 1959; 1962; 1965.

Kippenhahn, R., *see* Fricke, K. J., and Kippenhahn, R., 1972.

Klebesadel, R. W., and Strong, I. B. 1976, *Astrophys. Sp. Sci.*, **42,** 3 (1).

Klebesadel, R. W., *see* Strong, I. B., Klebesadel, R. W., and Evans, W. D., 1975.

Knigge, R., *see* Strohmeier, W., and Knigge, R., 1972.

Kobayashi, E., *see* Saio, H., Kobayashi, E., and Takeuti, M., 1977.

Kopal, Z. 1949, *Ap. J.*, **109,** 509 (17.9).

Korn, G. A., and Korn, T. M. 1968, *Mathematical Handbook for Scientists and Engineers* (N. Y.: McGraw-Hill) (6.1).

Korn, T. M., *see* Korn, G. A., and Korn, T. M., 1968.

Kovetz, A., *see* Kaniel, S., and Kovetz, A., 1967.

Kraft, R. P. 1961, *Ap. J.*, **134,** 616 (3.1).

Kraft, R. P. 1962, *Ap. J.*, **135,** 408 (3.2a).

Kraft, R. P. 1963, *Adv. Astron. Astrophys.*, **2,** 43 (3.2a).

Kruskal, M., Schwarzschild, M., and Härm, R. 1977, *Ap. J.*, **214,** 498 (18.2).

Krylov, M. M., and Bogoluibov, N. N. 1937, *Introduction to Nonlinear Mechanics* (Acad. Sci., Ukrainian S.S.R.).

Kukarkin, B. V. 1976, *Pulsating Stars* (transl. from Russian [1970]; N. Y.: Halsted Press, a division of John Wiley & Sons; IPST Astrophysics Library, Israel Program for Scientific Translations, Jerusalem) (1, 3).

Kukarkin, B. V., Kholopov, P. N., Efremov, Yu. N., Kukarkina, N. P., Kurochkin, N. E., Medvedeva, G. I., Perova, N. B., Fedorovitch, V. P., and Frolov, M. S. 1969, *General Catalogue of Variable Stars,* 3rd ed. (Moscow: Nauka) (1, 3.1).

Kukarkin, B. V., Kholopov, P. N., Efremov, Yu. N., Kukarkina, N. P., Kurochkin, N. E., Medvedeva, G. I., Perova, N. B., Pskovsky, Yu. P., Fedorovitch, V. P., and Frolov, M. S. 1974, *Second Supplement to the Third Edition of the General Catalogue of Variable Stars,* U.S.S.R. Acad. Sci., Moscow (3.1).

Kukarkin, B. V., Kholopov, P. N., Fedorovitch, V. P., Frolov, M. S., Kukarkina, N. P., Kurochkin, N. E., Medvedeva, G. I., Perova, N. B., and Pskovsky, Yu. P. 1976, *Third Supplement to the Third Edition of the General Catalogue of Variable Stars,* U.S.S.R. Acad. Sci., Moscow (3.1).

Kukarkin, B. V., and Parenago, P. P. 1963, in *Basic Astronomical Data,* ed. K. A. Strand (Chicago: University of Chicago Press), p. 328 (1, 3).

Kukarkina, N. P., *see* Kukarkin, B. V., Kholopov, P. N., Efremov, Yu. N., Kukarkin, N. P., Kurochkin, N. E., Medvedeva, G. I., Perova, N. B., Fedorovitch, V. P., and Frolov, M. S., 1969; Kukarkin, B. V., Kholopov, P. N., Efremov, Yu. N., Kukarkina, N. P., Kurochkin, N. E., Medvedeva, G., Perova, N. B., Pskovsky, Yu. P., Fedorovitch, V. P., and Frolov, M. S., 1974; Kukarkin, B. V., Kholopov, P. N., Fedorovitch, V. P., Frolov, M. S., Kukarkina, N. P., Kurochkin, N. E., Medvedeva, G. I., Perova, N. B., and Pskovsky, Yu. P., 1976.

Kunasz, P. B., *see* Mihalas, D., Kunasz, P. B., and Hummer, D. G., 1975.

Kurochkin, N. E., *see* Kukarkin, B. V., Kholopov, P. N., Fedorovitch, V. P., Frolov, M. S., Kukarkina, N. P., Kurochkin, N. E., Medvedeva, G. I., Perova, N. R., and Pskovsky, Yu. P., 1969; Kukarkin, B. V., Kholopov, P. N., Efremov, Yu. N., Kukarkina, N. P., Kurochkin, N. E., Medvedeva, G. I., Perova, N. B., Pskovsky, Yu. P., Fedorovitch, V. P., and Frolov, M. S., 1974; Kukarkin, B. V., Kholopov, P. N., Fedorovitch, V. P., Frolov, M. S., Kukarkina, N. P., Kurochkin, N. E., Medvedeva, G. I., Perova, N. B., and Pskovsky, Yu. P., 1976.

Kutter, G. S., and Sparks, W. M. 1972, *Ap. J.,* **175,** 407 (12.2).

Lamb, D. Q. 1974, *Ap. J.*, **192,** L129 (3.2a).
Lamb, D. Q., and Sorvari, J. M. 1972, IAU Circular No. 2442 (1).
Landau, L. D., and Lifshitz, E. M. 1959, *Fluid Mechanics* (London: Pergamon Press) (4, 5.5).
Langer, G. E. 1971, *M.N.R.A.S.*, **155,** 199 (9.2a, 10.2).
Latyshev, I. N. 1969, *Astrofizika*, **5,** 331 (3.1).
Lebovitz, N. R. 1965a, *Ap. J.*, **142,** 229 (17.8).
Lebovitz, N. R. 1965b, *Ap. J.*, **142,** 1257 (17.8).
Lebovitz, N. R. 1966, *Ap. J.*, **146,** 946 (17.8).
Lebovitz, N. R. 1967, *Ann. Rev. Astron. Astrophys.*, **5,** 465 (19.1a).
Lebovitz, N. R. 1970, *Ap. J.*, **160,** 701 (19.1a).
Lebovitz, N. R., Reid, W. H., and Vandervoort, P. O. 1978, eds., *Theoretical Principles in Astrophysics and Relativity* (Chicago and London: University of Chicago Press).
Lebovitz, N. R., *see* Chandrasekhar, S., and Lebovitz, N. R., 1964.
Ledoux, P. 1945, *Ap. J.*, **102,** 143 (19.1a).
Ledoux, P. 1951, *Ap. J.*, **114,** 373 (19.1c).
Ledoux, P. 1955, *Ann. d'Ap.*, **18,** 232 (8.12c).
Ledoux, P. 1958, *Handb. der Physik*, ed. S. Flügge (Berlin: Springer-Verlag), **51,** 605 (9.4, 19.4).
Ledoux, P. 1963, in *Star Evolution*, ed. L. Gratton (N. Y.: Academic Press), p. 394 (1, 9.3, 13.1, 19.4).
Ledoux, P. 1965, in *Stellar Interiors*, eds., L. H. Aller and D. B. McLaughlin (Chicago: University of Chicago Press), p. 499 (1, 19.2).
Ledoux, P. 1974, in *Stellar Instability and Evolution* (IAU Symp. No. 59), eds., P. Ledoux, A. Noels, and A. W. Rodgers (Dordrecht: Reidel), p. 135 (1, 17, 17.2, 17.8, 17.9, 18.2, 18.2a).
Ledoux, P. 1978, in *Theoretical Principles in Astrophysics and Relativity*, eds., N. R. Lebovitz, W. H. Reid, and P. O. Vandervoort (Chicago and London: University of Chicago Press) (1, 17, 17.9, 17.10).
Ledoux, P., Noels, A., and Rodgers, A. W., eds. 1974, *Stellar Instability and Evolution* (IAU Symp. No. 59) (Dordrecht: Reidel) (1).
Ledoux, P., and Pekeris, C. L. 1941, *Ap. J.*, **94,** 124 (8.15).
Ledoux, P., and Renson, R. 1966, *Ann. Rev. Astron. Astrophys.*, **4,** 293 (1, 3, 19.2).
Ledoux, P., Simon, R., and Bierlaire, J. 1955, *Ann. Astrophys.*, **18,** 65 (8.15, 10.1, 10.2).
Ledoux, P., and Smeyers, P. 1966, *C. R. Acad. Sci. Paris*, **262,** 841 (17.8, 17.9).
Ledoux, P., and Walraven, Th. 1958, *Handbuch der Physik*, ed. S. Flügge (Berlin: Springer-Verlag) **51,** 353 (1, 2.1, 3, 4, 4.2a, 4.2b, 4.2c, 5.1,

7.6, 8.4, 8.5, 8.9, 8.10, 8.11, 8.12c, 8.15, 9.2b, 12.1, 12.3, 16.2, 17.1, 17.5, 17.6b, 17.7, 17.8, 17.9, 17.11, 18.3, 19, 19.1a, 19.1b, 19.1c, 19.2).

Ledoux, P., and Whitney, C. A. 1961, in *Aerodynamic Phenomena in Stellar Atmospheres* (*Nuovo Cim.*, Suppl., **22,** 1) (1, 9.2a).

Leibacher, J., *see* Stein, R. F., and Leibacher, J., 1974.

Leighton, R. B., Noyes, R. W., and Simon, G. W. 1962, *Ap. J.*, **135,** 474 (18.3).

Lesh, J. R., *see* Aizenman, M. L., Cox, J. P., and Lesh, J. R., 1975; Fischel, D., Lesh, J. R., and Sparks, W. M., 1978.

Lewin, W. H. G. 1977, *Ann. N. Y. Acad. Sci.*, **302,** 210 (1).

Lewin, W. H. G., Hoffman, J. H., Doty, J., Clark, G. W., Swank, J. H., Becker, R. H., Pravdo, S. H., and Serlemitsos, P. J. 1977, *Nature*, **267,** 28 (1).

Lewin, W. H. G., and van Paradijs, J. 1979, *Sky and Telescope*, **57,** 446 (May, 1979) (1).

Lifshitz, E. M., *see* Landau, L. D., and Lifshitz, E. M., 1959.

Lighthill, J. 1978, *Waves in Fluids* (London: Cambridge University Press) (1).

Liller, W., *see* Forman, W., Jones, C. A., and Liller, W., 1972.

Linsky, J. L., *see* Ayres, T. R., Linsky, J. L., and Shine, R. A., 1975.

Love, A. E. H. 1889, *Phil. Mag.*, Ser. 5, **27,** 254 (19.1a).

Lovell, A. B. C. 1971, *Quart. J.R.A.S.*, **12,** 98 (3).

Lucy, L. 1976, *Ap. J.*, **206,** 499 (18.3, 19.7).

Luyten, W. J. 1971, ed., *White Dwarfs* (IAU Symp. No. 42) (Dordrecht: Reidel).

Lynden-Bell, D., and Ostriker, J. P. 1967, *M.N.R.A.S.*, **136,** 293 (5.3, 8.10, 9.3, 15.2, 15.3, 19.1a, 19.1b).

Maeder, A., and Rufener, F. 1972, *Astron. Astrophys.*, **20,** 437 (18.3).

Main, I. G. 1978, *Vibrations and Waves in Physics* (London: Cambridge University Press) (1).

Makarenko, E. N. 1972, *Astron. Cir.*, No. 673 (U.S.S.R. Acad. of Sciences) (reported in *Sky and Telescope*, **44,** 255) (3.1).

Manchester, R. N., *see* Taylor, J. H., and Manchester, R. N., 1977.

Markey, P., and Tayler, R. J. 1973, *M.N.R.A.S.*, **163,** 77 (19.2).

Markey, P., and Tayler, R. J. 1974, *M.N.R.A.S.*, **168,** 505 (19.2).

McCall, M. L., *see* Smith, M. A., and McCall, M. L., 1978.

McConnell, A. J. 1931, *Applications of the Absolute Differential Calculus* (London: Blackie and Sons Ltd.) (4.1).

McGraw, J. T. 1977, Ph.D. Dissertation, University of Texas (1, 3, 3.2a, 18.3, 19.7).

McGraw, J. T., and Robinson, E. L. 1975, *Ap. J.,* **200,** L89 (3.2a).

McGraw, J. T., and Robinson, E. L. 1976, *Ap. J.,* **205,** L155 (3.2a, 18.3).

McGraw, J. T., *see* Robinson, E. L., and McGraw, J. T., 1976a,b; Robinson, E. L., Nather, R. E., and McGraw, J. T., 1976.

McLaughlin, D. B., *see* Aller, L. H., and McLaughlin, D. B., 1965.

McNamara, D. H., and Feltz, K. A., Jr. 1978, *Publ. A.S.P.,* **90,** 275 (3).

Medvedeva, G. I., *see* Kukarkin, B. V., Kholopov, P. N., Efremov, Yu. N., Kukarkina, N. P., Kurochkin, N. E., Medvedeva, G. I., Perova, N. B., Fedorovitch, V. P., and Frolov, M. S., 1974; Kukarkin, B. V., Kholopov, P. N., Efremov, Yu. N., Kukarkina, N. P., Kurochkin, N. E., Medvedeva, G. I., Perova, N. B., Pskovsky, Yu. P., Fedorovitch, V. P., and Frolov, M. S., 1974; Kukarkin, B. V., Kholopov, P. N., Fedorovitch, V. P., Frolov, M. S., Kukarkina, N. P., Kurochkin, N. E., Medvedeva, G. I., Perova, N. B., and Pskovsky, Yu. P., 1976.

Meltzer, D. W., and Thorne, K. S. 1966, *Ap. J.,* **145,** 514 (8.5).

Meltzer, D. W., *see* Bardeen, J. M., Thorne, K. S., and Meltzer, D. W., 1966.

Melvin, P. J. 1975, unpublished manuscript (13.4).

Melvin, P. J. 1977, *SIAM J. Appl. Math.,* **33,** 161 (12.3).

Mestel, L. 1965, in *Stellar Interiors,* eds., L. H. Aller and D. B. McLaughlin (Chicago: University of Chicago Press) (19.1a).

Mestel, L. 1978, in *Theoretical Principles in Astrophysics and Relativity,* eds., N. R. Lebovitz, W. H. Reid, and P. O. Vandervoort (Chicago: University of Chicago Press) (19.2).

Middleditch, *see* Davidson, A., Henry, J. P., Middleditch, J., and Smith, H. E., 1972.

Mihalas, D. 1978, *Stellar Atmospheres,* 2nd ed. (San Francisco: W. H. Freeman) (3.4, 9.2).

Mihalas, D., Kunasz, P. B., and Hummer, D. G. 1975, *Ap. J.,* **203,** 265 (3.4).

Milne-Thomson, L. M. 1960, *Theoretical Hydrodynamics,* 4th ed. (N. Y.: MacMillan) (4, 4.2a, 4.2b).

Misner, C. W., Thorne, K. S., and Wheeler, J. A. 1973, *Gravitation* (San Francisco: W. H. Freeman) (1, 4, 19.5).

Mitropolsky, Yu. A. 1964, *Problems of the Asymptotic Theory of Non-Stationary Vibrations* (trans. from Russian; Jerusalem; Israel Program for Scientific Translations) (19.4).

Mitropolsky, Yu. A., *see* Bogoluibov, N. N., and Mitropolsky, Yu. A., 1961.

Mitton, S., *see* Eggleton, P., Mitton, S., and Whelan, J., 1976.

Mohan, C. 1972, *Publ. Astron. Soc. Japan,* **24,** 133 (12.3).

Monaghan, J. J. 1965, *M.N.R.A.S.*, **131,** 105 (19.2).
Monaghan, J. J. 1966a, *M.N.R.A.S,* **132,** 1 (19.2).
Monaghan, J. J. 1966b, *M.N.R.A.S,* **134,** 275 (19.2).
Monaghan, J. J. 1968a, *Z. Astrophys.*, **68,** 461 (19.2).
Monaghan, J. J. 1968b, *Z. Astrophys.*, **69,** 146 (19.2).
Monaghan, J. J. 1968c, *Z. Astrophys.*, **69,** 154 (19.2).
Monaghan, J. J., and Robson, K. W. 1971, *M.N.R.A.S.*, **155,** 231 (19.2).
Mook, D. E., *see* Hiltner, W. A., and Mook, D. E., 1970.
Moore, D. W., and Spiegel, E. A. 1966, *Ap. J.*, **143,** 871 (13, 18.3).
Morse, P. M. 1936, *Vibration and Sound* (N. Y.: McGraw-Hill) (1).
Morse, P. M., and Feshbach, H. 1953, *Methods of Theoretical Physics* (N. Y.: McGraw-Hill) (4.2b, 6.1).
Morton, K. W., *see* Richtmyer, R. D., and Morton, K. W., 1967.
Mumford, G. S. 1967, *Publ. A.S.P.*, **79,** 283 (1).
Murial, A., *see* Chiu, H.-Y., and Murial, A., 1973.
Murphy, J. O. 1968, *Australian J. Phys.* **21,** 465 (12.3).
Murphy, J. O., and Smith, A. C. 1970, *Proc. Astron. Soc. Australia,* **1,** 328 (12.3).
Murphy, J. O., *see* van der Borght, R., and Murphy, J. O., 1966.
Musman, S., and Rust, D. M. 1970, *Solar Phys.*, **13,** 261 (18.3).
Mutschlecner, J. P., *see* Keller, C. F., and Mutschlecner, J. P., 1971.

Nather, R. E. 1978, *Publ. A.S.P.*, **90,** 477 (1, 3, 3.1, 3.2a, 18.3).
Nather, R. E., *see* Patterson, J., Robinson, E. L., and Nather, R. E., 1977; Robinson, E. L., Nather, R. E., and McGraw, J. T., 1976.
Nelson, G. D., *see* Böhm-Vitense, E., and Nelson, G. D., 1976.
Newing, R. A., *see* Cowling, T. G., and Newing, R. A., 1949.
Ni, W. T. 1973, *Ap. J.*, **181,** 939 (19.5).
Nikolenko, L. D., *see* Feschenko, S. F., Shkil', N. I., and Nikolenko, L. D., 1967.
Nikolov, N., and Tsvetko, T. 1972, *Astrophys. Sp. Sci.*, **16,** 445 (3.1).
Noels, A., Boury, A., Scuflaire, R., and Gabriel, M. 1974, *Astron. Astrophys.*, **31,** 185 (7.6, 17.10).
Noels, A., *see* Ledoux, P., Noels, A., and Rodgers, A. W., 1974.
Nordlund, A. 1974, *Astron. Astrophys.*, **32,** 407 (19.3).
Nordlund, A. 1976, *Astron. Astrophys.*, **50,** 23 (19.3).
Novikov, I. D., *see* Zeldovich, Ya. B., and Novikov, I. D., 1971.
Noyes, R. W., *see* Leighton, R. B., Noyes, R. W., and Simon, G. W., 1962.

Okamoto, I. 1967, *Publ. Astron. Soc. Japan,* **19,** 384 (19.4).
Okamoto, I., and Unno, W. 1967, *Publ. Astron. Soc. Japan,* **19,** 154 (13).
Oke, J. B. 1961a, *Ap. J.*, **133,** 90 (3.3).

Oke, J. B. 1961b, *Ap. J.*, **134,** 214 (3.3).
Oke, J. B., and Searle, L. 1974, *Ann. Rev. Astron. Astrophys.*, **12,** 315 (1).
Olsen, K. H., *see* Cox, J. P., Cox, A. N., and Olsen, K. H., 1963; Cox, J. P., Cox, A. N., Olsen, K. H., King, D. S., and Eilers, D. D., 1966.
Öpik, E. J. 1951, *M.N.R.A.S.*, **111,** 278 (19.1a).
Osaki, Y. 1966, *Publ. Astron. Soc. Japan,* **18,** 7 (19.1a).
Osaki, Y. 1971, *Publ. Astron. Soc. Japan,* **23,** 405 (3.4).
Osaki, Y. 1972, *Publ. Astron. Soc. Japan,* **24,** 509 (19.1a).
Osaki, Y. 1975, *Publ. Astron. Soc. Japan,* **27,** 237 (17.9, 17.10, 17.11, 17.14, 18.2, 18.2a).
Osaki, Y. 1976, *Publ. Astron. Soc. Japan,* **28,** 105 (17.10, 17.14).
Osaki, Y. 1977, *Publ. Astron. Soc. Japan,* **29,** 235 (17.10, 18.3, 19.7).
Osaki, Y., and Hansen, C. J. 1973a, *Ap. J.*, **185,** 277 (3.2a, 17.5, 19.5).
Osaki, Y., and Hansen, C. J., 1973b, *Astron. Astrophys.*, **23,** 475 (5.1).
Osaki, Y., *see* Ando, H., and Osaki, Y., 1975; 1977; Shibahashi, H., and Osaki, Y., 1976a,b; Unno, W., Osaki, Y., Ando, H., and Shibahashi, H., 1979.
Ostriker, J. P. 1977, *Annals. N. Y. Acad. Sci.*, **302,** 229 (1).
Ostriker, J. P. 1978, in *Theoretical Principles in Astrophysics and Relativity,* eds., N. R. Lebovitz, W. H. Reid, and P. O. Vandervoort (Chicago and London: University of Chicago Press) (19.1).
Ostriker, J. P., *see* Herbst, W., Hesser, J. E., and Ostriker, J. P., 1974; Lynden-Bell, D., and Ostriker, J. P., 1967.
Owen, J. W. 1957, *M.N.R.A.S.*, **117,** 384 (17.9).

Papagionnis, M. D. 1977, ed., *Eighth Texas Symposium on Relativistic Astrophysics* (N. Y.: New York Acad. Sci.).
Papaloizou, J. C. B. 1973a, *M.N.R.A.S.*, **162,** 143 (12.3).
Papaloizou, J. C. B. 1973b, *M.N.R.A.S.*, **162,** 169 (12.2).
Papaloizou, J. C. B., and Pringle, J. E. 1978, *M.N.R.A.S.*, **182,** 423 (17.3, 19.1a).
Parenago, P. P., *see* Kukarkin, B. V., and Parenago, P. P., 1963.
Parsons, S. B. 1969, *Ap. J. Suppl.*, **18,** 127 (19.3).
Parsons, S. B. 1971, *Ap. J.*, **164,** 355 (3.3, 3.4).
Parsons, S. B. 1972, *Ap. J.*, **174,** 57, 192 (3.3, 3.4).
Parsons, S. B., and Bouw, G. D. 1971, *M.N.R.A.S.*, **153,** 133 (3.3).
Patterson, J., Robinson, E. L., and Nather, R. E. 1977, *Ap. J.*, **214,** 144 (3, 3.2a, 19.7).
Payne-Gaposchkin, C. 1951, in *Astrophysics: A Topical Symposium,* ed. J. A. Hynek (N. Y.: McGraw-Hill), p. 495 (1, 3, 3.1).
Payne-Gaposchkin, C. 1954, *Variable Stars and Galactic Structure* (London: Athlone Press) (1, 3).

Payne-Gaposchkin, C. 1957, *The Galactic Novae* (Amsterdam: North Holland) (1).

Payne-Gaposchkin, C. 1961, *Vistas in Astron.*, **8,** 184 (3.1).

Payne-Gaposchkin, C., and Gaposchkin, S. 1963, in *Basic Astronomical Data,* ed. K. A. Strand (Chicago: University of Chicago Press) (3, 3.1).

Payne-Gaposchkin, C., and Gaposchkin, S. 1965, *Smithsonian Astrophys. Obs. Rep.,* No. 9 (3.1, 13.3).

Payne-Gaposchkin, C., and Gaposchkin, S. 1966, *Vistas in Astron.,* **8,** 191 (3.1).

Payne-Gaposchkin, C., and Haramundanis, K. 1970, *Introduction to Astronomy,* 2nd ed. (Englewood Cliffs: Prentice-Hall) (3.1).

Payne-Gaposchkin, C., *see* Baade, W., 1963.

Pekeris, C. L. 1938, *Ap. J.,* **88,** 189 (17.7, 17.8).

Pekeris, C. L., *see* Ledoux, P., and Pekeris, C. L., 1941.

Pel, J. W. 1978, *Astron. Astrophys.,* **62,** 75 (3, 3.1).

Penrose, R. 1972, *Sci. Am.,* **226,** No. 5, p. 38 (1).

Percy, J. R. 1975, *Sci. Am.,* **232,** No. 6, p. 66 (1).

Perdang, J. 1968, *Astrophys. Sp. Sci.,* **1,** 355 (17.3).

Perdang, J., *see* Aizenman, M. L., and Perdang, J., 1971; 1972.

Perkins, F., *see* Axel, L., and Perkins, F., 1971.

Perova, N. B., *see* Kukarkin, B. V., Kholopov, P. N., Efremov, Yu. N., Kukarkin, N. P., Kuorchkin, N. E., Medvedeva, G. I., Perova, N. B., Fedorovitch, V. P., and Frolov, M. S., 1969; Kukarkin, B. V., Kholopov, P. N., Efremov, Yu. N., Kukarkina, N. P., Kurochkin, N. E., Medvedeva, G., Perova, N. B., Pskovsky, Yu. P., Fedorovitch, V. P., and Frolov, M. S., 1974; Kukarkin, B. V., Kholopov, P. N., Fedorovitch, V. P., Frolov, M. S., Kukarkina, N. P., Kurochkin, N. E., Medvedeva, G. I., Perova, N. B., and Pskovsky, Yu. P., 1976.

Perry, J. J., Burbidge, E. M., and Burbidge, G. R. 1978, *Publ. A.S.P.,* **90,** 337 (1).

Petersen, J. O. 1973, *Astron. Astrophys.,* **27,** 89 (3.2b).

Petersen, J. O. 1974, *Astron. Astrophys.,* **34,** 309 (3.2b).

Petersen, J. O. 1975, *Mem. Soc. R. Sci. Liège,* **8,** 299 (8.13).

Petersen, J. O. 1976, in *Multiple Periodic Stars,* ed. W. S. Fitch (Dordrecht: Reidel) (3, 3.1).

Petersen, J. O. 1978, *Astron. Astrophys.,* **62,** 205 (3.2b).

Philip, A. G. D. 1972, ed., *The Evolution of Population II Stars* (Dudley Observatory Report, No. 4) (1).

Phillips, H. B. 1933, *Vector Analysis* (N. Y.: John Wiley & Sons, Inc.) (4.2b, 4.2c, 17.3).

Pickering, E. C. 1912, *Harvard Circ.,* no. 173 (3.1).

Pollack, J., and Hansen, C. J. 1970, unpublished calculations (8.12c).

Pravdo, S. H., *see* Lewin, W. H. G., Hoffman, J. H., Doty, J., Clark, G. W., Swank, J. H., Becker, R. H., Pravdo, S. H., and Serlemitsos, P. J., 1977.

Press, W. H., *see* Eardley, D. M., and Press, W. H., 1975.

Price, R., and Thorne, K. S. 1969, *Ap. J.,* **155,** 163 (19.5).

Pringle, J. E., *see* Bath, G. T., Evans, W. D., and Pringle, J. E., 1974; Papaloizou, J. C. B., and Pringle, J. E., 1978.

Pskovsky, Yu. P., *see* Kukarkin, B. V., Kholopov, P. N., Efremov, Yu. N., Kukarkina, N. P., Kurochkin, N. E., Medvedeva, G., Perova, N., Pskovsky, Yu. P., Fedorovitch, V. P., and Frolov, M. S., 1974; Kukarkin, B. V., Kholopov, P. N., Fedorovitch, V. P., Frolov, M. S., Kukarkina, N. P., Kurochkin, N. E., Medvedeva, G. I., Perova, N. B., and Pskovsky, Yu. P., 1976.

Rabinowitz, I. 1957, *Ap. J.,* **126,** 386 (10.1, 10.2).

Rahe, J., *see* Kippenhahn, R., Rahe, J., and Strohmeier, W., 1977.

Ralston, A. 1965, *A First Course in Numerical Analysis* (N. Y.: McGraw-Hill) (8.12b).

Randers, G., *see* Rosseland, S., and Randers, G., 1938.

Reid, W. H., *see* Lebovitz, N. R., Reid, W. H., and Vandervoort, P. O., 1978.

Renson, R., *see* Ledoux, P., and Renson, R., 1966.

Rhodes, E. J., Ulrich, R. K., and Simon, G. W. 1977, *Ap. J.,* **218,** 901 (18.3).

Richer, H. B., and Ulrych, T. J. 1974, *Ap. J.,* **192,** 719 (3.2a).

Richtmyer, R. D., and Morton, K. W. 1967, *Difference Methods for Initial-Value Problems,* 2nd ed. (N. Y.: Interscience) (12.2).

Ricort, G., *see* Fossat, E., Ricort, G., Aime, C., and Roddier, F., 1974.

Robe, H. 1965, *Bull. Acad. Roy. Belge, Cl. 5c, 5^{e} series,* **51,** 595 (17.7).

Robe, H. 1968, *Ann. Astrophys.,* **31,** 475 (17.8, 17.9, 17.10).

Roberts, P. H., *see* Hurley, M., Roberts, P. H., and Wright, K., 1966.

Robinson, E. L. 1976, *Ann. Rev. Astron. Astrophys.,* **14,** 119 (1, 3, 19.7).

Robinson, E. L., and McGraw, J. T. 1976a, *Ap. J.,* **207,** L37 (1, 3.2a, 19.7).

Robinson, E. L., and McGraw, J. T. 1976b, in *Proc. Los Alamos Solar and Stellar Pulsation Conf.,* eds., A. N. Cox and R. G. Deupree, p. 98 (1, 3.2a, 19.7).

Robinson, E. L., Nather, R. E., and McGraw, J. T. 1976, *Ap. J.,* **210,** 211 (1, 3.2a, 19.7).

Robinson, E. L., *see* McGraw, J. T., and Robinson, E. L., 1975; 1976; Patterson, J., Robinson, E. L., and Nather, R. E., 1977; Warner, B., and Robinson, E. L., 1972.

Robson, K. W., *see* Monaghan, J. J., and Robson, K. W., 1971.

Roddier, F., *see* Fossat, E., Ricort, G., Aime, C., and Roddier, F., 1974.
Rodgers, A. W. 1957, *M.N.R.A.S.*, **117,** 85 (3.3).
Rodgers, A. W., and Gingold, A. R. 1973, *M.N.R.A.S.*, **161,** 23 (3.2b).
Rodgers, A. W., *see* Ledoux, P., Noels, A., and Rodgers, A. W., 1974.
Rojansky, V. 1938, *Introductory Quantum Mechanics* (N. Y.: Prentice-Hall) (17.3).
Rose, W. 1973, *Astrophysics* (N. Y.: Rinehart and Winston) (3).
Rosenberg, R. M., *see* Rudd, T. J., and Rosenberg, R. M., 1970.
Rosenbluth, M. N., and Bahcall, J. N. 1973, *Ap. J.*, **184,** 9 (19.6).
Rosendahl, J. D., and Snowden, M. S. 1971, *Ap. J.*, **169,** 281 (18.3).
Rosenwald, R. D., *see* Hill, H. A., Caudell, T. P., and Rosenwald, R. D., 1977; Hill, H. A., Rosenwald, R. D., and Caudell, T. P., 1977.
Ross, R. R., *see* Aizenman, M. L., Hansen, C. J., and Ross, R. R., 1975; Hansen, C. J., Aizenman, M. L., and Ross, R. R., 1976; King, D. S., Hansen, C. L., Ross, R. R., and Cox, J. P., 1975.
Rosseland, S. 1949, *The Pulsation Theory of Variable Stars* (Oxford: Clarendon Press; also available as a Dover paperback, 1964) (1, 4, 5.2, 8.12a, 8.15, 9.2b, 11.1, 12.3, 19.4).
Rosseland, S., and Randers, G. 1938, *Astrophys. Norvegica,* **3,** 71 (7.6).
Roxburgh, I. W. 1963, *M.N.R.A.S.*, **126,** 67 (19.2).
Roxburgh, I. W. 1966, *M.N.R.A.S,* **132,** 207 (19.2).
Roxburgh, I. W. 1970, in *Stellar Rotation,* ed. A. Slettebak (Dordrecht: Reidel) (19.1a).
Rudd, T. J., and Rosenberg, R. M. 1970, *Astron. Astrophys.*, **6,** 193 (13, 13.2, 13.4).
Ruderman, M. 1972, *Ann. Rev. Astron. Astrophys.*, **10,** 427 (1, 3).
Ruderman, M. 1975, *Ann. N. Y. Acad. Sci.*, **262,** 159 (1).
Rufener, F., *see* Maeder, A., and Rufener, F., 1972.
Ruffini, R., and Wheeler, J. A. 1971, *Phys. Today,* **24,** No. 1, p. 30 (1).
Rust, D. M., *see* Musman, S., and Rust, D. M., 1970.

Saio, H., and Cox, J. P. 1979a, in *Proceedings of the Workshop on Nonradial and Nonlinear Stellar Instabilities,* eds., H. Hill and W. Dziembowski (N. Y.: Springer-Verlag) (10, 19.7).
Saio, H., and Cox, J. P. 1979b, *Ap. J.*, in press. (18.1, 18.3).
Saio, H., Cox, J. P., Hansen, C. J., and Carroll, B. W. 1979, in *Proceedings of the Workshop on Nonradial and Nonlinear Stellar Instabilities,* eds., H. Hill and W. Dziembowski (N. Y.: Springer-Verlag) (19.1d).
Saio, H., Kobayashi, E., and Takeuti, M. 1977, *Sci. Rep. Tohoku University,* **51,** 144 (3.2b, 19.3).
Sandage, A. R. 1972, *Quart. J.R.A.S.*, **13,** 202 (3.1).

Sandage, A. R., and Tammann, G. A. 1968, *Ap. J.*, **151,** 531 (3.1).
Sandage, A. R., and Tammann, G. A. 1969, *Ap. J.*, **157,** 683 (3.1).
Sandage, A. R., and Tammann, G. A. 1971, *Ap. J.*, **167,** 293 (3.1).
Sandage, A. R., and Tammann, G. A. 1974, *Ap. J.*, **194,** 559 (3.1).
Sandage, A. R., and Tammann, G. A. 1976a, *Ap. J.*, **207,** L1 (3.1).
Sandage, A. R., and Tammann, G. A. 1976b, *Ap. J.*, **210,** 7 (3.1).
Sanders, J. H., and Wapstra, A. H. 1975, *Proc. Fifth International Conf. on Atomic Masses and Fundamental Constants* (N.Y.: Plenum Press) (1).
Sargent, W. L. W. 1964, *Ann. Rev. Astron. Astrophys.*, **2,** 297 (1, 19.2).
Sastri, V. K., and Simon, N. R. 1973, *Ap. J.*, **186,** 997 (19.4).
Sastri, V. K., *see* Simon, N. R., and Sastri, V. K., 1972.
Sauvenier-Goffin, E. 1951, *Bull. Soc. Roy. Sci. Liège,* **20,** 20 (17.7, 17.9).
Savedoff, M. P., *see* Van Horn, H. M., and Savedoff, M. P., 1976.
Schaltenbrand, R., and Tammann, G. A. 1970, *Astron. Astrophys.*, **7,** 289 (3.1).
Schatzman, E. 1953, *Ann. d'Ap.*, **16,** 162 (7.6).
Schatzman, E. 1956, *Ann. d'Ap.*, **19,** 45 (8.4).
Schatzman, E., *see* Thorne, K., 1967a.
Schiff, L. 1955, *Quantum Mechanics,* 2nd ed. (N. Y.: McGraw-Hill) (15.2).
Schmidt, E. G. 1971, *Ap. J.*, **165,** 335 (3.3).
Schmidt, E. G. 1974, *M.N.R.A.S.*, **167,** 613 (3.2b).
Schubert, G., *see* Goldreich, P., and Schubert, G., 1967.
Schutz, B. F. 1979, *Ap. J.*, **232,** 874 (19.1b).
Schutz, B. F., Jr. 1972, *Ap. J. Suppl.*, **24,** 319 (19.5).
Schwank, D. C. 1975, unpublished manuscript (17.8).
Schwank, D. C. 1976, *Astrophys. Sp. Sci.*, **43,** 459 (17.8, 17.13).
Schwarzschild, K. 1906, *Gesell. Wiss. Göttingen, Nachr., Math.-Phys. Klasse,* **1,** 41, (17.2, 18.2).
Schwarzschild, M. 1958, *Structure and Evolution of the Stars* (Princeton: Princeton University Press; also available as a Dover paperback, 1965) (7.5, 8.12a, 19.1a).
Schwarzschild, M. 1975, *Ap. J.*, **195,** 137 (18.3).
Schwarzschild, M., *see* Härm, R., and Schwarzschild, M., 1964; 1972; Kruskal, M., Schwarzschild, M., and Härm, R., 1974.
Scuflaire, R. 1974a, *Astron. Astrophys.*, **34,** 449 (17.8, 17.9, 17.10).
Scuflaire, R. 1974b, *Astron. Astrophys.*, **36,** 107 (17.9, 17.10, 17.11).
Scuflaire, R., *see* Noels, A., Boury, A., Scuflaire, R., and Gabriel, A., 1974.
Searle, L., *see* Oke, J. B., and Searle, L., 1974.
Sears, R. L., *see* Bahcall, J. N., and Sears, R. L., 1972.
Serkowski, K. 1970, *Ap. J.*, **160,** 1107 (18.3).

Serlemitsos, P. J., *see* Lewin, W. H. G., Hoffman, J. H., Doty, J., Clark, G. W., Swank, J. H., Becker, R. H., Pravdo, S. H., and Serlemitsos, P. J., 1977.

Shapley, H. 1914, *Ap. J.*, **40,** 448 (1).

Shawl, S. J. 1974, *Publ. A.S.P.*, **86,** 843 (18.3).

Shibahashi, H. 1979, *Publ. Astron. Soc. Japan,* **31,** 87 (Preface, III, 17, 17.9, 17.10, 17.12, 18, 18.3).

Shibahashi, H., and Osaki, Y. 1976a, *Publ. Astron. Soc. Japan,* **28,** 199 (17.10, 17.11, 17.14, 18.3).

Shibahashi, H., and Osaki, Y. 1976b, *Publ. Astron. Soc. Japan,* **28,** 533 (17.10, 17.14, 18.2, 18.3).

Shibahashi, H., *see* Unno, W., Osaki, Y., Ando, H., and Shibahashi, H., 1979.

Shine, R. A., *see* Ayres, T. R., Linsky, J. L., and Shine, R. A., 1975.

Shkil', N. I., *see* Feschenko, S. F., Shkil', N. I., and Nikolenko, L. D., 1967.

Shklovsky, I. S. 1968, *Supernovae* (N. Y. and London: Wiley-Interscience) (1).

Shortley, G. H., *see* Condon, E. U., and Shortley, G. H., 1935.

Sienkowicz, R., and Dziembowski, W. 1978, in *Nonstationary Evolution of Close Binaries,* ed., A. N. Żytkow (Warsaw: Polish Scientific Publishers) (18.3).

Simon, G. W., *see* Leighton, R. B., Noyes, R. W., and Simon, G. W., 1962; Rhodes, E. J., Ulrich, R. K., and Simon, G. W. 1977.

Simon, N. R. 1970, *Ap. J.,* **159,** 859 (12.3, 19.4).

Simon, N. R. 1971, *Ap. J.,* **164,** 331 (12.3, 19.4).

Simon, N. R. 1972a, *Astron. Astrophys.,* **21,** 45 (12.3).

Simon, N. R. 1972b, *Astron. Astrophys.,* **21,** 51 (12.3).

Simon, N. R. 1974, *Bull. Am. Astron. Soc.,* **6,** 469 (19.4).

Simon, N. R. 1976, in *Proc. Los Alamos Solar and Stellar Pulsation Conf.,* eds., A. N. Cox and R. G. Deupree, p. 173 (12.3).

Simon, N. R. 1977, *Astrophys. Sp. Sci.,* **51,** 205 (19.4).

Simon, N. R. 1979, *Astron. Astrophys.,* **75,** 140 (3.2b).

Simon, N. R., and Sastri, V. K. 1972, *Astron. Astrophys.,* **21,** 39 (12.3, 19.4).

Simon, N. R., *see* Sastri, V. K., and Simon, N. R., 1973.

Simon, R. 1969, *Astron. Astrophys.,* **2,** 390 (17.3, 19.1c).

Simon, R., *see* Ledoux, P., Simon, R., and Bierlaire, J., 1955.

Slettebak, A. 1969, ed., *Stellar Rotation* (N. Y.: Gordon and Breach) (19.1a).

Smeyers, P. 1966, *Acad. Roy. des Sci., des Lett. et des Beaux-Arts d. Belgique,* 5e Ser., **52,** 1126 (17.6a, 17.7, 17.8).

Smeyers, P. 1967, *Bull. Soc. Roy. Sci. Liège,* **36,** 357 (17.5, 17.5a, 17.8).

Smeyers, P. 1973, *Researche Astron.*, **8,** 359 (15.3).

Smeyers, P. 1976, in *Proc. Los Alamos Solar and Stellar Pulsation Conf.*, eds., A. N. Cox and R. G. Deupree, p. 140 (19.2).

Smeyers, P., and Denis, J. 1971, *Astron. Astrophys.*, **14,** 311 (19.1c).

Smeyers, P., *see* Aizenman, M. L., and Smeyers, P., 1977; Aizenman, M. L., Smeyers, P., and Weigert, A., 1977; Goossens, M., and Smeyers, P., 1974; Goossens, M., Smeyers, P., and Denis, J., 1976; Ledoux, P., and Smeyers, P., 1966.

Smith, A. C., *see* Murphy, J. O., and Smith, A. C., 1970.

Smith, E. V. P., and Jacobs, K. C. 1973, *Introductory Astronomy and Astrophysics* (Philadelphia: W. B. Saunders) (3).

Smith, F. G. 1977, *Pulsars* (Cambridge: Cambridge University Press) (1).

Smith, H. E. 1978, *Mercury,* **7,** 27 (Mar.-Apr. 1978) (1).

Smith, H. E., *see* Davidson, A., Henry, J. P., Middleditch, J., and Smith, H. E., 1972.

Smith, M. A. 1977, *Ap. J.*, **215,** 574 (1, 3.2c, 3.4, 19.7).

Smith, M. A. 1978, *Ap. J.*, **224,** 927 (1, 3.2c, 3.4).

Smith, M. A. 1979a, in *Proceedings of the Workshop on Nonradial and Nonlinear Stellar Instabilities,* eds., H. Hill and W. Dziembowski (N. Y.: Springer-Verlag) (1, 3.2c, 19.7).

Smith, M. A. 1979b, preprint (1, 3.2c).

Smith, M. A., Africano, S., and Worden, S. P. 1979, preprint (3.2c, 3.4).

Smith, M. A., and Buta, R. 1979, *Ap. J.*, **232,** L193 (3.2c, 8.4, 19.7).

Smith, M. A., and McCall, M. L. 1978, *Ap. J.*, **223,** 221 (1, 3.2c, 3.4, 19.7).

Smith, M. A., *see* Buta, R., and Smith, M. A., 1979.

Snowden, M. S., *see* Rosendahl, J. D., and Snowden, M. S., 1971.

Sobouti, Y. 1977, *Astron. Astrophys.*, **55,** 327 (17.8, 17.12).

Sorvari, J. M., *see* Lamb, D. Q., and Sorvari, J. M., 1972.

Sparks, W. M., *see* Fischel, D., and Sparks, W. M., 1975; Fischel, D., Lesh, J. R., and Sparks, W. M., 1978; Kutter, G. S., and Sparks, W. M., 1972.

Spiegel, E. A. 1963, *Ap. J.*, **138,** 216 (19.3).

Spiegel, E. A. 1971, *Ann. Rev. Astron. Astrophys.*, **9,** 323 (19.3).

Spiegel, E. A. 1972, *Ann. Rev. Astron. Astrophys.*, **10,** 261 (19.3).

Spiegel, E. A., and Veronis, G. 1960, *Ap. J.*, **131,** 442 (18.2).

Spiegel, E. A., *see* Moore, D. W., and Spiegel, E. A., 1966; Unno, W., and Spiegel, E. A., 1966.

Spillar, E. J., *see* Stiening, R. F., Hildebrand, E. H., and Spillar, E. J., 1979.

Stebbins, R. T., *see* Brown, T. M., Stebbins, R. T., and Hill, H. A., 1976; 1978; Hill, H. A., Stebbins, R. T., and Brown, T. M., 1975.

Stein, R. F., and Cameron, A. G. W. 1966, eds., *Stellar Evolution* (N. Y.: Plenum Press).

Stein, R. F., and Leibacher, J. 1974, *Ann. Rev. Astron. Astrophys.*, **12**, 407 (18.3).

Steinmetz, K. O., *see* Hillebrandt, W., and Steinmetz, K. O., 1976.

Stellingwerf, R. F. 1972, *Astron. Astrophys.*, **21**, 91 (13, 13.3).

Stellingwerf, R. F. 1974a, Ph.D. Dissertation, University of Colorado (5.1, 12.2, 13.4).

Stellingwerf, R. F. 1974b, *Ap. J.*, **192**, 139 (12.2, 13.4).

Stellingwerf, R. F. 1975, *Ap. J.*, **195**, 441 (12.2).

Stellingwerf, R. F. 1978, *A. J.*, **83**, 1184 (10, 10.1, 13.3, 19.7).

Stellingwerf, R. F. 1979, *Ap. J.*, **227**, 935 (10, 10.1, 13.1, 13.3).

Stellingwerf, R. F., *see* Cox, J. P., King, D. S., and Stellingwerf, R. F., 1972; Cox, J. P., and Stellingwerf, R. F., 1979.

Stewart, J. N., *see* Cox, A. N., and Stewart, J. N., 1970.

Stibbs, D. W. N., *see* Woolley, R. v. d. R., and Stibbs, D. W. N., 1953.

Stiening, R. F., Hildebrand, R. H., and Spillar, E. J. 1979, preprint (3, 3.2a, 19.7).

Stobie, R. S. 1969a, *M.N.R.A.S.*, **144**, 461 (12.2).

Stobie, R. S. 1969b, *M.N.R.A.S.*, **144**, 485 (12.2).

Stobie, R. S. 1969c, *M.N.R.A.S.*, **144**, 511 (12.2).

Stobie, R. S. 1970, *Observatory*, **90**, 20 (3.2b).

Stobie, R. S. 1972, *M.N.R.A.S.*, **157**, 167 (3.2b).

Stobie, R. S. 1977, *M.N.R.A.S.*, **189**, 631 (3.2b).

Stobie, R. S., and Hawardin, T. 1972, *M.N.R.A.S.*, **157**, 157 (3.2b).

Stobie, R. S., *see* Balona, J. M., and Stobie, R. S., 1979.

Stothers, R., and Frogel, J. A. 1967, *Ap. J.*, **148**, 305 (8.11, 8.12c).

Strittmatter, P. A. 1969, *Ann. Rev. Astron. Astrophys.*, **7**, 665 (19.1a).

Strohmeier, W., and Knigge, R. 1972, *Die Dr. Remeis-Sternwarte Bamberg und die "Veränderliche Sterne" als die Objekte ihrer Forschung* (Bamberg: Remeis-Sternwarte) (1, 3).

Strohmeier, W., *see* Kippenhahn, R., Rahe, J., and Strohmeier, W., 1977.

Strong, I., Klebesadel, R. W., and Evans, W. D. 1975, *Annals N. Y. Acad. Sci.*, **262**, 145 (1).

Strong, I. B., *see* Klebesadel, R. W., and Strong, I. B., 1976.

Swank, J. H., *see* Lewin, W. H. G., Hoffman, J. H., Doty, J., Clark, G. W., Swank, J. H., Becker, R. H., Pravdo, S. H., and Serlemitsos, P. J., 1977.

Sweet, P. A. 1950, *M.N.R.A.S.*, **110**, 548 (19.1a).

Swihart, T. L. 1968, *Astrophysics and Stellar Astronomy* (N. Y.: Wiley) (3).

Szeidl, B., *see* Fitch, W. S., and Szeidl, B., 1976.

Tabor, J. E., *see* Cox, A. N., King, D. S., and Tabor, J. E., 1973; Cox, A. N., and Tabor, J. E., 1976.

Tammann, G. A., *see* Sandage, A. R., and Tammann, G. A., 1968; 1969; 1971; 1974; 1976a,b; Schaltenbrand, R., and Tammann, G. A., 1970.

Takeuti, M., *see* Saio, H., Kobayashi, E., and Takeuti, M., 1977.

Tassoul, J.-L. 1979, *The Theory of Rotating Stars* (Princeton: Princeton University Press) (19.1).

Tassoul, J.-L., *see* Dedic, H., and Tassoul, J.-L., 1974.

Tayler, R. J. 1973a, *M.N.R.A.S.*, **161,** 365 (19.2).

Tayler, R. J. 1973b, *M.N.R.A.S.*, **162,** 17 (19.2).

Tayler, R. J., *see* Markey, P., and Tayler, R. J., 1973; 1974.

Taylor, D. J., *see* Cocke, W. J., Disney, M. J., and Taylor, D. J., 1969.

Taylor, J. H., and Huguenin, G. R. 1971, *Ap. J.*, **167,** 273 (2.1).

Taylor, J. H., and Manchester, R. N. 1971, *Pulsars* (San Francisco: W. H. Freeman and Company) (1).

Temple, G., and Bickley, W. G. 1956, *Rayleigh's Principle and Its Applications to Engineering* (N. Y.: Dover) (8.11).

Thomas, L. H. 1931, *M.N.R.A.S.*, **91,** 619 (19.4).

Thomson, W. 1863, *Phil. Trans. Roy. Soc. London,* **153,** 603 (17.7).

Thorne, K. S. 1967a, *High-Energy Astrophysics,* **3,** eds., C. DeWitt, E. Schatzman, and P. Veron (Les Houches Summer School of Theoretical Physics, 1966) (N. Y.: Gordon and Breach) (1, 4.2a, 19.5).

Thorne, K. S. 1967b, *Ap. J.*, **149,** 591 (19.5).

Thorne, K. S. 1969a, *Ap. J.*, **158,** 1 (19.5).

Thorne, K. S. 1969b, *Ap. J.*, **158,** 997 (19.5).

Thorne, K. S. 1977, *Ap. J.*, **212,** 825 (19.5).

Thorne, K. S. 1978, in *Theoretical Principles in Astrophysics and Relativity,* eds., N. R. Lebovitz, W. H. Reid, and P. O. Vandervoort (Chicago and London: University of Chicago Press) (19.5).

Thorne, K. S., and Żytkow, A. N. 1977, *Ap. J.*, **212,** 832 (19.5).

Thorne, K. S., *see* Bardeen, J. M., Thorne, K. S., and Meltzer, D. W., 1966; Campolattaro, A., and Thorne, K. S., 1970; Ipser, J. R., and Thorne, K. S., 1973; Meltzer, D. W., and Thorne, K. S., 1966; Misner, C. W., Thorne, K. S., and Wheeler, J. A., 1973; Price, R., and Thorne, K. S., 1969.

Tolstoy, I. 1963, *Rev. Mod. Phys.*, **35,** 207 (17.2, 17.9, 17.10).

Tolstoy, I. 1973, *Wave Propagation* (N. Y.: McGraw-Hill) (1, 5.4b, 17.2, 17.9, 17.10).

Tralli, N., *see* Goertzel, G., and Tralli, N., 1960.

Tsvetko, T., *see* Nikolov, N., and Tsvetko, T., 1972.

Tuggle, R. S., and Iben, I., Jr. 1973, *Ap. J.*, **186,** 593 (10.3, 19.3).

Ulrich, R. K. 1970a, *Astrophys. Sp. Sci.,* **7,** 71 (19.3).
Ulrich, R. K. 1970b, *Astrophys. Sp. Sci.,* **7,** 183 (19.3).
Ulrich, R. K. 1970c, *Astrophys. Sp. Sci.,* **9,** 80 (19.3).
Ulrich, R. K. 1972, *Ap. J.,* **172,** 165 (18.2a).
Ulrich, R. K. 1976, *Ap. J.,* **207,** 564 (19.3).
Ulrich, R. K., *see* Rhodes, E. J., Ulrich, R. K., and Simon, G. W., 1977.
Ulrych, T. J., *see* Richer, H. B., and Ulrych, T. J., 1974.
Unno, W. 1965, *Publ. Astron. Soc. Japan,* **17,** 205 (4.3, 7.5, 8.3, 8.4, 9.2a, 18.3).
Unno, W. 1967, *Publ. Astron. Soc. Japan,* **19,** 1210 (13, 19.3, 19.4).
Unno, W. 1968, *Publ. Astron. Soc. Japan,* **20,** 356 (19.4).
Unno, W. 1975, *Publ. Astron. Soc. Japan,* **27,** 81 (17.10).
Unno, W., and Kamijo, F. 1966, *Publ. Astron. Soc. Japan,* **18,** 23 (13).
Unno, W., Osaki, Y., Ando, H., and Shibahashi, H. 1979, *Nonradial Oscillations of Stars* (Tokyo: Tokyo University Press) (Preface, 1, III, 17, 17.6b, 17.10, 17.12, 17.14, 18.3).
Unno, W., and Spiegel, E. A. 1966, *Publ. Astron. Soc. Japan,* **18,** 85 (18.3).
Unno, W., *see* Kato, S., and Unno, W., 1967; Okamoto, I., and Unno, W., 1967.
Unsöld, A. 1977, *The New Cosmos,* 2nd ed. (Berlin: Springer-Verlag) (3).
Usher, P. D., and Whitney, C. A. 1968, *Ap. J.,* **154,** 203 (13, 13.2, 13.4).

van der Borght, R. 1968, *Proc. Australian Soc. Astron.,* **1,** 87 (12.3).
van der Borght, R. 1969, *Australian J. Phys.,* **22,** 497 (12.3).
van der Borght, R. 1970, *Proc. Australian Soc. Astron.,* **1,** 325 (12.3).
van der Borght, R., and Murphy, J. O. 1966, *M.N.R.A.S.,* **131,** 225 (12.3).
van Genderin, A. M. 1970, *Astron. Astrophys.,* **7,** 244 (3.1).
Van Horn, H. M. 1973, in *Physics of Dense Matter,* ed. C. J. Hansen (Dordrecht: Reidel) (19.12).
Van Horn, H. M. 1976, in *Multiple Periodic Variable Stars,* ed. W. S. Fitch (Dordrecht: Reidel) (18.3).
Van Horn, H. M. 1978, in *Current Problems in Stellar Pulsation Instabilities,* eds., D. Fischel, J. R. Lesh, and W. M. Sparks (3, 3.2a, 18.3).
Van Horn, H. M., and Savedoff, M. P. 1976, in *Proc. Los Alamos Solar Stellar Pulsation Conf.,* eds., A. N. Cox and R. G. Deupree, p. 109 (19).
Van Horn, H. M., Wesemael, F., and Winget, D. E. 1980, *Ap. J.,* **253,** L143 (19.7).

Van Horn, H. M., *see* Hansen, C. J., Cox, J. P., and Van Horn, H. M., 1977; Hansen, C. J., and Van Horn, H. M., 1979.

Vandervoort, P. O., *see* Lebovitz, N. R., Reid, W. H., and Vandervoort, P. O., 1978.

van Paradijs, J., *see* Lewin, W. H. G., and van Paradijs, J., 1979.

Vauclair, G. 1971, in *White Dwarfs* (IAU Symp. No. 42), ed. W. J. Luyten (Dordrecht: Reidel) (18.3).

Vaughan, G. J. 1972, *M.N.R.A.S.*, **159,** 375 (8.12a).

Vemury, S. K. 1978, *Ap. J.*, **221,** 258 (19.4).

Veron, P., *see* Thorne, K. S., 1967a.

Veronis, G., *see* Spiegel, E. A., and Veronis, G., 1960.

Vitense, E. 1953, *Zs. f. Ap.*, **32,** 135 (19.3).

von Sengbusch, K. 1973, "*Mitteilungen der Astronomischen Gesellschaft,*" No. 32, p. 228 (12.2).

von Sengbusch, K., *see* Baker, N., and von Sengbusch, K., 1969; 1970.

von Zeipel, H. 1924, *M.N.R.A.S.*, **84,** 665 (19.1a).

Wagoner, R. V. 1969, *Ann. Rev. Astron. Astrophys.*, **7,** 553 (1).

Walraven, Th., *see* Ledoux, P., and Walraven, Th., 1958.

Wapstra, A. H., *see* Sanders, J. H., and Wapstra, A. H., 1975.

Warner, B. 1976a, in *Structure and Evolution of Close Binary Systems* (IAU Symp. No. 73), eds., P. Eggleton, S. Mitton, and J. Whelan (Dordrecht, Boston: Reidel) (1, 3, 3.2a).

Warner, B. 1976b, in *Multiple Periodic Stars,* ed. W. S. Fitch (Dordrecht: Reidel) (3, 3.2a).

Warner, B., and Brickhill, A. J. 1974, *M.N.R.A.S.*, **164,** 673 (3.2a).

Warner, B., and Robinson, E. L. 1972, *Nature Phys. Sci.*, **239,** 2 (3, 3.2a).

Watts, J. W., Jr., *see* Fishman, G. J., Watts, J. W., Jr., and Derrickson, J. H., 1978.

Weinberg, S. 1972, *Gravitation and Cosmology* (N. Y.: Wiley) (19.5).

Weigert, A., *see* Aizenman, M. L., Smeyers, P., and Weigert, A., 1977.

Wesemael, F., *see* Van Horn, H. M. Wesemael, F., and Winget, D. E., 1979.

Wesselink, A. J. 1946, *B. A. N.*, **10,** 91 (3.3).

Wesselink, A. J. 1947, *B. A. N.*, **10,** 252 (3.3).

Wheeler, J. A., *see* Misner, C. W., Thorne, K. S., and Wheeler, J. A., 1973; Ruffini, R., and Wheeler, J. A., 1971.

Wheeler, J. C., *see* King, D. S., Wheeler, J. C., Cox, J. P., and Hodson, S. W., 1978; King, D. S., Wheeler, J. C., Cox, J. P., Cox, A. N., and Hodson, S. W., 1979.

Whelan, J., *see* Eggleton, P., Mitton, S., and Whelan, J., 1976.

White, H. E., *see* Jenkins, F. A., and White, H. E., 1957.
White, O. R. 1977, ed., *The Solar Output and Its Variation* (Boulder: Colorado Associated University Press) (1).
Whitney, C. A., *see* Cox, J. P., and Whitney, C. A., 1958; Ledoux, P., and Whitney, C. A., 1961; Usher, P. D., and Whitney, C. A., 1968.
Will, C. M. 1974, *Ap. J.,* **190,** 403 (19.5).
Willson, L. A., *see* Hill, S. J., and Willson, L. A., 1979.
Wilson, O. C., and Bappu, M. K. V. 1957, *Ap. J.,* **125,** 661 (18.3).
Winget, D. E., *see* Van Horn, H. M., Wesemael, F., and Winget, D. E., 1979.
Wolf, E., *see* Born, M., and Wolf, E., 1959.
Wolff, S. C. 1976, in *Multiple Periodic Variable Stars* (IAU Colloq. No. 29), ed. W. S. Fitch (Dordrecht: Reidel) (19.2).
Wolff, C. L. 1977, *Ap. J.,* **216,** 784 (19.1c).
Wolff, C. L. 1979, *Ap. J.,* **227,** 943 (Preface, III, 17, 17.9, 17.10, 17.11, 18, 18.2, 18.3).
Wood, P. R. 1973, Ph.D. Dissertation, Australian National University (19.3).
Wood, P. R. 1976, *M.N.R.A.S.,* **174,** 531 (9.2c).
Woolley, R. v. d. R., and Stibbs, D. W. N. 1953, *The Outer Layers of a Star* (Oxford: Clarendon Press) (3.4, 9.2a).
Worden, S. P., *see* Smith, M. A., Africano, S., and Worden, S. P., 1979.
Wright, G. A. E. 1969, *M.N.R.A.S.,* **146,** 197 (19.2).
Wright, G. A. E. 1973, *M.N.R.A.S.,* **162,** 339 (19.2).
Wright, K., *see* Hurley, M., Roberts, P. H., and Wright, K., 1966.
Wrubel, M. 1958, in *Handb. d. Phys.,* ed. S. Flügge (Berlin: Springer-Verlag) **51,** 38 (19.1a).

Zahn, J. P. 1968, *Astrophys. Lett.,* **1,** 209 (13).
Zahn, J. P. 1977, ed., *Problems of Stellar Convection* (IAU Colloq. No. 38) (Dordrecht: Reidel).
Zeldovich, Ya. B., and Novikov, I. D. 1971, *Relativistic Astrophysics,* eds., K. S. Thorne and W. D. Arnett (Chicago: University of Chicago Press) (1, 19.5).
Zheleznyakov, V. V., *see* Ginzburg, V. L., and Zheleznyakov, V. V., 1975.
Zhevakin, S. A. 1953, *Russian A. J.,* **30,** 161 (10.2).
Zhevakin, S. A. 1954a, *Russian A. J.,* **31,** 141 (10.2).
Zhevakin, S. A. 1954b, *Russian A. J.,* **31,** 335 (10.2).
Zhevakin, S. A. 1963, *Ann. Rev. Astron. and Astrophys.,* **1,** 367 (1).
Ziebarth, K. 1970a, Ph.D. Dissertation, University of Colorado (9.2b).
Ziebarth, K. 1970b, *Ap. J.,* **162,** 947 (9.2b).
Żytkow, A. N., 1978, ed., *Nonstationary Evolution of Close Binaries* (Warsaw: Polish Scientific Publishers) (18.3).
Żytkow, A. N., *see* Thorne, K. S., and Żytkow, A. N., 1977.

List of Symbols

Below is a list of symbols that have been used extensively throughout this book. The list is not necessarily complete or comprehensive. Generally, each symbol in the list has been used more than once; however, in some cases its importance has been the deciding factor, even though the symbols may only have been used once. The number after each symbol gives the section in which the symbol was first used or defined. If two numbers appear, the second one gives the section in which the symbol is especially heavily used.

English Symbols

Symbol	Description	Section
$\mathbf{a} = (a_1, a_2, a_3)$	identification parameters in the Lagrangian description of fluid behavior	4.1
a	radiation pressure constant ($a = 4\sigma/c$, where σ = Stefan-Boltzmann constant)	9.2b
a	quasi-adiabatic quantity (subscript)	10.1
a	oscillatory moment of inertia	19.1b
ad	adiabatic quantity (subscript)	4.2c
$\mathbf{A}$	$(\equiv \rho^{-1}\nabla\rho - [\Gamma_1 P]^{-1}\nabla P)$	17.1
A	radial component of $\mathbf{A}$; convective stability criterion	17.1
A	"acoustic" region	17.10
$\mathbf{A}(\boldsymbol{\xi})$	$(\equiv \boldsymbol{\xi})$	19.1b
$A_{2\nu}$	expansion coefficient for A/r	17.6a
b	quantity defined by eq. (19.6)	19.1b
$B(T)$	integrated Planck function	4.3
B	$(\equiv 3\Gamma_1 - 4)$	13.1
$\mathbf{B}(\boldsymbol{\xi})$	$(\equiv 2iM\boldsymbol{\xi})$	19.1b
$B_{2\nu}$	expansion coefficient for Ag/r^2	17.6a
$B - V$	color index	3.1
c	velocity of light in vacuo	2.4
c	value at stellar center (subscript)	8.12a
c	quantity defined by eq. (19.6)	19.1b
c_P	specific heat per unit mass at constant pressure	4.2c
c_V	specific heat per unit mass at constant volume	4.2c
c_1	$(\equiv [r/R]^3 M/m)$	17.5b

Symbol	Meaning	Section
C	"work integral"	9.1
C	quantity determining rotational splitting of frequencies	19.1c
$\mathbf{C}(\xi)$	$(\equiv M^2\xi + \mathcal{P}[\xi] + \mathcal{V}[\xi])$	19.1b
C_i	imaginary part of work integral	9.1
C_r	real part of work integral	9.1
$C_{2\nu}$	expansion coefficient for $\rho/(\Gamma_1 P)$	17.6a
d	days (subscript or superscript)	2.1
$d\tau$	volume element	4.2a
$\mathcal{D}M_i$	"effective" mass of ith interface in mass zoning scheme	8.12b
D	constant used in Baker one-zone model	13.1
D	thermal imbalance integral	19.4
D/Dt	Stokes derivative	4.1
$D_{2\nu}$	expansion coefficient for ρ	17.6a
$\mathbf{e}_{1,2,3}$	unit basis vectors in a Cartesian coordinate system	4.2b
$\mathbf{e}_{r,\theta,\phi}$	unit basis vectors in a spherical coordinate system	6.1
$\mathbf{e}_{\varpi,\phi,z}$	unit basis vectors in a cylindrical coordinate system	19.1c
E	internal energy per unit mass	4.2c
$\mathcal{E}$	$(\equiv [(\nabla \cdot \mathbf{F})/\rho]_0)$	17.14
E_{grav}	total gravitational energy	19.2
E_{rot}	rotational kinetic energy	19.2
E_{mag}	total magnetic energy	19.2
$E_{2\nu}$	expansion coefficient for $\rho g/(r\Gamma_1 P)$	17.6a
$\mathbf{f}$	total "body" force per unit mass	4.2b
f	denotes an f mode, or an f-mode quantity	17.7
f_t	"tangential component" of a general vector $\mathbf{f}$	17.3
$f(\mathbf{r},t)$	arbitrary function of $\mathbf{r}$ and t	4.1
$\mathbf{f}(\mathbf{r},t)$	gravitational force per unit mass	4.5
$\mathbf{F}$	total vector flux of heat due to all transport mechanisms	4.3
F_ν	radiant flux per unit frequency interval	3.3
$\mathbf{g}$	gravitational acceleration	4.5
g	magnitude of gravitational acceleration $(= Gm[r]/r^2)$	6.1
g_e	effective gravitational acceleration	11.3
g	denotes a g mode or a g-mode quantity (sometimes a subscript)	17.7
$g(\omega)$	$[\equiv f(\omega)/x_1(\omega)]$	9.2b

Symbol	Meaning	Section
G	gravitational constant	1
$G_{1,2}$	linear operators	9.2b
$G1, G2$	matrices	9.2b
$G_{1,3}$	thermal imbalance integrals	19.4
G	"gravity" region	17.10
h	specific enthalpy ($\equiv E + P/\rho$)	11.3
h	hours	3.1
H	magnetic field strength	19.2
$\mathbf{I}$	unit tensor	4.2b
$\mathcal{I}$	unit matrix	8.12b
I	moment of inertia	8.8
I	action ($\equiv \eta^2\omega$)	19.4
$\mathcal{J}_E$	energy flux vector	4.2c
J	number of zones in mass zoning scheme	8.12b
J_k	oscillatory moment of inertia for the kth mode	8.8
$J(\mathbf{r}[\mathbf{a},t])$	Jacobian ($\equiv \lvert \partial x_j/\partial a_k \rvert$)	4.2a
k	coefficient in limb darkening function $\phi(\theta)$	3.4
$\mathbf{k}$	propagation vector	5.5
k	denotes the kth mode ($k = 0, 1, 2 \ldots$ for radial oscillations; $k = 1, 2, \ldots$ for nonradial oscillations) (subscript)	8.5
k_r	radial part of $\mathbf{k}$ (r sometimes omitted)	8.4, 17.10
k_H	"horizontal" component of $\mathbf{k}$	18.2
k_T	total wave number	18.2
$K_{1,2}$	linear operators	9.2b
$K1, K2$	matrices	9.2b
K	nonadiabaticity parameter used in Baker one-zone model	13.1
K	"radiative conductivity"	17.14
K	thermal conductivity	18.2a
l	order of the spherical harmonic $Y_m^l(\theta,\phi)$	17.3
L	stellar luminosity	2.3
L_ν	stellar luminosity per unit frequency interval	3.3
$\mathcal{L}$	linear Hermitian operator used in the LAWE	8.8
L_r	interior luminosity ($\equiv 4\pi r^2 F$)	6.1
L_1	luminosity at the bottom of the hydrogen ionization zone	11.3
L_2	luminosity at the top of the hydrogen ionization zone	11.3

L^2	"Legendrian"	17.3
m	magnitude (usually a superscript)	3.1
m	the quantity used in Rudd-Rosenberg one-zone model (defined by eq. [13.28]), and the constant used in the Stellingwerf model, §13.3	13.2
m	azimuthal spherical harmonic index	17.3
$m(r)$	mass interior to a sphere of radius r (r dependence sometimes not shown explicitly)	6.1
M	stellar mass	1.0
$\mathcal{M}$	mass of gas in the zone of the Rudd-Rosenberg one-zone model	13.2
M	operator ($\equiv \mathbf{v}_0 \cdot \nabla$)	19.1b
M_{bol}	bolometric absolute magnitude	3.1
M_V	visual absolute magnitude	3.1
$\mathbf{n}$	unit vector pointing along the outward normal to a surface	4.2b
n	exponent of ρ in the opacity law	7.5
n	exponent of P in the opacity law	11.3
n	polytropic index	8.12c
n	mode index (subscript)	17.7
n	($\equiv m\Gamma_1 - 2$)	13.2
na	nonadiabatic (subscript)	19.4
n_0	($\equiv r_0/r_c$)	13.2
N	Brunt-Väisälä frequency	17.2
0	unperturbed or equilibrium value (subscript)—often omitted	5.3
$O(\cdots)$	order symbol ("of the order of")	9.3
p	correction factor for observed radial velocity	3.3
p	denotes a p mode or a p-mode quantity (sometimes a subscript)	17.12
P	total pressure	2.1
P	total pressure tensor	4.2b
$\mathcal{P}$	linear vector operator	15.3
P_g	gas pressure	8.3
P_r	radiation pressure	8.3
q	heat content per unit mass	4.2c
q	ratio of interior mass to total stellar mass ($\equiv m/M$)	8.12a
q	($\equiv \sigma/4[n + 1]$)	11.3

Q	pulsation constant	2.1
r	radial coordinate, measured from center of star, in a spherical coordinate system	2.1
$\mathbf{r}$	position vector	4.1
r	radial component (subscript)	17.3
r_c	radius of rigid stellar core	13.2
R	stellar radius	2.1
R	surface value (subscript)	8.12c
R_s	Schwarzschild radius	1
Re	real part of [e.g., Re(ζ)]	8.2
s	seconds	3
s	specific entropy	5.4c
s	exponent of T^{-1} in the opacity law	7.5
s	eigenfrequency	13.1
$\mathfrak{s}$	complex angular frequency	13.1, 18.2
$\mathfrak{s}_0$	$\mathfrak{s}$ taken in the adiabatic limit	18.2a
sp	space part (subscript)	9.1
S	surface of an enclosed volume	4.2b
S_l	"critical acoustical frequency"	17.10
$\mathcal{S}$	quantity defined by eq. (18.15)	18.2a
t	time	2.1
t'	$(\equiv\omega_0 t)$	13.4
$\tilde{t}$	$(\equiv\epsilon t)$	13.4
t_{ff}	"free-fall" or dynamical time scale	2.2
t_K	Kelvin time scale	2.3
t_{nuc}	"nuclear" time scale	2.4
t_s	time characteristic of evolution, or "slow" change, of "static" stellar model	19.4
T	temperature	4.1
$\mathcal{T}$	total kinetic energy	4.6
T_e	effective temperature	3.1
TR	denotes a "transition region" value (subscript)	10.1
$\mathbf{u}$	arbitrary "sufficiently regular" vector function	15.2
u	arbitrary "sufficiently regular" function	8.8
u	$(\equiv r^2\delta r)$	17.5a
$\mathbf{u}_k$	k^{th} eigenfunction of L	15.2
$\mathbf{u}(\mathbf{r})$	space part of $\delta\mathbf{r}(\mathbf{r},t)$	15.2
U	$(\equiv d\ln m/d\ln r)$	17.5b
$\mathcal{U}_\nu$	expansion coefficient for u	17.6a
$\mathbf{v}$	fluid velocity	4.1

v	$(\equiv uP^{1/\Gamma_1})$ $(=r^2\delta rP^{1/\Gamma_1})$	17.9
v_S	adiabatic (or Laplacian) sound speed	5.5
v_M	velocity in the meridional plane	19.1c
v_{gr}	group velocity	17.12
v_{ph}	phase velocity	17.12
V	stellar volume	2.3
V	actual velocity of stellar surface $(=\dot{R})$	3.4
$V(t)$	observed radial velocity of stellar surface	3.3
V	specific volume $(\equiv 1/\rho)$	4.2a
V	$(\equiv -d\ln P/d\ln r)$	8.12a
$\mathcal{V}$	linear vector operator	15.3
w	$(\equiv y\rho P^{-1/\Gamma_1})$ $(=P'P^{-1/\Gamma_1})$	17.9
x	Cartesian coordinate	4.1
$\mathbf{x}$	position vector	4.5
x	$(\equiv r_0/R_0)$	7.3
$x_{1,2,3}$	components of $\mathbf{r}$ in a Cartesian coordinate system	4.1
x_i	fractional mass abundance of nuclear species i	5.2
X	hydrogen mass fraction	18.2a
X_i	column matrix used in mass zoning scheme $[\equiv(\mathcal{D}M_i)^{1/2}\delta r_i]$	8.12b
y	Cartesian coordinate	4.1
y	$(\equiv P'/\rho)$	17.5a
y_1	$(\equiv\delta r/r)$	17.5b
y_2	$[\equiv(gr)^{-1}(P'/\rho+\psi')]$	17.5b
y_3	$[\equiv(gr)^{-1}\psi']$	17.5b
y_4	$[\equiv g^{-1}d\psi'/dr]$	17.5b
Y	helium mass fraction	10.3
$Y_l^m(\theta,\phi)$	spherical harmonic	17.3
Y_ν	expansion coefficient for y	17.6a
z	Cartesian coordinate	4.1
Z	mass fraction of elements heavier than helium	10.3

Greek symbols

α	$(\equiv\nabla\cdot\boldsymbol{\delta r})$	17.3
α	$[\equiv(\partial\ln\epsilon/\partial\ln X)_{\rho,T}]$	18.2a
$\alpha(r)$	$(\equiv w^{-1}dv/dr)$	17.10
β	ratio of gas pressure to total pressure $[\equiv P_g/(P_g+P_r)]$	8.3

$\beta(r)$	$(\equiv v^{-1}dw/dr)$	17.10
γ	ratio of specific heats $(\equiv c_P/c_V)$	4.2c
$\Gamma_{1,2,3}$	adiabatic exponents	4.2c
δ	Lagrangian variation (e.g., $\delta\mathbf{r}$)	5.3
δ	$(\equiv \lvert C/J\Sigma^3 \rvert)$	9.3
δ_{k1}	Kronecker delta	8.8
δt	"tangential component" of $\delta\mathbf{r}$	17.3
$\delta t_{\theta,\phi}$	transverse components of $\delta\mathbf{r}$	17.3
$\delta\Phi$	"effective potential energy" of a star in its perturbed configuration	8.11
$\delta\Psi$	total (kinetic and potential) pulsation energy $(\equiv \delta\mathcal{T} + \delta\Phi)$	9.1
Δ	arbitrary, small variation (e.g., Δu)	8.10
$\Delta m(r)$	mass lying above the radius r $(\equiv M - m[r])$	10.1
ϵ	a small, pure number	13.4
ϵ	total rate of heat per unit mass gained from energy sources (usually taken to be thermonuclear reactions)	4.3
ζ	relative Lagrangian variation of $r(\equiv \delta r/r_0)$	7.1
$\boldsymbol{\zeta}$	$(\equiv \delta\mathbf{r})$	15.2
ζ	quantity defined by eq. (17.93b), a measure of the transverse component of $\delta\mathbf{r}$	17.11
ζ	nonadiabaticity parameter used in Stellingwerf model	13.3
η	growth rate for pulsation amplitude	18.3
θ	polar angle in a spherical coordinate system	6.1
Θ	$[\equiv L_0/(\sigma\chi\Delta m_0)]$	11.3
κ	opacity	7.5
κ	stability coefficient $(\lvert\kappa\rvert \equiv 1/\tau_d)$	9.1
κ_{I}	stability coefficient for the relative variation $\delta r/r$	19.4
κ_{II}	stability coefficient for the absolute variation δr	19.4
κ_E	stability coefficient for total pulsational energy	19.4
λ	exponent of ρ in the thermonuclear energy generation rate relation	7.6
λ	wavelength of a disturbance $(=2\pi/k)$	2.1
λ_P	local pressure scale height $(\equiv -dr/d\ln P)$	7.3

Λ	$\{\equiv 4(\Gamma_3 - 1) + [s(\Gamma_3 - 1) - n] - 4/3\}$	13.1
Λ'	$\{\equiv 4(\Gamma_3 - 1) + [s(\Gamma_3 - 1) - n] - 4/m\}$	13.3
μ	mean molecular weight	17.2
ν	exponent of T in the thermonuclear energy generation rate relation	7.6
ν	polytropic index	17.8
ν_e	effective polytropic index	17.9
$\xi(r)$	spatial part of $\zeta(r,t)$	8.2
Π	pulsation period (subscripts 0,1,2, . . . usually mean, respectively, fundamental, first overtone, second overtone, . . .)	2.1
ρ	mass density	2.1
ρ_c	central density	8.12a
$\bar{\rho}(r)$	mean density interior to $r(=m[r]/\{[4/3]\pi r^3\})$	8.9
σ	angular pulsation frequency	2.1
σ'_{rot}	that part of ω due to rotational effects ("*rot*" subscript sometimes omitted)	19.1c
Σ^2	real number defined by eq. (8.41)	8.10
Σ^2	real number defined by eq. (15.16)	19.2
$\phi(\theta)$	limb darkening function	3.4
ϕ	azimuthal angle in a spherical coordinate system	6.1
ϕ	phase of $(\delta L/L)_2$ relative to $(\delta L/L)_1$	11.3
ϕ	phase angle	13.4
Φ	total gravitational potential energy	4.6
χ	$(\equiv P'/\rho + \psi')$	17.3
χ_T	$(\equiv [\partial \ln P/\partial \ln T]_\rho)$	4.2c
χ_μ	$(\equiv [\partial \ln P/\partial \ln \mu]_{\rho,T})$	17.2
χ_ρ	$(\equiv [\partial \ln P/\partial \ln \rho]_T)$	4.2c
$\psi(\mathbf{r},t)$	gravitational potential	4.5
Ψ	"total" energy $(\equiv \mathcal{T} + \mathcal{U} + \Phi)$	4.6
Ψ	total pulsation energy $(=\mathcal{T} + \delta\Phi)$	8.11
Ψ_ν	expansion coefficient for ψ'	17.6a
ω	angular pulsation frequency	8.12b
ϖ	perpendicular distance from axis of rotation (cylindrical coordinates)	19.1a
$\omega_{1,2,3}$	solutions to $i\omega(\omega^2 - \Sigma^2) = C/J$	9.3
Ω	stellar gravitational potential energy	2.3
Ω	dimensionless angular frequency $(\equiv\sigma^2 R^3/GM)$	8.6
Ω	angular rotation velocity	19.1a

Miscellaneous Symbols and Notations

$\cdot$ (dot)	time (Stokes) derivative (e.g., $\dot{r}$)	2.1, 4.2a
$^{-}$ (bar)	average quantity (e.g., $\bar{\rho}$)	2.1
$\odot$	solar value (e.g., $\rho_{\odot}$)	2.1
$\langle\ \rangle$ (angular bracket)	average quantity (e.g., $\langle\rho\rangle$)	3.1
∇ (del)	gradient operator	4.1
: (double dot)	double dot product in dyadic notation [e.g., $\mathsf{P}:(\nabla\mathbf{v})$]	4.2c
∇^2 (del squared)	the Laplacian operator ($\equiv\nabla\cdot\nabla$)	4.5
$\lvert\ \rvert$	magnitude of a vector (e.g., $\lvert\mathbf{v}\rvert$)	5.2
$\lvert\ \rvert$	absolute value	7.1
$\lvert\ \rvert$	determinant of a matrix	8.12b
$'$ (prime)	Eulerian variation (e.g., f')	5.3
δ	Lagrangian variation, departure from equilibrium value (e.g., δP)	2.1
* (asterisk)	complex conjugate	8.8
$'$ (prime)	differentiation with respect to $x(\equiv r_0/R_0)$	8.12a
∇_x	gradient with respect to $\mathbf{x}$	15.1
∇	logarithmic temperature gradient ($\equiv[d\ln T/d\ln P]$)	17.2
∇_{rad}	($\equiv[\Gamma_2-1]/\Gamma_2$)	17.2
$\propto$	is proportional to	17.8

Index

This is a subject index only. Reference section also serves as author index.

Library of Congress Cataloging in Publication Data

Cox, John P
Theory of stellar pulsation.

Includes bibliographical references.
1. Pulsating stars. I. Title.
QB838.C69 523.8′4425 79-3198
ISBN 0-691-08252-9
ISBN 0-691-08253-7 pbk.